PROSPECTS FOR POLAR TOURISM

PROSPECTS FOR POLAR TOURISM

Edited by

J.M. Snyder

Strategic Studies Inc., 1789 E, Otero Avenue, Centennial, Colorado 80122, USA

and

B. Stonehouse

Scott Polar Research Institute, University of Cambridge, Lensfield Road, Cambridge, CB2 1ER, UK

www.cabi.org

CABI is a trading name of CAB International

CABI Head Office
Nosworthy Way
Wallingford
Oxon OX10 8DE
UK

Tel: +44 (0)1491 832111
Fax: +44 (0)1491 833508
E-mail: cabi@cabi.org
Website: www.cabi.org

CABI North American Office
875 Massachusetts Avenue
7th Floor
Cambridge, MA 02139
USA

Tel: +1 617 395 4056
Fax: +1 617 354 6875
E-mail: cabi-nao@cabi.org

A catalogue record for this book is available from the British Library, London, UK.

A catalogue record for this book is available from the Library of Congress, Washington, DC.

ISBN-13: 978 1 84593 247 3

The paper used for text pages in this book is FSC certified. The FSC (Forest Stewardship Council) is an international network to promote responsible management of the world's forests.

Typeset by Columns Design Ltd, Reading, UK
Printed and bound in the UK by Cromwell Press, Trowbridge

Contents

Contributors   vii

Editor's Introduction   ix

Foreword   xi

Part I   Tourism and the Polar Environment   1
Bernard Stonehouse

1   The Growing Significance of Polar Tourism   3
John M. Snyder and Bernard Stonehouse

2   Pioneers of Polar Tourism and Their Legacy   15
John M. Snyder

3   Polar Tourism in Changing Environments   32
Bernard Stonehouse and John M. Snyder

Part II   Economic Roles of Polar Tourism   49
John M. Snyder

4   The Polar Tourism Markets   51
John M. Snyder

5   Tourism in Rural Alaska   71
*Henry Huntington, Mike Freeman, Bill Lucey, Grant Spearman
and Alex Whiting*

6   Development of Tourism in Arctic Canada   84
Mike Robbins

7   The Economic Role of Arctic Tourism   102
John M. Snyder

8   Gateway Ports in the Development of Antarctic Tourism   123
     Esther Bertram, Shona Muir and Bernard Stonehouse

Part III   Developments in Antarctic Tourism   147
     Bernard Stonehouse

9   Antarctic Ship-borne Tourism: an Expanding Industry   149
     Esther Bertram

10   Antarctic Adventure Tourism and Private Expeditions   170
     Machiel Lamers, Jan H. Stel and Bas Amelung

11   Antarctic Scenic Overflights   188
     Thomas Bauer

12   Antarctic Tourism: What are the Limits?   197
     Denise Landau and John Splettstoesser

13   Antarctic Tourism Research: the First Half-Century   210
     Bernard Stonehouse and Kim Crosbie

Part IV   Managing the New Realities   229
     John M. Snyder and Bernard Stonehouse

14   Managing Polar Tourism: Issues and Approaches   231
     John M. Snyder

15   Tourism on South Georgia: a Case for Multiple
     Resource Management   247
     John M. Snyder and Bernard Stonehouse

16   Tourism Management on the Southern Oceanic Islands   263
     Phillip Tracey

17   Tourism Management for Antarctica   285
     Esther Bertram and Bernard Stonehouse

Index   311

Contributors

Amelung, Bas, *International Centre for Integrated Assessment and Sustainable Development, Maastricht University, Postbox 616, 6200 MD Maastricht, The Netherlands.*

Bauer, Thomas, *School of Hotel and Tourism Management, Hong Kong Polytechnic University, Hung Hom, Kowloon, Hong Kong SAR, China.*

Bertram, Esther, *Department of Geography, Royal Holloway, University of London, Egham, Surrey, UK.*

Crosbie, Kim, *International Association of Antarctica Tour Operators, PO Box 2178, Basalt, CO 81621, USA.*

Freeman, Mike, *Alaska Department of Fish and Game, PO Box 63, Yakutat, AK 99689, USA.*

Hall, C. Michael, *Department of Management, University of Canterbury, Christchurch, New Zealand.*

Huntington, Henry, *23834 The Clearing Drive, Eagle River, AK 99577, USA.*

Lamers, Machiel, *International Centre for Integrated Assessment and Sustainable Development, Maastricht University, Postbox 616, 6200 MD Maastricht, The Netherlands.*

Landau, Denise, *International Association of Antarctica Tour Operators, PO Box 2178, Basalt, CO 81621, USA.*

Lucey, Bill, *Yakutat Salmon Board, PO Box 160, Yakutat, AK 99689, USA.*

Muir, Shona, *Institute of Antarctic and Southern Ocean Studies, University of Tasmania, Private Bag 77, Hobart 7001, Tasmania, Australia.*

Robbins, Mike J., *The Tourism Company, 146 Laird Drive, Suite 102, Toronto, Ontario, Canada M4G 3V7.*

Snyder, John M., *Strategic Studies, Inc., 1789 E. Otero Avenue, Centennial, CO 80122, USA.*

Spearman, Grant, *Simon Paneak Memorial Museum, PO Box 21085, Anaktuvik Pass, AK 99721, USA.*

Splettstoesser, John, *PO Box 515, Waconia, MN 55387, USA.*

Stel, Jan H., *International Centre for Integrated Assessment and Sustainable Development, Maastricht University, Postbox 616, 6200 MD Maastricht, The Netherlands.*

Stonehouse, Bernard, *Scott Polar Research Institute, University of Cambridge, Lensfield Road, Cambridge CB2 1ER, UK.*

Tracy, Phillip, *Australian Government Antarctic Division, Department of Environment and Heritage, Hobart, Tasmania, Australia.*

Whiting, Alex, *Native Village of Kotzebue, PO Box 296, Kotzebue, AK 99757, USA.*

Editors' Introduction

Tourism or travel for pleasure results from a spreading of prosperity that is relatively recent in the history of mankind. Early travellers crossed oceans and continents to seek wealth, to colonize new lands, to exploit human populations and resources, or to right real or imagined wrongs. Travel for its own sake, to satisfy curiosity or to learn, required surplus wealth, time and motivation, and was rare before the 18th century. Among the earliest tourists were the scions of wealthy families, a tiny minority of the population, seeking culture in the discomforts and dangers of the European Grand Tour. During the 19th and early 20th centuries, steamships and expanding railroad networks vigorously promoted commercial tourism. Entrepreneurs exploited these new transport modes, allowing a broader spectrum of people affordable access to a wide variety of attractions.

Following World War II, international prosperity, accompanied by commercial air travel and the omnipresent automobile, provided both the financial means and the transport modes for millions of people to travel. Commercial enterprises created resorts, theme parks, golf courses and other amusements, while simultaneously governments established more national parks and protected areas. Collectively, these diverse attractions now comprise a huge industry, reputedly the world's largest, supplying tourist experiences on a grand scale, virtually unlimited by distance.

Polar regions are the newest tourist destinations. Remoteness, lack of access and public perceptions that they were cold, distant and unattractive, delayed their acceptance: who could possibly want a holiday in a polar region? The earliest recreational tours to the Arctic began during the mid-19th century, to the Antarctic in 1958. From small beginnings, within the span of a human generation, polar tourism has become a distinct sector within the tourism industry. Still small, but now rapidly growing and diversifying, it caters to rising numbers of tourists who, in the spirit of the Grand Tour, seek unique experiences in remote places.

Polar regions are still not everyman's ideal for holiday-making – a fact that has safeguarded essentially wilderness areas from massive human incursions. But, as individual chapters in this volume testify, polar tourism is expanding both north and south. Tourism does not merely occur; it transforms the regions within which it occurs, catalysing changes in environment, economics and culture, and affecting decisions regarding their use and management. Host countries and communities dedicate considerable financial, human and political resources to tourism, creating and implementing environmental management, economic development and cultural conservation plans. Simultaneously the environments in which these activities occur experience significant change.

The keynote textbook *Polar Tourism*, edited in 1995 by Colin Michael Hall and Margaret E. Johnson, first spelled out some of the issues and problems presented by this small but burgeoning corner of the industry. The editors' approach was largely geographical, and most chapters covered important issues occurring throughout wide areas. At a time when polar research in the north was concerned mainly with social and political issues, and in the south with the physical and biological sciences, Hall and Johnston's book was a timely reminder that those who were managing either polar region faced an emergent industry with far from negligible impacts.

Since the publication of *Polar Tourism*, both the magnitude of tourism impacts and the contexts within which they are evaluated have changed considerably. Tourists now overwhelmingly outnumber residents at most polar destinations. Tourism's role in polar economies has grown from minor contributor to prominent force. Indigenous Arctic peoples are exerting more decisive roles in tourism, and insisting that their voices be heard and views respected.

Polar tourism now thrives on a remarkable combination of human-induced and natural events. Its magnitude and growth will inevitably produce changes, but changes need not be for the worse. That tourism has arrived late in polar regions means that people, jurisdictions and organizations responsible for stewardship can benefit from a wealth of research and management experience gained in other regions, selectively adapting to meet unique polar circumstances. Contributors to this book, all professionally involved in different aspects of polar management, examine polar tourism in relation to its environmental, economic and cultural settings, and explore resource management techniques to fashion appropriate responses.

John M. Snyder
Bernard Stonehouse

Foreword

The End of a Frontier or Last Chance to See?

Since the early 18th century leisured tourists have travelled outwards from the metropolitan tourist-generating regions of Europe and North America and, more recently, East Asia, to the polar regions. Fuelled by curiosity at nature's wonders, and able to travel ever quicker and further thanks to improvements in transport technology, wave after wave of tourists has advanced over the globe, consuming landscapes and the many perfect images of the tourism marketer. In the search for 'authentic', 'unique' and/or fashionable experiences, no part of the globe is now untouched by that outsider of the modern era – the tourist. For a long time the very isolation of the polar regions from metropolitan centres meant that they were relatively immune from the effects of a 'pure' form of leisure tourism, although scientific and commercial curiosity was in itself a form of temporary mobility into the polar regions. However, as Johnston and Hall (1995: 309) observed, 'the isolation which has long served the polar regions is now the very feature that attracts tourists to visit these places'.

Polar regions, particularly the non-road-accessible areas, are still among the world's last tourism frontiers. But those frontiers are now fading as a result of advances in transport technology, and corresponding changes in the tourism market that are enabling greater numbers of people to visit polar destinations. Regular scheduled and charter flights and sailings in both polar regions are testimony to the extent to which these areas have now been drawn into the mental map of the tourist industry. In the sub-Arctic, places compete to be the home of Santa Claus, while Japanese tourists pay to travel under the northern lights. Vicarious ecotourism experiences on *Animal Planet* and the *Discovery Channel*, whether with whales, polar bears or penguins, perhaps make the polar environment seem all the more accessible and more desirable to visit (Gössling and Hultman, 2006). Yet at the same time as the Arctic and Antarctic have seemingly become more accessible, their fragility to human impact becomes all the more pronounced.

In 1995, in the conclusion of *Polar Tourism*, Johnston and Hall (1995) made several predictions:

1. That tourism would continue to grow in both polar regions, but that growth in the Arctic would prove to be greater in absolute terms than in the south.

2. There would be considerable expansion of cruise tourism in the Arctic and sub-Arctic, including the gradual opening up of the Russian Arctic regions to marine tourism.

3. Cultural tourism related to indigenous peoples, and historic tourism based on Arctic exploration and industrial development, would increase. This would be an important element of deseasonalization, particularly in the northern Nordic regions.

4. Visitor codes of conduct would continue to be an extremely important management tool, but there would be increased demands from some stakeholders for improved regulatory regimes.

5. There would be increased concerns over the environmental impacts of tourism in polar regions while there would simultaneously be greater government interest in the economic dimensions of such tourism, including its contribution to conservation practice.

To a great extent, as revealed by chapters in the present volume, these predictions have all come to pass. However, the growth that has occurred with respect to tourism in the polar regions has now become all the more problematic as a result of concerns over global environmental change (Gössling and Hall, 2006; Johnston, 2006).

Unfortunately, governments and industry continue to have an almost paradoxical attitude towards tourism in the polar regions, particularly in the northern hemisphere. On the one hand, tourism is regarded as an important element in the diversification of peripheral economies, whether in Greenland, northern Finland, Siberia or Alaska. Yet at the same time tourism is simultaneously criticized, not only for its potential immediate local environmental impact, but increasingly because of its contribution on a global scale to greenhouse gas emissions and subsequent global warming effects. This tension between the various impacts of tourism highlights the importance of better understanding the nature, patterns and dimensions of tourism in polar regions, and the relative contribution it makes to sustainable development practice.

Arguably there is perhaps no better example of the complexity of the issues facing polar regions with respect to tourism than the potential effects of climate change. In Alaska, Western Canada and Eastern Russia, average winter temperatures have increased by as much as 3–4°C (4–7°F) in the past 50 years and are projected to rise by 4–7°C (7–13°F) over the next 100 years (ACIA, 2004). Warming in polar regions has already had an impact on polar ecology and geography. Arctic summer sea ice is projected to decline by at least 50% by the end of this century, with some models showing its near-complete disappearance (ACIA, 2004). Such changes are extremely likely to have devastating consequences for some Arctic animal species such as ice-living seals, bird species, the iconic polar bears and caribou, and also for local peoples for whom some of these animals are a primary food source. However, at the same time, reduced sea ice extent particularly during the summer navigation period is likely to increase marine access to some of the region's resources which would actually serve to make some of the areas more accessible for tourists, even though some of the attractions may already be suffering severe environmental stress as a result of climate change. In fact, tourism's embeddedness in the complexity of polar resource management does not end there, as economic utilization through such mechanisms as tourism becomes one means of exercising sovereignty and jurisdiction in an increasingly ice-free maritime environment.

Although the above examples mainly relate to the northern polar regions, the retreat of Antarctic Peninsula glaciers, changes to shelf ice as well as greater melting of sea ice are also raising alarm at the impacts of climate change on the Antarctic environment. Again, concerns over increased accessibility for tourism in newly ice-free areas in summer, particularly from private yachts, increases biosecurity risks such as the introduction of weeds or diseases. And even in the transnational space of the Antarctic there is continued expression of the use of tourism by some national governments as a means to support their Antarctic presence.

As Johnston and Hall (1995) concluded over a decade ago, it is almost impossible to halt the growth in tourist visitation to the polar regions. Even though restrictions may be enabled in some locations of high scientific or economic interest, other areas of access are increasingly becoming available. Therefore, given the potential impacts of global environmental change, there is an increased urgency in formulating sustainable approaches to tourism at the polar regions. Sustainable tourism in the polar context means conserving the productive basis of the physical environment by preserving the integrity of the biota, ecological processes and cultural values, and at the same time producing tourism commodities without destroying other aspects of resource use, such as indigenous and local people's activities.

Although it might make ecological sense to reduce tourism substantially or to prohibit it completely in all polar regions, or at least in some selected areas, this is generally unrealistic given the ever-increasing public demand and the impossibility of halting access in most cases. It is also unrealistic given national and stakeholder interests in both regions and the economic expectations of indigenous peoples and local residents in northern latitudes. Additionally, and paradoxically, by allowing people to visit polar areas, we may continue to encourage an interest in polar conservation by the public, who then might attempt to persuade policy-makers and governments to maintain or designate protected area status. The polar bear is likely to become as much an emblem of international conservation efforts with respect to climate change over the coming decades as the panda or elephant has been with respect to land-use change in recent years. Vicarious appreciation through books and documentaries is important, but it is not necessarily sufficient to create a groundswell of public support for good conservation practice.

The development of a tourism industry also provides a useful economic argument in helping to ensure that the polar areas are maintained in their relatively pristine state for future generations, rather than being exploited completely for other natural resource uses. Nevertheless, now that isolation is less a factor in limiting visitor arrivals, it is essential that appropriate management regimes are put in place to regulate tourist activity and that environmental considerations occur at all stages of the trip, not just at the polar destination. Indeed, a key future issue for polar resource management is that visitor and operator codes of conduct are not enough, while in a global age neither are national regulatory frameworks. Instead, the challenge for the polar regions in the coming years will be to develop comprehensive international regulatory and governance regimes that can manage not only the effects of global environmental change but also flows of people. Clearly, such a goal requires not just political will and stakeholder interest but also a strong scientific and research base, and it is here that the present volume will undoubtedly make a substantial further contribution to our understanding of tourism in polar areas.

C. Michael Hall
Department of Management
University of Canterbury
Christchurch
New Zealand

References

ACIA (2004) *Impacts of a Warming Arctic; Arctic Climate Impact Assessment.* Cambridge University Press, Cambridge, UK.

Gössling, S. and Hall, C.M. (eds) (2006) *Tourism and Global Environmental Change: Ecological, Social, Economic and Political Interrelationships.* Routledge, London.

Gössling, S. and Hultman, J. (eds) (2006) *Ecotourism in Scandinavia.* CABI, Wallingford, UK.

Johnston, M.E. (2006) Impacts of global environmental change on tourism in the polar regions. In: Gössling, S. and Hall, C.M. (eds) *Tourism and Global Environmental Change: Ecological, Social, Economic and Political Interrelationships.* Routledge, London, pp. 37–53.

Johnston, M.E. and Hall, C.M. (1995) Visitor management and the future of tourism in polar regions. In: Hall, C.M. and Johnston, M.E. (eds) *Polar Tourism: Tourism in the Arctic and Antarctic Regions.* John Wiley & Sons, Chichester, UK, pp. 297–314.

I

Tourism and the Polar Environment: Introduction

BERNARD STONEHOUSE

Little more than a generation ago the phrase 'polar tourism' would have been considered a contradiction in terms. Not perhaps by the knowledgeable few who had already discovered the Arctic, but certainly by the masses whose tolerance for travel stopped short of the unknown (and reputedly uncomfortable) ends of the earth. Only a few years ago, when polar tourism began to make its mark, few travel agents could have helped the casual enquirer with information about polar tours, still less with a booking – in particular to that remotest of all destinations, Antarctica.

'Times, they are a-changing', as the three chapters in this first section, 'Tourism and the Polar Environment', testify. Tourism to polar regions is today no more or less remarkable than tourism to African safari parks, Caribbean holiday islands or European capitals, though it may prove more expensive, and intending travellers may need to book longer ahead. With websites and colourful brochures showing polar bears and penguins (mercifully separate), icebergs, red anoraks, jaunty small ships and majestic liners – always in splendid scenery and perfect weather – polar tourism now offers familiar patterns of vacation travel, to different but no less intriguing destinations.

Both editors of *Prospects for Polar Tourism* have many years' experience of tourism in polar regions, John Snyder mainly in the north, Bernard Stonehouse mainly in the south. In Chapter 1 they combine to establish the main contention of the book – that polar tourism is a rapidly expanding industry at both ends of the world, with prospects that demand thought from its managers. Like any other large and growing human development, polar tourism needs careful, positive and continuing management to maintain the integrity of environments in which it operates. Though slow in starting, polar tourism is currently expanding in every

© CAB International 2007. *Prospects for Polar Tourism*
(eds J.M. Snyder and B. Stonehouse)

measurable aspect. Within a generation it has become, in the authors' words, 'the single largest human activity in both polar regions', and is now 'old enough to have established recognizable patterns of procedure, and mature enough ... to have earned respectful consideration by those who aspire to manage it'.

In Chapter 2 John Snyder outlines more fully the history of polar tourism, from its tentative Arctic beginnings in the early 19th century to the most recent developments in both polar regions. Though polar recreational travel has always appealed to a minority, the 20th century advent of mass travel by sea and air opened up new opportunities to which polar boundaries were no barrier. The author draws attention to the curious fact that, however much polar tourists are attracted by remoteness, wilderness, scenic beauty, wildlife and other obvious characteristics of the regions, the theme of discovery continues to permeate tourism experience in both polar regions. No other kind of commercial tourism so strongly and consistently relies on historical traditions to sustain its business. He concludes that 'to a very considerable extent, the prospects for polar tourism will be a reflection of its past'.

In Chapter 3 the editors again combine to consider 19th and 20th century changes in polar environments, from consequences of economic exploitation at both ends of the world, to more recent relics of the Cold War in the Arctic and of scientific and technological exploration in the Antarctic. Superimposed on changes due to direct human intrusion is a dramatic secular shift in climate, for which the responsibility of man is still being debated, but of which the consequences must clearly affect both short-term and long-term prospects for polar tourism. The authors conclude that in planning for the future, tour operators and environmental managers alike would be well-advised to take these changes into account.

1

The Growing Significance of Polar Tourism

JOHN M. SNYDER[1] AND BERNARD STONEHOUSE[2]

[1]Strategic Studies, Inc., 1789 E. Otero Avenue, Centennial, CO 80122, USA; [2]Scott Polar Research Institute, University of Cambridge, Lensfield Road, Cambridge CB2 1ER, UK

Introduction: Polar Tourism in a Dynamic Setting

> Tourism is the discovery of the well-known, whereas travel is the discovery of the ill-known, and exploration is the discovery of the unknown.
>
> (Brendon, 1991)

Until the late 19th century the polar regions were virtually unknown to the general public, and poorly understood. Following a long history that characterized them as remote, inaccessible and austere, 20th century exploration brought the regions from obscurity to world prominence. Though still remote, they are now readily accessible and, at least in summer, far less austere than most people imagined. They are also of considerable political significance, scientific interest and natural beauty, with engaging wildlife. Not surprisingly, they are of growing interest to tourists and tour operators. Both north and south, polar regions today host expanding tourism industries that are diversifying in range of recreational activities, variety and size of transport modes, extension of seasons, and penetration into new geographical areas.

The environmental, economic and cultural settings of polar tourism change constantly and at accelerating rates. Biological, climatic and oceanographic indicators agree that polar and sub-polar environments are undergoing changes, both natural and human-induced, of which climatic warming is but one example (ACIA, 2004). As mining, fisheries and other industries based on natural resources decline, and military presence in the Arctic diminishes, indigenous human populations are forced to look elsewhere to maintain their economies and social well-being. Tourism has become the Arctic-wide answer. Throughout annual tourist seasons of steadily increasing length, numbers of tourists greatly exceed numbers of residents in all popular Arctic venues (Fig. 1.1). Similarly the scientists and support staff who for a generation or more were dominant in the Antarctic are now far outnumbered by summer tourists (Fig. 1.2). This chapter concerns the current role and future implications of tourism in the changing polar world.

Polar Tourism Activities

To understand the impacts of tourism, both beneficial and otherwise, on polar environments, we need to know not only how many tourists visit, but where, when and how they make their marks. For example, the several thousand cruise-ship passengers who passively view the Arctic from offshore, and occasionally invade land-based souvenir shops, affect the region in ways that differ from the smaller numbers who are active in river rafting, wildlife photography, mountaineering and sport fishing. By accurately identifying the full array of tourist activities and their behavioural patterns, and placing that information within the context of their natural and human resource settings, we can begin to understand their relationships.

To explore probable future effects of tourism on the environmental, economic and cultural resources of polar regions, we need to draw on many different sources of information. Apart from numbers, essential data include accurate profiling of tourists, analysis of their behaviour and an understanding of the economic, jurisdictional and social conditions that facilitate tourism in any setting. That daunting responsibility is faced by policy-makers and managers throughout both polar regions.

Polar regions currently offer tourism experiences ranging from very passive to extremely active recreational activities. Of these, cruise-ship travel and recreational hunting and fishing have the longest histories. Cruise ships

Fig. 1.1. Penguin meets tourists: Antarctic Peninsula. (Photo: J.M. Snyder.)

Fig. 1.2. Cruise ship *Cunard Princess* and float plane attend tourists at Juneau, Alaska. (Photo: J.M. Snyder.)

represent the largest mass tourism activity operating in either polar region. Tourists are essentially passive while aboard, marginally more active should they undertake excursions ashore. At the other end of the scale, anglers and hunters attracted to high-latitude rivers, forests, oceans and tundra are more active and make greater demands on the environment – incidentally supporting one of the largest and most lucrative branches of the polar tourism market.

Smaller-scale and more individualistic tourism may involve outdoor adventure and personal challenges; for example, river rafting, sea and river kayaking, mountaineering, snow-shoe treks, ski-trekking, backpacking, camping, wildlife viewing and nature photography. All these activities are catered for in the Arctic, and several are being added to Antarctic itineraries too. Those seeking contemplative experiences in remoteness and solitude can find them in Arctic Wilderness Areas, National Parks and World Heritage Sites. There are no such areas in Antarctica, where individual or small-group travel is less easily arranged: for good sailors, cruises on small yachts offer the best possibilities. Tourists who seek to visit, and participate in the lives and art of indigenous cultures, now find this possible among the Saami of the Scandinavian Arctic, native Alaskans and the Nunavut people of Canada. These examples illustrate how multi-faceted polar tourism has become, catering for a wide range of interests among its clients. The rapid diversification of the polar tourism market is reviewed in several chapters of this volume.

Where Does Polar Tourism Occur?

'Polar' is a relative term, indicating vaguely the northern and southern ends of the world. 'Polar tourism' is equally vague, signifying recreational visits to lands and communities in what travellers from warmer countries regard as primitive or wilderness areas of the far north or south. Thus a week in the highly civilized city of Tromsö, north of the Arctic Circle in Norway, has much in common with a week in any other city, while ski-trekking in nearby wilderness areas more closely matches the polar image.

In the Arctic, tourism began in the early 19th century as daring visits to the northern extremities of inhabited countries – for example, to northern Canada, Alaska and Scandinavia, where indigenous peoples pursued lifestyles completely different from those in the south. Often travellers were seeking opportunities to escape briefly from civilization, to find the frontier of immediate forebears, to experience wilderness, live precariously at subsistence level, catch fish, shoot and bring back trophy heads and skins – for all of which outfitters proved eager to cater (Auer, 1916; Murphy, 1983; Viken and Jorgensen, 1998). Those who sought more passive and comfortable experiences of wilderness were well-served by ship-borne and railway package tours. Both categories of tourists continue to be served in every country throughout the Arctic, as far south as the edge of civilization; for a review of polar tourism's historic development see Chapter 2 of this volume, and for perspectives on its several roles in the Arctic see Part II.

Antarctic tourism began a whole century later, with overflights and ship-borne cruises from South America to the Antarctic Peninsula and neighbouring islands of the maritime Antarctic. More remote than the Arctic, with a harsher climate and no indigenous human population, Antarctica up to the mid-20th century was regarded as the preserve of heroic explorers and scientists. Once overcome, this tradition served as a spur to ship-borne tourism: anyone could now visit this extraordinary place, and experience some of its hardships in reasonable comfort and safety. Antarctic tourism covers the continent and nearby islands, and extends to many of the cold, remote islands in the southern oceans, for example South Georgia, which share some of its characteristics (Snyder and Shackleton, 2001). Part III of this volume discusses the expansion of diversification of Antarctic tourism.

Barriers to Entry

Throughout most of its history, the viability of polar tourism has been challenged by obstacles that economists refer to as 'barriers to entry'. The concept suggests that the extent to which these barriers are increased, reduced, altered or eliminated directly controls the amount, geographic distribution, seasonal duration and types of tourism likely to occur in polar regions. Many different agents of change can affect barriers to entry – natural, man-made or combinations of both. The characteristics of these agents range from obvious and readily understood, to subtle and difficult to verify.

The most obvious barriers affecting polar tourism include difficulty of access, environmental conditions (both real and imagined), high costs resulting from remoteness and lack of infrastructure, the short season in which travel is feasible, and jurisdictional conditions that act as constraints on the industry. Individual and collective assessment of these factors and their interrelationships offers insights into how the growing tourist industry may ultimately affect the regions in which it is operating (Clawson and Knetsch, 1966; Walsh, 1986). The factors are introduced below, and discussed further in subsequent chapters.

Access

The ability to travel to and throughout polar regions is heavily influenced by both natural conditions and human events. Access by non-native people was inhibited by the sheer size of the physical barriers, insufficient geographic knowledge to accomplish navigation, and inadequate or inappropriate transportation technology, making these some of the latest areas of the world to be discovered by man.

The most obvious physical barrier to access on both sea and land is ice – sea ice temporarily or permanently preventing access by canoe or ship, and land ice preventing easy passage both along the shore (often the preferred route for migrant or expanding populations) or inland. Both forms of ice inhibited movement not only of indigenous populations, but also of explorers and entrepreneurs from the south, intent on discovering new routes and exploiting natural resources.

The skills and persistence of early navigators and their more recent successors provided the accurate maps and marine charts that make Arctic polar tourism possible. For much of the Arctic, topographic and cadastral land surveys were either not conducted or not published until the late 19th and early 20th centuries. Evidence of the lack of good mapping appears, for example, in government hearings on 19th century boundary disputes between the USA and Canada, and 20th century contests for possession of Svalbard (US Senate, 1900). The first reliable hydrographic and coastal surveys of the North Pacific and Arctic oceans were not published until the late 19th century. For centuries valuable nautical information on this area and the islands of the southern oceans remained hidden in the logbooks and personal records of whalers, sealers and commercial fishermen. Antarctica itself remained a suspected but virtually unknown continent until 1820, and was not properly mapped and charted to modern standards until after World War II (Baughman, 1994; Gurney, 1997).

Whatever protection against intrusion may in the past have been afforded by the presence of sea and land ice, its effects in the short term are likely to diminish. Both polar regions are currently warming, the Arctic more rapidly than the Antarctic, and sea ice appears to be diminishing in extent and thickness (Chapter 3 of this volume). One probable effect is increased access by cruise ships to parts of both regions that are currently closed to tourism.

Another may be dramatic changes in flora, fauna and scenery, which discerning tourists might wish to see for themselves.

Polar climates have indeed in the past provided a second barrier, impeding every effort of man (a temperate-climate and tropical species) to explore, map and even inhabit polar lands. During the past millennium the Arctic has been anomalously cold, with particularly cold spells during the 14th and 15th centuries, followed by a slow and erratic shift towards warming which is currently accelerating. A closely related third barrier affecting human usage was seasonal lack of food, which permitted only thinly distributed, nomadic and mainly coastal or riparian human populations in the high Arctic, and marginal, unreliable agriculture and stock-rearing on the periphery.

In modern terms relating to tourism, these three barriers are still extant. Seasonal sea ice still restricts the movements of tourist ships, and landings are restricted to the relatively small areas that are free of ice in summer – in the case of Antarctica, to less than 2% of the continent. Cold and inclement weather remain constant constraints on the enjoyment of polar regions, and tour operators must be prepared to bring in almost every item of food that their clients will require, often at great expense.

A fourth barrier has been the lack not only of good maps, but also of the many other forms of geographical information that we find readily for other areas. Only since the mid-20th century has a flurry of scientific and technical activity, for example that accomplished during International Geophysical Year (1957/8), established a comprehensive foundation for understanding Antarctica. Since then numerous international operations have added substantially to our knowledge of the biology, geology, climates and other parameters of both polar regions, and remote imaging and other mapping techniques have contributed to our knowledge of land forms and oceanographic conditions. Nevertheless, these areas remain the least known places on earth. The north geographic pole was not verifiably reached until Wally Herbert attained it in 1964, and to date the entire Northwest Passage has been successfully navigated only six times (Brigham and Ellis, 2005).

The invention of diverse fast and safe modes of transport, and their adaptation to extreme cold, has enormously facilitated polar travel. The mid-to-late 19th century saw the introduction of steamships and steam locomotives, which opened the door to popular travel throughout the world, including the Arctic (Runte, 1984; Brendon, 1991). Steam-powered icebreakers challenged the inhibiting presence of sea ice from 1871. Cruise ships carrying tourists began operating in North American waters in 1867 (Twain, 1869). Automobiles toured in northern Canada and Alaska during the early 20th century, very soon after their invention. From the late 1920s aircraft fitted with wheels, skis or pontoons began to provide the ultimate access in both polar regions, proving especially valuable over the vast northern expanses of North America and Siberia (Glines, 1964; Van Doren, 1993).

Tourism has been quick to take advantage of each new method of travel. In recent decades a remarkable diversity of motorized transport has directly contributed to the growth of polar tourism. A seemingly endless variety of ships, trains, airplanes, helicopters, automobiles, motor coaches, snow machines (snowmobiles), snow buggies and tractors regularly deliver groups of people to and around the Arctic with efficiency and comfort. In the category of non-motorized transport, traditional dog sleds, kayaks (baidarkas), skis and snow-shoes have been re-designed and manufactured with new materials to the demanding specifications both of individuals and of commercial guide services. The newest forms of non-motorized transport, particularly river rafts and mountain bikes, have become extremely popular forms of backcountry transport in the polar regions. Even sled dogs are being especially conditioned and trained to take part in such widely publicized tourist events as the Alaskan Iditorad.

The collapse of the Soviet Union in 1991 resulted in the sudden availability of Russian ice-strengthened ships and icebreakers for commercial cruise ships, providing tourist access to new areas in both the Arctic and Antarctic, and extending the length of the travel season in both hemispheres. Small ice-strengthened scientific research vessels became cheap and efficient cruise ships for use in ice-strewn waters. Small icebreakers with on-board helicopters allowed Antarctic tourists to land in out-of-the-way emperor penguin colonies. Powerful Russian Yamal-class nuclear icebreakers, designed to keep open the shipping lanes of the Northeast Passage, take tourist passengers on regular scheduled runs to the North Pole (Armstrong, 1991). Soviet submersibles made available for chartering have been used to carry tourists to a variety of destinations including the North Pole itself.

The creation of new transport technologies, and personal motivations to visit new areas, has overwhelmingly demonstrated that no parts of the globe, including the polar regions, are beyond tourist access.

Environmental conditions – real and perceived

Tourists are frequently attracted to unusual environmental settings and unique wildlife viewing opportunities. But those attractions are also inherently unpredictable and visiting them may, on occasion, be uncomfortable and dangerous. While experiencing a particular attraction, the tourist may encounter difficult terrain, the vagaries of weather, immediate proximity to wild animals, strong ocean currents and exposure to similar authentic environmental conditions, which set limits on their enjoyment and ability to travel in the natural world. Such conditions can produce exhilaration, or generate concerns extending from discomfort to fear. The tourism and recreation industries tend to flourish when they can avoid, dispel or minimize the inherent discomfort and danger of natural environmental conditions. Both the real and the perceived dangers are formidable constraints that must be diminished to lower this particular barrier to entry.

Authentic environmental conditions result from the combination of two kinds of events: (i) naturally occurring environmental events associated with the ecology of an area; and (ii) human intervention. The range of human intervention can extend from aggressive exploitation to sustainable environmental management practices, or indeed to benign neglect. The interplay of these dynamics creates the authentic environmental conditions that the traveller experiences.

An example of this interplay is provided by a current situation in the Antarctic Peninsula area, where relatively ice-free conditions, splendid scenery and spectacular wildlife provide the continent's most popular area for Antarctic tourism. The area is, however, becoming more popular with Antarctic fur seals (*Arctocephalus gazella*), which were previously hunted almost to extinction but have recently recovered and returned in considerable numbers. Many areas that, 10 years ago, were devoid of this species, now support large summer populations. Under Antarctic Treaty regulations, fur seals are completely protected and cannot be controlled. At several sites of great tourist interest, visitors are at serious risk of bodily injury from the aggressive behaviour of the fur seals. So the naturally occurring situation, of ideal conditions for tourism, has recently altered and will require human intervention – management – if it is to retain its attraction. There appears to be no question of killing or other drastic interference with the seals, which are incidentally destroying vegetation as well as the tourists' peace of mind. Nevertheless management intervention is required that responds to the new environmental conditions.

Tourism is also directly affected by the extent to which harsh environmental conditions, especially weather, can be offset. Advances in clothing and equipment technologies, especially since World War II, have diminished the discomfort and dangers that have historically been the hallmarks of polar regions. Clothing now provides warmth, wind protection and water repellency that early explorers would have envied. Similarly modern kayaks with fibreglass shells are more comfortable, durable and safe for recreational use than those made by native peoples from bones and hide. Modern communication technologies, and such way-finding equipment as hand-held Global Positioning Systems, are vastly more reliable than earlier radio and navigation technologies. Despite numerous advances, there is no suggestion that the harsh environmental conditions in the polar regions have been overcome. They have merely been mitigated by increased knowledge of principles, modern clothing and equipment, allowing tourists more comfortably and safely to visit these regions.

Public perceptions greatly influence people's desire both to travel and to be selective in their choice of destinations. For most of their history, the polar regions have conjured forebodings among the public in general, and dread among mariners in particular. Early descriptions of polar environments contributed to a sizeable lexicon of dismay. The considerable loss of life and fortune associated with efforts to explore the polar regions substantiated and embellished this reputation. The perception that polar regions were cold, bleak and inhospitable constituted a formidable barrier to tourism.

But as our knowledge of places changes, so do our perceptions. Photographs and popular articles in illustrated magazines provided newer and on the whole truer accounts of the polar regions, later reinforced by television programmes and feature films that showed them in the most favourable light, and tourists returning from early visits with glowing reports. After more than a century of recreational travel to the Arctic, and half as long to the Antarctic, more balanced and positive views prevail. Opinion of the desirability of travel to the polar regions continues to evolve: they are increasingly perceived as places where, in an overcrowded and over-busy world, serenity and solitude can still be found.

Most recently they are considered as among the safest destinations, least likely to attract terrorists and other dangerous or undesirable elements. The tourism industry acts quickly to avoid these risks. In October 1985 a terrorist attack on the cruise ship *Achille Lauro*, in the Mediterranean Sea, resulted in the murder of a wheelchair-bound retiree and a US Navy medic. In 1986 an explosion in the Chernobyl nuclear reactor spread contamination over a wide area of western and southern Europe. Fearing that their ships might either 'blow or glow', leaders of the cruise industry set courses for polar destinations that they perceived to be safe, some expanding existing operations in polar regions, others entering polar cruising for the first time (R. Tuegas, Vice President, Holland America, personal communication, 1987). They were all economically successful, and none has so far seen reason to withdraw their polar itineraries.

Costs of travel

Historically, cost has always provided an effective barrier to participation in tourism. Prior to the mid-19th century, travel to remote destinations was virtually restricted to society's wealthy elite. By the latter half of the century, aggressive competition among railroads and shipping companies provided cheaper mass travel to a wider range of destinations, including remote regions. Popular tourism, as offered by Thomas Cook and other pioneers, became economically viable. Continuous advances in transportation technologies, especially automobiles and commercial airliners, and their continuing popularity in a massive market, further reduced travel costs. Concurrently the evolution from agrarian to post-industrial economies helped to raise household incomes and spread wealth to growing numbers of people, giving them money to travel for pleasure (Feifer, 1985).

Initially the more accessible destinations, with well-established infrastructure, enjoyed a substantial competitive cost advantage over such remote and less accessible locations as the polar regions. But, following the economic development model for the rest of the world, polar tourism grew as a result of initiatives in both transportation and economic measures. The private sector providing the planning and means of transport improved in efficiency, and the public sector invested – not exclusively for the benefit of tourism, but very much to tourism's advantage – in improvements to ports, airports and roads. These measures have steadily reduced costs of polar travel and continuously narrowed the competitive cost disadvantage. These trends are anticipated to continue, so any concomitant increase in personal wealth will most probably result in increasing travel to polar regions.

It is worth noting that a staunch segment of the polar tourism market dedicates considerable effort to surmounting the cost barrier. These are clients who are strongly motivated to participate in high-quality outdoor recreational experiences, and will devote whatever time is needed to save for the unique experiences that are offered by polar tourism. Their commitment to saving provides them with opportunities to patronize expensive sport fishing lodges, rafting expeditions, mountaineering, wildlife viewing and photography, expedition ship cruising and other types of unique recreation activities.

Time for travel

Tourists' selections of acceptable destinations have always been influenced by the amount of time available to them, and the efficiency of transportation. Polar regions are literally the ends of the earth, and the time needed to reach them from inhabited regions has always been a serious consideration. To an average tourist with a limited vacation heading for an interesting destination, time spent in getting there is simply time wasted; the longer the journey, the less attractive the whole operation becomes. This is particularly important for travel to Antarctica from North America or Europe – the starting points for most Antarctic voyagers – who are likely to spend 4 days out of an advertised 12-day vacation in travel. This is one reason why many travellers to polar destinations are retired or otherwise leisured. The facts that an enormous number of the world's population is nearing retirement age, leisure time has consistently increased throughout the workforce, and transportation progressively achieves greater efficiencies, make it appear likely that the time constraint associated with travel to the polar regions will be reduced.

A further time-based barrier is the need to take polar vacations during a limited season, normally the 3 or 4 months of summer at either end of the world. Not everyone who is still working can get away in November to February to see Antarctic penguins, or in May to August to see Arctic tundra flowers and polar bears. Again this constraint favours the leisured classes with time (and money) on their hands.

Jurisdictional constraints

Jurisdictions have the authority to define allowable uses for polar lands and coastal marine zones, and where necessary enforce their decisions. They may allow unlimited or limited use, or prohibit use altogether. They may facilitate tourism by establishing development zones, providing economic incentives and investments, and creating attractions such as national parks. They may inhibit it by establishing conservation reserves, prohibiting commercial development in attractive areas, limiting visits to particular seasons and types of use, or banning visits altogether.

Polar regions function within a remarkable assortment of jurisdictions that individually and collectively influence the location, type and magnitude of tourism activities. These change radically from time to time. Several Arctic lands especially have recently undergone shifts in sovereignty and jurisdictional relationships that provide new authorities, with different approaches to economic development and natural resource management. Prominent among these changes are the attainment of sovereignty by indigenous Arctic peoples of Canada and Siberia, most of whose new governing bodies favour controlled tourism and are setting forth legislation that is designed to achieve it. The Antarctic, which in pre-tourism days was managed loosely and somewhat absent-mindedly by seven claimant nations, has since 1961 been governed under the international Antarctic Treaty System, which promotes Antarctica as a continent of peace and freedom of scientific investigation, but does not aspire towards promotion of tourism.

Jurisdictional circumstances surrounding 'allowable uses' in polar regions in general, and tourism in particular, are further challenged by legitimacy to govern and the need to manage 'the commons'. The central common feature of the Arctic region is the Arctic Ocean, which is under no single jurisdictional control or management authority. The central feature of the Antarctic is the continent itself, that seven nations continue to regard as their property (though currently under international governance) and ultimately their responsibility, while others regard it as the world's vastest 'commons'.

Arctic jurisdictions and Antarctic Treaty parties are confronted with a variety of management issues that uniquely characterize polar regions. Management techniques currently employed and potentially available to the jurisdictions and treaty parties are presented in Part IV of this volume.

Summary and Conclusions

Polar tourism, though relatively new in the history of the tourism industry, is old enough to have established recognizable patterns of procedure, and mature enough, even in its newest venue Antarctica, to have earned respectful consideration by those who aspire to manage it. The constraints under which it operates are severe, but have nevertheless proved surmountable. Polar tourism was slow to start, but is now a popular and rapidly growing industry that is expanding in terms of tourists, tour operators, diverse recreational pursuits, geographic scope and seasons of use. Arctic economies have seen it evolve from an incidental activity to a vital sector upon which they increasingly rely. This has been particularly true for newly enfranchised indigenous peoples of the Arctic seeking self-sufficiency and for gateway cities in the southern hemisphere eager to realize the economic benefits of Antarctic tourism.

Tourism now constitutes the single largest human activity in the polar regions. For anyone seriously interested in the well-being of those regions, the mere acknowledgement of that fact is not sufficient. Rather, a comprehensive look at the role of polar tourism, and the context within which it operates, warrants thoughtful attention. The content of this book provides multiple perspectives intended to advance our understanding of the role of tourism in the polar world.

References

ACIA (2004) *Impacts of a Warming Arctic; Arctic Climate Impact Assessment.* Cambridge University Press, Cambridge, UK.

Armstrong, T.E. (1991) Tourist visits to the North Pole, 1990. *Polar Record* 27(161), 130.

Auer, H.A. (1916) *Camp Fires in the Yukon.* Stewart & Kidd Co., Cincinnati, Ohio.

Baughman, T.H. (1994) *Before the Heroes Came: Antarctica in the 1890s.* University of Nebraska Press, Lincoln, Nebraska.

Brendon, P. (1991) *Thomas Cook: 150 Years of Popular Tourism.* Secker & Warburg, London.

Brigham, L. and Ellis, B. (eds) (2005) *Arctic Marine Transport Workshop, 28–30 September 2004.* Northern Printing, Anchorage, Alaska.

Clawson, M. and Knetsch, J.L. (1966) *Economics of Outdoor Recreation.* Johns Hopkins Press, Baltimore, Maryland.

Feifer, M. (1985) *Tourism in History from Imperial Rome to the Present.* Stein and Day, New York, New York.

Glines, C.V. (1964) *Polar Aviation.* Franklin Watts, Inc., New York.

Gurney, A. (1997) *Below the Convergence – Voyages toward Antarctica 1699–1839.* W.W. Norton and Co., New York.

Murphy, P.E. (ed.) (1983) *Tourism in Canada: Selected Issues and Options.* Western Geographical Series Vol. 21. University of Victoria, Victoria, British Columbia.

Runte, A. (1984) *Trains of Discovery – Western Railroads and the National Parks.* Northland Press, Flagstaff, Arizona.

Snyder, J. and Shackleton, K. (2001) *Ship in the Wilderness: Voyages of the MS 'Explorer' through the Last Wild Places on Earth.* Gaia Books Ltd, London.

Twain, M. (1869) *The Innocents Abroad.* American Publishing Co., Hartford, Connecticut.

US Senate (1900) *Report Number 1023: Compilation of Narratives of Explorations in Alaska.* Government Printing Office, Washington, DC.

Van Doren, C.S. (1993) PanAm's legacy to world tourism. *Journal of Travel Research* Summer, 3–12.

Viken, A. and Jorgensen, F. (1998) Tourism on Svalbard. *Polar Record* 34(189), 123–128.

Walsh, R.G. (1986) *Recreation Economic Decisions: Comparing Benefits and Costs.* Ventura Publishing, Inc., State College, Pennsylvania.

2 Pioneers of Polar Tourism and Their Legacy

JOHN M. SNYDER

Strategic Studies, Inc., 1789 E. Otero Avenue, Centennial, CO 80122, USA

Introduction

Since the early 1800s polar regions have witnessed the birth and development of two distinct forms of tourism. The first, independent travel, attracted individuals with personal curiosity and wanderlust, and later those with enthusiasm for a diversity of outdoor recreation activities and backcountry adventures. The second, mass tourism, provides a variety of experiences to be shared by groups of people travelling together. Experiences associated with the two forms of tourism are quite different, and appeal to different types of client. Most significantly, each form has firmly established distinct patterns of resource uses, human behaviour and economic dependencies that influence the polar regions.

The first polar tourists consisted of a very few curious and intrepid persons who travelled to the Arctic during the early 1800s. Their journals describe adventures and remark upon strange environments, unusual animals that were sometimes killed for trophies, and encounters with people who were very different from themselves. Published journals provided very personal accounts that occasionally served as guides for future travellers; see for example John Lainige's 1807 travel journal *A Journey to Spitzbergen*. In the late 1800s the Arctic increasingly drew anglers and hunters attracted to abundant fish and wildlife, and adventurers who revelled in the wilderness that offered seemingly unlimited recreational opportunities. During the late 19th and early 20th centuries, the Arctic's sporting attractions achieved European notoriety from aristocrats, who touted the chase of trophy game, the collection of curios and their personal explorations. Americans were similarly attracted to the Arctic by their long tradition of fishing and hunting. America's leaders of that time, for example President Theodore Roosevelt, strongly encouraged hunting and outdoor recreation as an essential part of a healthy life. Sporting attractions were further popularized by vivid descriptions in new publications, such as *Field and Stream* and *Recreation*. By igniting the imagination and unleashing the recreation expenditures of anglers, hunters and outdoor adventurers, these popular publications played a vital role in establishing an Arctic tourism market.

By the 1890s commercial guiding enterprises and equipment manufacturers were satisfying the independent Arctic traveller's recreation demands. Providing guide services, specialized recreation and sports equipment, transport logistics and backcountry facilities best suited for Arctic conditions soon constituted a flourishing business. The Primus Oil Stove was not only available for the Arctic sportsmen, but its original advertisements were personally endorsed by Fridtjof Nansen. David Abercrombie offered waterproof tents, aluminium cooking outfits, sleeping bags and other recreation equipment that would eventually make his firm (developed with a fellow by the name of Fitch) one of the world's great purveyors of sports equipment. No outdoor adventure in the Arctic was complete without capturing the angling and hunting trophy by new inventions called the 'Kodak' by Eastman and 'The Tourist Hawk Eye' by the Blair Camera Company.

Polar Tourism: Success of an Improbable Idea

Polar outdoor recreation and adventure tourism continue today, although changed by social and environmental values. Individuals still enjoy catching fish, but most anglers now practise catch-and-release techniques – or are admonished to do so. More pursue wildlife with cameras rather than guns: polar bears that were once killed for furs in Svalbard, Norway and Churchill, Canada are now sought for photo opportunities. However, highly regulated hunting remains popular: in all locations and for all seasons throughout the Arctic, wildlife managers now decide the fate of the fish, birds and mammals. But wildlife regulations have not in any way diminished the ardour of anglers and hunters, and in fact have assured the sustainability of those recreation resources. The historical progression of angling and hunting in the Arctic has, within the last century, evolved to become a lucrative segment of today's polar tourism economy.

Adventure tourism, originally the unregulated domain of intensely independent recreationists, has also witnessed a history of increased control and management. Early adventurers included passengers aboard private yachts who visited the arctic whaling grounds at Point Barrow, Alaska in 1891 and others who boated down the Yukon River during the 1890s. These unfettered experiences have been replaced by well-orchestrated and regulated Arctic charter-boat operations and river rafting trips. Mountaineering, kayaking, dog sledding, backcountry hiking and wilderness camping – once the arduous means for accomplishing polar exploration and pioneer settlement – have all become popular tourist activities. Some are offered by commercial enterprises, others by permits obtained from natural resource agencies, but all are subject to rigorous regulatory enforcement by a variety of jurisdictions.

Independent travel to the polar regions has evolved from the curious few to an enormous market comprised of individuals seeking personal challenges and involvement with nature. Most importantly, the impacts that they have on the polar world are disproportionate to their numbers. Their direct contact with polar environments, native cultures and economies can produce significant impacts. Their needs for emergency services, communication infrastructure and hospitality services place considerable demands on the professional skills and financial resources of Arctic communities. And their pursuit of wildlife and adventure tourism directly affects the traditional resource uses and fishing and hunting cultures of the indigenous Arctic peoples.

Explorers, Excursionists and Entrepreneurs

Pioneers of popular mass tourism to the Arctic in the mid-1800s created successful visitor experiences and commercial delivery systems that continue today. Entrepreneurs audaciously suggested that their guests should personally experience the harsh conditions, remoteness and hazards of uncharted waters and wilderness conditions in absolute comfort. This bravado, enhanced by employing distinguished polar explorers and scientists as guides and on-board lecturers, started a tradition that endures to the present.

Publicity surrounding polar exploration and gold discoveries popularized the first attractions for Arctic 'excursionists'. This new demand, and its potential revenues, did not escape the attention of the nascent mass tourism industry. In the mid-1840s a new breed of entrepreneur called the tour manager created a business that combined the public's desire to see attractions they had previously only read about, with recently invented transportation modes that accommodated mass travel. Serving in this catalytic role, tourism managers actively sought opportunities to promote Arctic tourism. In 1850 Thomas Bennett established a Norwegian tour agency to facilitate steamship tours to coastal fjords and the North Cape. His business risks were justified when the North Cape, the Svartisen Glacier and Spitzbergen attained international popularity in the 1870s. Between 1861 and 1869 a New York artist, William Bradford, sponsored seven expeditions to the Arctic, of which the last achieved fame by involving a prominent explorer with a flair for publicity.

The Hayes/Bradford expedition

Dr Isaac I. Hayes, who had previously achieved international publicity as a participant in the Second Grinnell Expedition, 1853–1855, later as leader of an 1860 Arctic Ocean exploratory expedition, in 1869 led a

Bradford-sponsored summer-long pleasure cruise to Greenland and the upper waters of Baffin Bay, subsequently reporting it in an article, 'Across the Arctic Circle', that featured prominently on the front page of *Harper's Weekly* (Hayes, 1871). Aware of the precedent being set by this journey, Hayes wrote:

> The steamship Panther crossed the arctic circle July 31, 1869, bound for the waters of upper Baffin Bay. She was not bound upon a voyage of discovery, nor did she belong to the whaling fleet which for the past three centuries has annually visited the icy regions; nor was she in pursuit of the codfish, salmon, and halibut which abound in the Greenland seas and lakes; but she simply bore a party of excursionists, who had resolved to make a summer trip to the regions of the arctic circle.

The expedition is noteworthy both for its timing and for the many precedents established during the voyage. The fact that a pleasure cruise would be planned, conducted and highly publicized during an era when polar exploration was characterized by tragedy and danger was itself remarkable. In addition, several of its features still characterize polar cruises today:

- On-board lecturing by famous polar explorers and scientists.
- Shore excursions as a vital part of the tourist experience.
- On-board competitions to sight Arctic features and wildlife.
- Purchasing souvenirs from indigenous people.
- Capturing images by camera and on canvas.
- Consuming outrageous amounts of food and beverages.

Fig. 2.1. *Panther*, ship of the Hayes/Bradford tourist expedition, crossing the Arctic Circle, 31 July 1869, as illustrated in *Harper's Weekly*.

Despite hitting an unchartered rock in Melville Bay, the Hayes/Bradford pleasure excursion was an enormous success. The event was summarized as 'Socially the day was one perpetual lunch, and the night an endless dinner'. Numerous shore excursions explored glaciers, fauna and local people. Tourists made friends with the indigenous Greenlanders and generously shared gifts with them. Sport fishing was popular and provided additional entrees to the already plentiful menu. Some among the passengers hunted birds, but most aimed only cameras at the wildlife. Besides providing a moment of fame to the pioneering excursionists, Dr Hayes reinforced his credentials as a polar explorer by publishing articles about Greenland in two issues of *The Atlantic Monthly*. The significance of his polar research may be debated, but his efforts to fortify his reputation as polar expert are undeniable. Photographs, paintings and woodcuts preserved memories of the trip, ultimately serving as illustrations for the magazine articles. Both *Harper's Weekly* and *The Atlantic Monthly* enjoyed an international audience, and thus helped validate the idea of pleasure travel to the Arctic.

Managed tours and guidebooks

The Hayes/Bradford cruise made a favourable start to polar tourism. Public interest in Greenland and the Arctic was sustained when such explorers as Otto Nordenskjold, A.W. Greeley, Fridtjof Nansen, Robert Peary and Frederick Cook published both their discoveries and disasters. The precedent set by Hayes was followed by other polar explorers: Dr Frederick Cook is reported to have 'made ends meet as a tour operator, taking his well-heeled clientele from New York to Greenland and Ellesmere Island' (*The Explorers Journal*, 2004).

A later contribution to the success of Arctic tourism was the creation of guidebooks, generally acknowledged to have been initiated by John Murray's *Handbook for Travellers* series, first published in 1836 (Murray IV, 1919). By the late 1880s Arctic tourism received a promotional boost from the publication of Paul du Chaillu's *The Land of the Midnight Sun* and handbooks for Iceland, Scandinavia and Lapland (Dufferin, 1873; du Chaillu, 1881; Coles, 1882; Murray, 1893).

Once the popularity of Arctic attractions and routes was verified, tour operators quickly expanded the geographic scope of their operations. For example, Thomas Cook's publication, appropriately entitled the *Excursionist*, advertised tours to Scandinavia in 1875, escapes to Norway – 'The Land of the Midnight Sun' – in 1879, an Arctic cruise to Iceland and Greenland in 1881, and popular tours to Iceland in 1883 (Brendon, 1991). Popular tourism to Greenland, the American Arctic and the Canadian Arctic aboard steamships and railroads were all well established by the late 1800s. The first tours to visit the Russian Arctic were offered in 1899, using the recently completed trans-Siberian Railway between St Petersburg and Vladivostok. When those courageous tourists disembarked at Vladivostok in 1899, popular tourism could rightfully claim that its presence extended throughout the entire

Fig. 2.2. A lady tourist of the 1880s.

Arctic. But, again, the significance of polar tourism is not its age, but the indelible patterns of resource use, visitor behaviour and economic development it created throughout the polar world.

From its inception, popular polar tourism established key precedents, specific visitor expectations, operational techniques and infrastructure that have been steadfastly employed throughout its entire history. The birth and evolution of commercial operations and tourist experiences in the polar regions demonstrate how remarkably 'history has repeated itself'. The remainder of this chapter offers examples of the historical development of polar tourism, to illustrate how it was both created and replicated throughout the polar world.

Alaska and American Arctic Tourism Pioneers

In October 1879 John Muir and two companions paddled into an ice-filled Alaskan bay of spectacular beauty that for hundreds of years the Hoonah Tlingit, native people, had called Sitadakay (Ice Bay). Muir returned to California to publicize the discoveries in a series of lectures he called the 'Fairweather Glaciers', and later in articles written for the San Francisco *Evening Bulletin*. During the next summer he continued his glacial research, while a hydrographic reconnaissance by Captain L.A. Beardslee of the US Navy chartered Muir's discovery and officially named the region Glacier Bay.

No time was lost in transforming Muir's discovery into a tourism opportunity. Before sending his chart to the hydrographic office, Beardslee provided a tracing of it to Captain James Carroll of the Pacific Coast Steamship Company. Sailing instructions into the bay were provided to Captain Carroll by Muir himself, who described safe transit to the largest tidewater glaciers. Less than a year after its discovery, a commercial operator had obtained sufficient information from the original explorers to conduct tours to Glacier Bay, Alaska (Beardslee, 1882).

Muir's lectures, and the several articles and books he subsequently published (Muir, 1915, 1981), were rhapsodies of words that espoused a glacial gospel – a gospel that, as one editor of his publications concluded, expressed the belief that wilderness ultimately provides a place for journeys of the spirit. Muir loved the Alaskan wilderness and urged people to travel north. 'Go', he said, 'go and see'. People soon heeded his words. Captain Carroll too understood how to promote new business opportunities through the popular press and political connections. In 1881 the Pacific Coast Steamship Company leveraged a US Government mail delivery contract to begin its commercial travel to Alaska. It was obvious that commercial diversification was needed to offset the perils and costs of steamship travel to Alaska. Captain Carroll's idea of collecting revenues from tourism was an appealing prospect and he was granted the authority to use the mail steamship *Idaho* for that purpose.

In July 1883, the first tourists cruised into Glacier Bay. Among the passengers was the Honorary Associate Editor of *National Geographic Magazine*, Eliza Ruhamah Scidmore, whom Carroll had expressly invited aboard. Alaska's inaugural tourism event included the naming of known glaciers, the discovery of new ones, numerous shore excursions, and the collection of souvenirs, photographs and native Alaskan crafts. Ms Scidmore wrote of the 'unparalleled scenic grandeur' of the glaciers and immense Alaskan wilderness, instantly establishing a new and highly desirable tourism destination (Scidmore, 1885).

Not content with only one Alaskan destination, based on five subsequent visits Scidmore produced *Appleton's Guide Book to Alaska and the Northwest Coast* (Scidmore, 1896) which provided tourists with all the information they would need to enjoy attractions from Glacier Bay to the Arctic Ocean. During the 1880s and 1890s the Pacific Coast Steamship Company and the Alaska Commercial Company regularly transported thousands of tourists through southeastern Alaska and the Arctic, respectively. Their shipping schedules advertised multiple tourism departures from April through October. Their commercial tourism endeavours were very successful and by 1885 they reported that:

> The demands of business have induced the Pacific Coast Steamship Company to double their service to Alaska, and steamers now run twice a month instead of monthly, as heretofore. The sagacity of this movement is indicated by the fact that the summer excursion lists are rapidly filled months in advance of the days of sailing.

By 1890 international travel agencies were booking large groups of excursionists aboard steamships departing for Alaska.

Travel writers fuelled promotional campaigns with assurances that:

The tourist who makes the voyage from Tacoma to Glacier Bay through the inland sea has the opportunity of beholding some of the grandest scenery and natural phenomena on the globe.

(Ballou, 1890)

During the same year, Alaska's Territorial Governor reported, with some amazement:

A large number of summer vacation travelers have visited Southeastern Alaska during the present season. These people have manifested much interest in the country, and it is thought the knowledge they acquire may prove of service to this Territory.

The mutual benefits derived from a partnership between commercial tour operators, explorers and scientists continued, as they do to this day. What started as a marriage of convenience between scientists and the polar tourism industry has been sustained to the present. John Muir, for example, continued his scientific studies throughout the Arctic from Glacier Bay to Wrangell and Heard Islands. In 1890 he returned to Glacier Bay aboard the tourist steamship *Queen* and off-loaded a pre-cut cabin that he built near his namesake glacier. This structure was a precursor of the numerous polar scientific stations built with the support of commercial tourism companies. Muir's relationship with the Pacific Coast Steamship Company exemplified the emergence of a mutually supportive partnership between scientists and polar tourism. The relationship created opportunities for scientists to pursue their research while, simultaneously, the tourist experience was enriched by knowledgeable information.

As a prominent historian of Glacier Bay wrote about the events of 1890: 'Many of the tourists who visited Glacier Bay that season had the unique experience of learning their natural science from John Muir himself' (Bohn, 1967). Muir's initial research was continued by Dr H.F. Reid, H.P. Cushing and other renowned scientists. Their scientific accomplishments were made possible by the free transport, accommodation, food and beverage provided by the Pacific Coast Steamship Company. The culmination of this unique partnership occurred in 1899 when Captain Carroll took the Harriman Alaska Expedition on a pleasure cruise to the Arctic. The scientists and artists on that cruise comprised a who's who of distinguished persons. John Burroughs, one of the ornithologists, wrote: 'The expedition was known as the Harriman Alaska Expedition, and its object was to combine pleasure with scientific research and exploration' (Burroughs, 1904).

The Harriman Alaska Expedition, as chronicled by C. Hart Merriam, is remarkable in the sense that it established themes that have continued to characterize polar tourism for more than a century. Specifically:

Among the unusual features which contributed to the success of the Expedition, three are worthy of special mention:
(1) The ship had no business other than to convey the party withersoever it desired to go.
(2) The scientific staff represented varied interests and was made up of men trained in special lines of research.

(3) The equipment was comprehensive, including naptha launches, small boats and canoes, camping outfits, stenographers, photographers, and extra men for oarsmen and helpers. The launches were of utmost service, landing large parties quickly and safely, and conveying men and supplies out of reach of the ship.

The first polar tourists, guided by explorers such as Dr Hayes and scientists such as John Muir, would find a lot in common with today's polar tours led by prestigious explorers and prominent scientists. Although contemporary discovery no longer includes filling in vast blank spaces of the globe, the remoteness and dynamics of polar regions continue to provide attractions that satisfy the polar tourist's personal desire to explore.

Gold: Tourism Joins the Rush

Arctic tourism received a tremendous boost from the Gold Rush years that began in Alaska in the late 1800s and then stampeded to the Klondike, Yukon Territory of Canada in the summer of 1897. Herculean construction projects suddenly provided access to a seemingly impenetrable wilderness extending from the Stickeen River of Ketchikan, Alaska in the south, then spreading across Canada's Yukon Territory, and finally reaching the beaches of Nome, Alaska in the north (Wharton, 1972). New sea ports provided marine access, railroads established land routes, and steamboats provided transport along the rivers of this Arctic region.

International publicity devoted to the gold rushes was remarkable by any standard. These sometimes-frenzied accounts were soon accompanied by a genre of extremely popular literature exemplified by the writing of Jack London (London, 1903) and the poetry of Robert Service (Service, 1907). The attraction and curiosity surrounding gold, the availability of transport and the intensity of publicity all contributed to the demand for tours to these Arctic regions. The steamship companies of the era were able to quickly profit from this tourism opportunity. Advertisements for tours of 'The Gold Fields' soon appeared in a variety of promotional materials.

Although the Alaska and Klondike gold rushes did not endure, their fame and the transport access they financed made significant, long-lasting contributions to the development of Arctic tourism. By example, access to an enormous Canadian Arctic region was opened when the White Pass & Yukon Railroad Company completed its 110.4-mile route from Skagway, Alaska to Whitehorse in Canada's Yukon Territory. Passengers then embarked on river steamboats that provided travel along 460 miles of the Yukon River. By 1900 the Klondike Gold Rush was over and prospectors abandoned that region to stampede the gold-laden sands on the beaches of Nome, Alaska (Cohen, 2002). But tourist demand for travel along that historic route is enormous and growing. In 2006 the railroad set an all-time passenger record, carrying 431,249 tourists along its scenic route in refurbished antique railway carriages. Based on the expansion of both cruise-ship arrivals and increased

Fig. 2.3. Steamship entering Five Finger Rapids, Yukon, Canada.

numbers of passenger cars, the railroad is forecasting an even larger total for the 2006 season (White Pass & Yukon Route Railroad, 2006).

Tourism Promoters and Their Transport Partners

As mass tourism began to flourish in the late 1800s, it is important to recognize that most entrepreneurial tourism companies were essentially agents of or partners with steamship and railroad companies. While steamship companies were pioneering routes through the world's oceans, railway companies were aggressively building vast transportation networks across continents. Fierce competition among transport companies focused on reducing travel times, obtaining efficient routes, attracting the greatest amount of freight and number of passengers, and leveraging whatever location advantages could be secured. Entrepreneurial travel agents convincingly showed that tourism produced commercial benefits that included increased numbers of passengers, economic value for scenic routes and destinations, and competitive advantages resulting from visitor comfort and safety.

The steamship and railroad companies of the 1800s not only provided transportation services to previously inaccessible regions, they also enabled all classes of people to travel. With revenues dependent upon both numbers of passengers and volume of freight, steamship and railroad companies soon realized the profitability of mass tourism, and quickly became staunch investors as well as leading proponents. Steamship companies, often with polar explorers near the helm, pioneered cruise-ship tourism throughout the Arctic. Railroads, especially those in the USA and Canada, not only facilitated travel to the Arctic, but were instrumental in its creation and development. During the late 19th and early 20th centuries, railroads, environmental organizations and national governments struck a curious but effective alliance to establish National Parks and Forest Reserves. Correctly perceiving tourism's market opportunities, railroads strongly supported the government's designation of national parks and monuments. American and Canadian railroads constructed elegant accommodations along scenic routes and within the parks. By 1900 railroads became the world's leading advertisers of vacation travel to national parks (Runte, 1984). The National Park concept and associated tourism developments were soon implemented by Scandinavian countries further expanding Arctic tourism opportunities.

The Arctic became increasingly accessible as land transport technologies improved. During the early 20th century, trains found themselves sharing land transport roles with automobiles and by 1920 the Arctic regions of Scandinavia, Alaska and Canada were enticing tourists with hundreds of miles of roadway and diverse services (Burr, 1919; Rand McNally, 1922). These advances enabled tour companies to expand geographically, operate more efficiently, accommodate more persons and offer more safety and comfort.

Individual cruise companies sometimes thrived or failed, but the industry never lost interest in exploratory cruising in the Arctic. For example, in 1931, the Soviet Union's state-owned tourism company Intourist promoted Arctic tourism aboard the icebreaker *Malgyin*, stating that the esteemed polar explorer 'Professor V. Yu Vize, is intending to use the voyage for scientific work, [and] has agreed to undertake overall leadership of the expedition'. Vize was a member of G.L. Sedov's 1912–1914 expedition that sighted the Novaya Zemlya Islands, the last major land mass to be discovered anywhere on Earth. Vize was joined by an impressive team that included General Umberto Nobile. The cruise accomplished shore excursions to historic sites; made the scientific discovery that 'Arthur Island' and 'Alfred Harmsworth Island' were one and the same; successfully rendezvoused with the German airship *Graf Zeppelin*; and was occasionally terrified by ice, fog and shoals (Barr, 1980). All in all, the Soviet's Arctic tour experiences were quite similar to previous polar tourism ventures.

Intourist re-established tourist cruising to the Soviet Arctic in the 1960s and 70s. Vessels carried Soviet and Eastern European vacationers to a variety of locations in the Arctic Ocean, Barents and Kara seas. When the USSR collapsed, Russians were pleased to discover that their fleet of nuclear icebreakers could be profitably employed as tourist cruise ships. In 1990 the icebreaker *Rossiya* took tourists to the North Pole during an 18-day voyage. Emboldened by the commercial success of that enterprise, Russia fully engaged their nuclear icebreaker fleet for polar tourism. In 1991, Quark Expeditions conducted a tour to the North Pole using the nuclear-powered

Yamal, and offered the trip again in 1992 (Armstrong, 1991). Since that time the North Pole has been visited many times by Russian icebreakers and all tours were advertised as opportunities to re-discover this remote site in the company of distinguished polar explorers and scientists. Russian icebreakers are now used for polar tourism throughout both the north and south polar regions.

The legacy of exploratory cruising in the Arctic continues to this day. These voyages are perhaps best exemplified by the journeys of the *MS Lindblad Explorer*. Built for the sole purpose of exploratory tourism cruising by Lars-Eric Lindblad, the ice-strengthened *Lindblad Explorer* was launched in 1969, coincidentally 100 years after Dr Hayes's excursionist cruise to the Arctic Ocean. The *Lindblad Explorer*'s first transits through the Arctic Ocean occurred in 1972, and on that journey set a 'Farthest North' record for a passenger ship by achieving latitude 82°12'N. Other achievements in the Canadian Arctic included successful transits of the Northwest Passage in 1984 and 1988. With Sir Wally Herbert aboard as staff lecturer, the *Lindblad Explorer* celebrated the 30th anniversary of his historic first crossing of the North Pole by dog sled in 1998 (Snyder and Shackleton, 2001). The *Lindblad Explorer* and many other polar cruise ships have contributed to a better understanding of both the Arctic and Antarctic through the production of hydrographic charts, climate observations, wildlife surveys, and logistical support for numerous scientific research endeavours.

Flight represents the most recent and significant transportation event in the polar regions. Polar aviation provides a multitude of tourism opportunities that even a few decades ago were considered impossible. Not only were polar regions made more accessible, but reduced travel time, competitive pricing and flexible scheduling created an entirely new set of tourism development opportunities. Polar aviation allows huge groups of tourists to travel on regularly scheduled commercial airlines; small groups of tourists and individuals to travel by charters; general aviation offers polar entry for the private pilot; and helicopter services provide quick access to a diversity of remote and previously inaccessible sites. Aviation expanded tourist seasons and, in many locations, this has meant year-round access. The allocation of the tourist's time also changed. Shorter travel times to polar destinations allow the tourist more time on site to pursue recreational activities. By significantly altering travel time and producing virtually unlimited access, aviation forever affected the delivery of polar tourism experiences.

Polar aviation also challenges the definition of wilderness and its recreational uses. Planes equipped with wheels, pontoons or skis can land on almost any type of surface in the polar regions. Anglers, hunters and backcountry adventurers now have virtually unlimited access to areas that are either *de facto* or officially designated wilderness. Polar aviation services provide the means for establishing, marketing and supplying backcountry lodges and campsites. From another perspective, flight-seeing may be the least intrusive and damaging way for tourists to enjoy and appreciate the polar wilderness. The dichotomies of polar aviation are plentiful. While polar aviation improves rural Arctic communities' access to emergency services, it simultaneously permits those communities to become tourism destinations.

Polar Tourism Heads South

With more than 35 years' experience providing commercial tours to four continents, Thomas Cook and Sons began advertising tours to Australia and New Zealand in 1879. By the 1900s the southern hemisphere was a well-established tourism market served by a variety of companies. Realizing the economic benefits of positive publicity, Cook and Sons were attracted by the enormous press coverage being devoted to British Antarctic exploration during the early 1900s. At that time a crescendo of favourable British publicity reported the heroic exploits of Robert Falcon Scott's 1901–1904 British National Expedition, the William Bruce Scottish Expedition of 1902–1904, and Earnest Shackleton's 1907–1909 *Nimrod* Expedition. Employing their successful business model that leveraged favourable publicity to promote tours to recently discovered regions of the world, the company advertised a tour to McMurdo Sound, Antarctica in 1910. For reasons that remain undocumented, the tour never occurred. But the intriguing possibility of popular tours to the Antarctic was seriously suggested by a veteran tour operator.

Heroic endeavours to reach the South Pole continued to fuel the popular press, and by 1912 the world knew that the feat was accomplished triumphantly by Roald Amundsen's Norwegian team and tragically by Robert Falcon Scott's British team. Whatever lustre Antarctic tourism may have possessed in 1910 was soon lost by the tragic tales that emerged from the South Pole in 1912. If tourism to Antarctica were ever to happen it would take extraordinary accomplishments and extremely positive publicity to overcome the world's dismal perception of the southern continent. Amazingly, Britain's next Antarctic expedition, Sir Ernest Shackleton's Trans Antarctic Expedition of 1914–1917, provided both the deeds and the publicity needed to reverse public opinion.

Although Sir Ernest Shackleton had a well-deserved talent for self-promotion, it was never his intention to promote Antarctic tourism. Ironically, Shackleton's expedition would, from its planning stages through to its legacy, play a significant role in the history of Antarctic tourism. The irony begins with the Belgian polar explorer Adrien de Gerlache, leader of the Belgica Expedition. Gerlache barely survived history's first overwintering in the Antarctic, but by 1913 he and a Norwegian shipbuilder named Lars Christianson were planning tourist voyages to Greenland and Spitsbergen. Circumstances prevented them from achieving their plans, so they sold their ice-strengthened, ten-cabin tourist ship called *Polaris* to Shackleton, who promptly renamed it *Endurance* and then, as they say, sailed into history.

Key personnel associated with Shackleton's expedition played prominent roles in the establishment of popular tourism to the Antarctic. In 1919, when the world could divert its attention from World War I, Frank Hurley, the expedition's photographer and cinematographer, released a film called 'South' and published a stunning collection of Antarctic photographs. In contrast to the recent horrors of World War I, Hurley's images offered inspiring examples of triumph and survival. The public was once again introduced to Antarctica and its image, although austere, mirrored victory. But Antarctica's improved public perception did not immediately hasten tourism. Despite the visits of a few individual tourists who travelled aboard mail steamers transiting the Southern Ocean during the 1920s, it took the notoriety of determined individuals to initiate popular tourism to the southern continent.

None other than Sir Ernest Shackleton's two famous boat captains, Captain J.R. Stenhouse and Commander Frank Worsley, personally attempted to start popular tourism in the Antarctic. Advertised by the Holland America Line as nothing less than 'The Most Wonderful Voyage Ever Planned', the 'Antarctic World Cruise' was scheduled to depart New York on 15 December 1931 returning 18–19 April 1932. The advertisement emphasized that this tourism enterprise was:

> Under the personal direction of Lieut. Com. J.R. Stenhouse, D.S.O., O.B.E., D.S.C., the most famous Antarctic navigator, who commanded the 'Aurora' (Sir Ernest Shackletons Expedition), and the Royal Research Ship 'Discovery'.

In addition, Captain Stenhouse persuaded Shackleton's famous navigator and captain of the *Endurance*, Commander Frank Worsley, to accompany him as an on-board guide. Unfortunately the trip never occurred, most probably because of the Great Depression's severe economic conditions. Holland America Line proposed another cruise to the Antarctic for the 1932/3 sailing season, but economic conditions again scuttled this trip. Although neither trip occurred, widespread publicity raised public awareness that Antarctica was a potential tourist destination and continued the tradition of famed polar explorers serving as tourist guides.

The heroic legacy of Shackleton's expedition remains a potent force in the history of Antarctic tourism. Interest is sustained by popular films, books and articles, the republication of Hurley's photographs, and the popularity of travelling exhibits, especially those that display the lifeboat *James Caird*. Shackleton's continued popularity is evidenced in the design of tourist itineraries, promotional campaigns and the sale of Antarctic tours that not only feature his feats, but also honour his memory with graveside toasts in Grytviken, South Georgia.

In 1933 Antarctica was viewed by a few tourists travelling aboard the Argentine ship *Pampa*. The vessel was a merchant ship rather than a cruise ship, but most significantly, the cruise demonstrated the suitability of South America's Patagonia region as a 'tourism gateway' to the Antarctic. This was the first time that the Argentine ports of Buenos Aires and Ushuaia were used by Antarctic tourists (Capdevila, 1984).

The tradition of Antarctic explorer as tourism promoter resumed in 1936 when Douglas Mawson, the Australian explorer, gave his presidential speech to the Australian and New Zealand Association for the Advancement of

Science. He proposed that the southern continent offered 'prospects for economic development' and mused that 'a winter sports ground for diversion in summer, Antarctica would be a thrill to Australians'. And like other polar explorers who nearly perished from their harrowing polar experiences, Mawson saw 'no reason to delay the dispatch from our ports modern liners on summer pleasure cruises amongst the pack ice'.

The modern era of continuous visits to the southern continent began in the mid-1950s. The birth of Antarctica's modern tourism industry was brought about by the unique combination of favourable worldwide publicity associated with the International Geophysical Year (IGY), the availability of modern transportation technologies, a pent-up demand for recreation and new personal wealth. The announcement and implementation of the IGY and the first commercial jet and cruise-ship transport of Antarctic tourists all occurred between 1956 and 1958.

The IGY initiated scientific exploration of the continent by such internationally renowned people as Admiral Richard Byrd, Sir Vivian Fuchs and Sir Edmund Hillary. The critically important mapping of the continent by aerial surveys and the publication and international distribution of those maps brought favourable attention to the continent. The National Geographic Society, for example, distributed 2,270,000 new maps to its international membership. The IGY achieved unprecedented cooperation among the 46 nations that established scientific stations throughout Antarctica. The construction of housing, storage facilities and infrastructure associated with each of those stations resulted in the first, large-scale development of Antarctica. Especially significant were the construction of huge runways for wheeled aircraft and the installation of radar facilities for air navigation. Once again polar exploration and science led the way in capturing favourable public interest to visit the polar regions.

Entrepreneurial tour operators seized the opportunity to initiate Antarctica's modern tourism industry. In 1956 the first tourist flight over Antarctica was flown by a Chilean operator. In 1957, a Pan American Airways Stratocruiser flew from Christchurch to land tourists briefly at the 6000 foot runway built buy the US Navy at McMurdo Sound. Mawson's suggestion that Antarctic might become a 'winter sports ground' was echoed in the September 1957 issue of *National Geographic Magazine* that declared 'the first tourist flight [is] heralding the day when the airplane may make the white continent a winter sports playground'.

In 1958 the first tourist cruise ship, the Argentine vessel *les Eclaireurs*, transported 200 persons during two cruises to the Antarctic Peninsula and the South Shetland Islands. The next year the Argentine vessel *Yapeyu* and the Chilean vessel *Navarino* each provided tourist voyages to the Antarctic. The commercial tourism opportunities originally envisioned by Cook, Stenhouse, Worsley and Mawson were now realized. As Sir Edmund Hillary wrote:

> When we built Scott Base on Ross Island in 1957, we could not possibly have imagined that tourist icebreakers would be regularly visiting it by the end of the

second millennium. Now that it is more readily accessible, I recommend an
Antarctic trip to anyone.

(McGonigal and Woodworth, 2001)

Much like the Arctic tourism that preceded it, Antarctica's tourism
experience is predominantly characterized by exploration and discovery. In
1966 Lars Eric Lindblad applied the Arctic's successful tourism formula of
polar exploration led by famous explorers to create a visitation experience of
exploratory cruising and shore excursions in Antarctica. The unique visitor
experiences created by that formula have been replicated ever since. Antarctic
tourism perpetuates a tradition of polar tourism that, from its inception in the
1860s, made scientific inquiry a vital part of the polar tourism experience.

Tourism in the Antarctic is increasingly popular. Visitation has grown as a
result of extensive publicity, improved transport technologies and the
discovery of new attractions. Shackleton's legacy continues to capture media
attention, and although it has no resident population, the Antarctic receives
an inordinate amount of publicity due to the world's interest in global
warming. Russian icebreakers, commercial cruise ships and charter boats now
provide marine-based Antarctic tours. Air-borne tourists obtain views from
high-altitude overflights, short-term visits via helicopter journeys, and
backcountry adventures made possible by the 'blue-ice runways' courageously
pioneered by Antarctic Network International. Modern land, sea and air
transport to the Antarctic increasingly provides more access, new visitor
experiences and larger tourism capacities.

Summary and Conclusions

The features that uniquely characterize the polar tourism experience remain
steadfastly rooted in two centuries of history. Those characteristics include the
employment of famous explorers and scientists as guides, exploratory shore
excursions, educational lectures provided in transit and ashore, and
extraordinary visitor comfort amid some of the Earth's most remote
wilderness. In summary, the theme of discovery continues to permeate both
the Arctic and Antarctic tourism experience, and perhaps no other type of
commercial tourism so strongly and consistently relies upon historical
traditions to grow and sustain its business. To a very considerable extent, the
prospects for polar tourism will be a reflection of its past.

References

Armstrong, T.E. (1991) Tourist visits to the North Pole, 1990. *Polar Record* 27(161), 130.
Ballou, M.M. (1890) *The New Eldorado: A Summer Journey to Alaska*. Houghton Mifflin and
     Co., The Riverside Press, Cambridge, Massachusetts.
Barr, W. (1980) The first tourist cruise in the Soviet Arctic. *Arctic* 33(4), 671–685.

Beardslee, L.A. (1882) *Reports of Captain Beardslee, US Navy, Relative to Affairs in Alaska and Operations of the USS Jamestown Under his Command While in the Waters of that Territory.* Senate Document Number 71. US Government Printing Office, Washington, DC.

Bohn, D. (1967) *Glacier Bay; The Land and The Silence.* Sierra Club, San Francisco, California.

Brendon, P. (1991) *Thomas Cook: 150 Years of Popular Tourism.* Secker & Warburg, London.

Burr, A.R. (1919) *Alaska – Our Beautiful Northland of Opportunity.* The Page Company, Boston, Massachusetts.

Burroughs, J. (1904) *Far and Near.* Houghton Mifflin and Co., New York, New York.

Capedevila, R. (1984) Las primeras mujeres que estuvieron en las Orcadas fueron Argentina, *Artartida* 13, 34.

Cohen, S. (2002) *The White Pass and Yukon Route: A Pictorial History.* Pictorial Histories Publishing Company, Missoula, Montana.

Coles, J. (1882) *Summer Travelling in Iceland.* John Murray, London.

du Chaillu, P.B. (1881) *The Land of the Midnight Sun – Summer and Winter Journeys Through Sweden, Norway, Lapland and Northern Finland*, 2 vols. Harper & Brothers, New York, New York.

Dufferin, Lord (1873) *A Yacht Voyage Letters from High Latitudes: Being some account of a voyage, in 1856, in the schooner yacht 'Foam', to Iceland, Jan Mayen, and Spitzbergen.* John Murray, London/Lovell, Adam, Wesson & Co., New York, New York.

The Explorers Journal (Spring 2004) Out of the Arctic a club is born. Personal communications with the Archivist of The Explorers Club.

Hayes, I.I. (1871) Across the Arctic Circle. *Harper's Weekly Supplement* 7 January, 19–22.

London, J. (1903) *The Call of the Wild.* Grossett & Dunlop, New York, New York.

McGonigal, D. and Woodworth, L. (2001) *Antarctica and the Arctic: The Complete Encyclopedia.* Firefly Books, Willowdale, Ontario.

Muir, J. (1915) *Travels in Alaska.* Houghton Mifflin, Boston, Massachusetts.

Muir, J. (1981) *The Discovery of Glacier Bay.* 1895 Century Magazine Reprint. Outbooks, Golden, Colorado.

Murray, J. III (1893) *Handbook for Travellers in Denmark, with Schleswig and Holstein and Iceland.* John Murray, London.

Murray, J. IV (1919) *John Murray III.* John Murray, London.

Rand McNally (1922) *Guide to Alaska and Yukon.* Rand McNally & Co., Chicago, Illinois.

Runte, A. (1984) *Trains of Discovery – Western Railroads and the National Parks.* Northland Press, Flagstaff, Arizona.

Scidmore, E.R. (1885) *Alaska: Its Southern Coast and the Sitkan Archipelago.* D. Lothrop and Co., Boston, Massachusetts.

Scidmore, E.R. (1896) *Appleton's Guide Book to Alaska and the Northwest Coast.* D. Appleton and Co., New York, New York.

Service, R. (1907) *The Spell of the Yukon and Other Verses.* Barse and Hopkins, New York, New York.

Snyder, J. and Shackleton, K. (2001) *Ship in the Wilderness: Voyages of the MS Explorer Through the Last Wild Places on Earth.* Gaia Books Ltd, London.

Wharton, D. (1972) *The Alaska Gold Rush.* Indiana University Press, Bloomington, Indiana.

White Pass & Yukon Route Railroad (2006) Press releases of 26 September 2005 and 19 June 2006. White Pass & Yukon Route Railroad, Skagway, Alaska.

3 Polar Tourism in Changing Environments

BERNARD STONEHOUSE[1] AND JOHN M. SNYDER[2]

[1]*Scott Polar Research Institute, University of Cambridge, Lensfield Road, Cambridge CB2 1ER, UK*; [2]*Strategic Studies, Inc., 1789 E. Otero Avenue, Centennial, CO 80122, USA*

Introduction

Environmental change is a familiar concept in our understanding of earth processes. Some changes are slow, others rapid: some recent changes have occurred more rapidly than earlier stasis led us to expect. This chapter defines the polar regions geographically and discusses recent changes in polar ecosystems, both man-made and natural, that have occurred within the past human generation. We describe specific ways in which changes are likely to affect aspects of polar tourism, including lengths of season, problems faced by wildlife, management issues in protected areas, changing perceptions of the tourists themselves, and impacts of change on the native communities that are a major attraction for Arctic tourists. Environmental changes cannot be ignored: the future of polar tourism depends largely on how its managers will respond to recent and ongoing changes in both polar regions.

Polar Boundaries; Polar Regions

Polar regions are defined for different purposes by different circumpolar boundaries:

- Geographers and administrators tend to favour the polar circles, at approximately 66°32' north and south.
- Climatologists favour isotherms (lines joining points of equal mean temperature). In Köppen's climatic classification, climates in which the mean temperature of the warmest month does not exceed 10°C (50°F) are polar: thus the 10°C isotherm for the warmest month (normally July in the northern hemisphere, January in the south) limits the polar climatic regions.

- Ecologists prefer boundaries between recognizable ecosystems or prominent plant or animal communities. For the Arctic, often-used boundaries are the tree-line (the northern limit beyond which trees are stunted or absent) and the closely linked transitional forest–tundra zone. For the Antarctic, the Antarctic Convergence or Polar Front is generally used – a circumpolar line in the oceans where cold Antarctic surface waters meet and sink below slightly warmer and more saline sub-Antarctic waters.

These boundaries appear in Figs 3.1 and 3.2; for fuller discussion of their origins and significance *see* Stonehouse (1989: 9–11).

The polar circles have no direct climatic or ecological significance, but are valuable for comparative purposes. Equidistant (2606 km, 1619 miles) from their poles, they enclose equal areas (40.3 million sq km, 15.8 million sq miles) of the Earth's surface. The Antarctic Circle is mainly maritime, bisecting the Antarctic Peninsula but otherwise passing almost entirely through fringing

Fig. 3.1. The Arctic region. (Source: Stonehouse, 1989.)

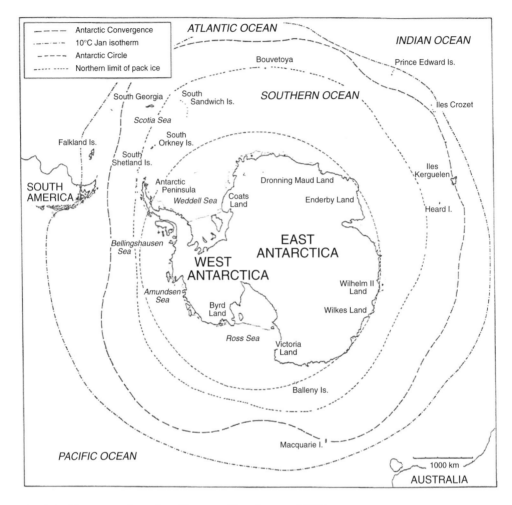

Fig. 3.2. The Antarctic region. (Source: Stonehouse, 1989.)

polar seas. Within it lies an ice-bound desert continent with no trees, shrubs, ground cover or permanent human habitation. The Arctic Circle by contrast is mostly land-based, ringing forests, tundra, farmlands, cities, industrial complexes and settled human populations of approaching four million inhabitants (Stonehouse, 1990: 18).

In the present context of changing polar climates and ecosystems, climatic and ecological boundaries provide more relevant bases for comparison. The 10°C summer isotherm for the Antarctic encloses a much larger region than that for the Arctic: latitude for latitude the northern hemisphere is warmer than the southern. In the north the limiting isotherm follows reasonably closely the tree-line, in the south the Polar Front. Isotherms too are better indicators of latitudinal variations in temperature, showing for example the influence of the North Atlantic Drift in bringing anomalously warm conditions to northern Scandinavia, and the immense pool of winter cold characterizing central Siberia.

However defined, the two polar regions agree in being cold, windy and (except in coastal regions) generally arid. Snow is more plentiful than rain: water surfaces and ground remain frozen for much or all of the year. Coasts are generally milder than their hinterlands, prompting climatologists (e.g. Shear, 1964: 310) to distinguish marine and continental categories within Köppen's polar category.

The fact that polar regions are currently much colder than sub-polar and temperate regions is an anomaly in world history. We live in an ice age in which both polar regions support persistent ice caps, both terrestrial and marine, which expand during periods of world cooling and contract during warmer spells. Present time is a period of warming, affecting the world in general and polar regions in particular. The warming may have been initiated, and has almost certainly been intensified and accelerated, by human-induced liberation of carbon dioxide into the atmosphere from the mid-19th century onwards.

Polar cold and its consequences have protected both polar regions from the spread of humanity. Humans seem physiologically better adapted for warmer rather than colder climates; few human populations are indigenous to the Arctic, and relatively few entrepreneurs have sought their fortunes within polar boundaries. However, currently several thousands of recreational travellers seek polar regions each year for their holidays; tourism is growing in both the Arctic and the Antarctic. If polar environments are indeed changing, what are the changes, and how are they likely to affect the tourist industry?

What Kinds of Change?

Two kinds of environmental changes are considered here: (i) those unequivocally induced by man, exemplified by despoliation of animal and plant communities through hunting, mining and other commercial activities; and (ii) those due primarily to cosmic events, exemplified by radical climatic changes, possibly triggered and intensified by human activities.

Human influences include tourism, which has gained an unfortunate reputation for despoliation throughout the world and cannot be excluded as an actual or potential source of damage to polar regions. The degree to which polar regions have been affected by tourism is discussed below.

Changes due to human activities

Human intrusions and exploitation in polar regions have included:

- Long-term use of the Arctic by indigenous populations.
- Fur trapping for non-indigenous markets throughout the Arctic tundra and sub-Arctic forest regions.

- Whaling and sealing for oil, baleen and skins, including walrus hunting for ivory.
- Commercial fishing.
- Extraction of minerals, including ores and hydrocarbons.
- Establishment of military and scientific stations.

While the first of these involved only local areas and populations in the Arctic, the rest have been due to intrusions from outside the indigenous populations. Fur trapping for southern markets was the original motive for colonization of much of the Arctic by southern cultures. Commercial whaling and sealing drew on maritime ecosystems at both ends of the world (Fig. 3.3), from which some stocks have never fully recovered. Commercial fishing, both controlled and clandestine, continues to do so today. Mineral extraction has so far been limited to the Arctic, being specifically proscribed in Antarctica under the Antarctic Treaty System. Both regions have been affected by the presence of intrusive long-term military and scientific stations, particularly during the second half of the 20th century when the Cold War dominated the north and both politics and science invaded the south.

Exploitation by small indigenous Arctic populations appears to have been sustainable. Exploitation for massive southern markets was not. Fur trapping,

Fig. 3.3. Commercial whaling and sealing in the 18th to 20th centuries devastated stocks in both polar regions. (Source: Palmer archive, Cambridge, UK.)

whether by settlers or by native hunters working for them, was market-led. So in turn were southern sealing, whaling and fishing, with little or no reference to sustainability.

No less damaging have been more recent forms of exploitation, particularly in the former Soviet Arctic. Perhaps its most drastic change was diversion of water resources. Some 80–85% of Siberian river waters flow northwards into the Arctic and North Pacific oceans, away from the main Soviet centres of population and industry. River diversion from the 1930s onwards redirected immense quantities of fresh water to arid regions in the south, where it is used in agriculture, mining and manufacturing. Goldman (1972) quoted the contention of Hubert Lamb, a pioneer investigator into causes of climate change, that large-scale diversion could lead to an overall warming of the northern hemisphere and of the world as a whole. In the early 1980s Lewis (1982) calculated that the three major Siberian rivers Ob, Yenesi and Pechora together contributed over 80% of the total river discharge into the Arctic basin, and Lamb (1962; cited in Goldman, 1972), quoting his own earlier views, expressed fears that:

> Soviet proposals to divert Siberian rivers for irrigation in central Asia may remove the freshwater layer on the surface of the Arctic Ocean and so remove much of the pack ice cover, especially in the Atlantic – European – Kara Sea (north-west Siberian sector).

Whether for this reason, or as a symptom of more radical global warming, Lamb's fears appear to have materialized. Perennial pack ice is diminishing annually in the Arctic basin.

Radical alteration of Siberian river systems has contributed also to loss of habitat for Arctic animals. As Pryde (1991) notes:

> ...a good example of the loss of wintering habitat can be seen in the case of the red breasted goose (*Branta ruficollis*), which breeds along the Arctic Coast of the USSR. In between their tundra breeding seasons, as many as 25,000 of these geese wintered in the wetlands around the Araks and Kura rivers in Transcaucasia as recently as the early 1960s. But as the river became dammed and diverted in the interests of irrigated agriculture, by 1970, the wintering birds numbered only several hundred, and by the mid-1970's only a few dozen (Vinokurov, 1986). The result has been that its total number have declined so sharply in the USSR that it is now listed by Russia as a threatened species.

A further disastrous change induced by the Soviet economy is that the remaining northward-flowing rivers now carry lethal mixtures of hazardous materials to the Arctic. Feshbach and Friendly (1992) record that the Angara River, flowing from the southern end of Lake Baikal to the Arctic Ocean across a distance of 2500 miles, has become:

> ...an aqueduct for poisons. Yearly it carries 257,000 tons of chlorides, 140,000 tons of sulfates, over 30,000 tons of organic wastes, and 10,000 tons of nitrates from factories built in the 1960s and 1970s along its banks.

Similarly Zelikman (1989) reported that 'most of the rivers flowing into the Barents Sea are catastrophically polluted due to the operation of ore and chemical industries'. The presence in Siberia of large industrial, mineral and energy enterprises, together with military installations and testing grounds, produces water-borne contaminants that either flow directly into the Arctic and Pacific oceans, or are leached into the soil. Given the collapse of the Soviet Union and Russia's dedication of resources to other purposes, these problems seem unlikely to be solved in the near future.

Similar but lesser and more local problems in the North American Arctic are associated with mining and Cold War military installations, now long abandoned and deteriorating to rubbish piles. Attempts to remove them are complicated and made more expensive by the presence of asbestos, fuel oils and other contaminants. Extraction of hydrocarbons continues, again with relatively minor and localized environmental effects.

A further man-made change of environmental significance is Arctic haze – a persistent form of atmospheric pollution that occurs over much of the Arctic basin. First noted in 1956, it is present throughout the year in the lowest 5 km of atmosphere, intensifying every summer to reduce visibility, absorb solar radiation, and leave measurable deposits of aerosol chemicals and particulate pollutants on snowfields. Its origins have been traced to smoke emissions from industrial plants in circumpolar temperate and Arctic latitudes; for further details see individual papers in Stonehouse (1986).

Antarctica has suffered less from industrial and other contaminants originating outside the area, though distance from the rest of the world has not ensured its complete immunity. Chemical insecticides, soil-dwelling, air-borne and sea-borne pathogens, and other pollutants have long been known to be present in Antarctic organisms. Almost every corner of the continent and neighbouring islands has now been visited and to some degree contaminated by man. It would be difficult to sustain, in any technical sense, claims frequently made in official publications that Antarctica remains 'pristine'. It remains only relatively free from man-made pollution and damage.

Changes due to climatic warming

The Arctic Council (a consortium of the eight Arctic nations) commissions reports on changing conditions within the Arctic region. The Arctic Climate Impact Assessment, a recent comprehensive report on climate changes affecting the Arctic and the rest of the world, is available in overview form in both print (ACIA, 2004) and from the website (www.acia.uaf.edu); for a summary see Corell et al. (2004). The report makes the following points: worldwide climatic warming is particularly intense in the Arctic, where mean temperatures have recently risen twice as fast as in the rest of the world. This trend is likely to accelerate during the current century, due to accumulation of greenhouse gases in the atmosphere. The Arctic also receives increased ultraviolet radiation, due to depletion of atmospheric ozone. Warming is evidenced in widespread melting of glaciers, reductions in extent and persistence of sea ice, and of snow and ice cover on land, increasing precipitation, and shorter and warmer winters. Melting of land ice results in rises in global sea level, and may slow oceanic circulation that carries tropical heat poleward.

Likely consequences of warming in the Arctic, generally regarded as deleterious both to the environment and to wildlife and human populations, include:

- Contraction of the region, manifest in poleward migration of the tree-line, with consequent loss of tundra and diminution of cold polar waters.
- Flooding of parts of the tundra due to enhanced river flow, drying-out of other parts, with consequent redistribution of tundra plants and animals, and possible invasions of competitive alien species and pathogens.
- Changes in coasts and coastal features, including increased erosion and loss of traditional terrestrial and inshore marine feeding grounds.
- Retreating sea ice, with consequent environmental challenges to ice-dependent marine mammals (seals, polar bears) and cold-water stocks of whales, birds, fish and planktonic organisms.
- Challenges to indigenous human populations from flooding rivers and thawing permafrost, including disruption of buildings and communications.
- Loss of traditional hunting and fishing grounds on land, in rivers, on pack ice and in the sea, on some or all of which indigenous human communities are at least part-dependent.

Not all the changes are spread evenly throughout the Arctic: the report considers slightly differing scenarios sector by sector. Overall it stresses that many of these changes are already detectable, and all will be considerable before the end of the 21st century.

A second relevant Arctic Council study, the Arctic Marine Strategic Plan (AMSP, 2004), embodies the Arctic Council's intentions to ensure 'a healthy and productive Arctic Ocean and coasts that support environmental, economic and socio-cultural values for current and future generations'. From the plan has arisen an ongoing Arctic Marine Shipping Assessment (AMSA) and a Protection of the Arctic Marine Environment initiative (PAME, 2006), which seeks to quantify current levels of shipping and related environmental impacts and levels projected for the years 2020 and 2050.

There is no southern equivalent to the Arctic Council concerned to generate equivalent studies for the Antarctic, where human activities are on a smaller scale and less critically affected by consequences of warming. Environmentally, Antarctic cold is generally more intense, and evidence of warming, though present, is patchily distributed. The most striking manifestations are retreat of glaciers on peripheral islands (notably South Georgia), break-up of pericontinental ice shelves, particularly in the Peninsula area, and a general diminution or thinning of annual sea ice.

Possible Consequences for Tourism

The main attractions of polar tourism are the differences between polar environments and those of the inhabited, everyday world, coupled with the relative ease with which they can now be reached. Differences of scenery, wildlife, culture and history all play their part in attracting tourists in tens of thousands annually to the Antarctic, and in hundreds of thousands to the Arctic. While prolonged warming may ultimately reduce these differences, the warming predicted within the next two or three generations seems likely to affect only certain popular venues adversely, while enhancing others and opening new opportunities to the tourist industry.

Enhanced tourist access; longer seasons

Historically, polar regions have been among the globe's least accessible regions, due mainly to the presence of sea ice and severe weather. Significant reductions in sea ice, and the moderation of climate conditions, particularly in the Arctic, have helped towards the recent general expansion of tourism operations by: (i) increasing the number of destinations; (ii) lengthening the tourism season; and (iii) allowing longer tourist visits. Improved access now enables tourist ships regular transit of the Northwest Passage, cruises to the North Pole and access to other previously difficult venues in both polar regions. Seasons in which travel is possible start earlier and end later in the year, and winter visits are becoming increasingly feasible. The cumulative impacts of reduced sea ice and a moderating climate will probably result in larger numbers of polar tourists spending more time in more locations.

Wildlife attractions: more or less?

Ecological changes in the Arctic, particularly those associated with diminished sea ice and relatively warmer weather, bring benefits to some polar species and problems for others. Geographical effects include an increase in terrestrial ice-free areas, allowing the polar extension of tundra and polar desert vegetation, and a poleward shift of the tree-line, affecting the viability and distribution of many terrestrial and freshwater plant and animal species. Similar changes may expected at sea, manifest in redistribution of water masses and – most notably – redistribution and reduction of sea ice.

Changes on land are likely to result in the northwards spread of such resident species as snow buntings, hares and musk oxen, and enlarge areas of tundra and forest for migratory reindeer, caribou, moose, brown bears and the many species of migrant wildfowl. Shorelines and coastal areas freed from ice will provide more habitats for both wildlife and plant communities, and contribute more nutrients to inshore waters, encouraging greater local diversity of marine wildlife. As both inland and coastal areas provide more sustenance, larger populations of resident and migratory wildlife may be sustained, with corresponding changes of wildlife migration routes. A greater abundance of wildlife and increased certainty of experiencing these Arctic species could contribute to the growth of both nature-based tourism and sport hunting.

Warming will affect not only recreational opportunities but also the comfort and interest of visitors. Accelerated plant succession resulting from the recession of glaciers and reduction of seasonal ice fields will offer tourists more plants to view, and a dynamic colonization process to witness. Soil and plant cover will warm the environment and provide tolerable conditions over a wider area. To this extent more of the Arctic will welcome visitors. However, damp soil and standing water in place of semi-permanent snow and ice will provide more breeding habitat for the Arctic's notorious biting flies, and drier conditions over the tundra as a whole will offer more opportunities for wildfire in Arctic regions. Thawing permafrost will not only alter vegetative regimes and hydrology (incidentally releasing stored 'greenhouse gases' into the atmosphere), but also make roads more hazardous for both tourists and resource managers. The emergence of these new environmental events will create the need to re-think traditional resource management practices and calculations of potential hazards, in particular wildlife viewing, sport fishing, hunting, wildlife photography, backpacking, kayaking and river rafting recreational activities.

Changes in ocean currents and water masses affect the distribution of plankton, fish, whales, seals and seabirds, which are all interdependent and to some degree influenced by the annual cycle of sea ice. Radical changes in sea ice distribution will affect the breeding of ice-dependent seals and Arctic foxes, the feeding of whales, and the annual movements of fish stocks and other species. Some species will become more vulnerable, some may alter migration routes, others may thrive on increased nutrients and prey species. These modifications will be reflected in related human activities, including subsistence hunting of native peoples, commercial harvesting and tourism opportunities.

Polar bears (Fig. 3.4), which depend on inshore sea ice for their winter survival, are one species most apparently at risk from warming. According to Stirling's extensive research on the polar bear and the ecological integrity of its Canadian Arctic habitat:

> There is a significant positive relationship between the time of breakup and the condition of adult males and females (i.e., the earlier the breakup, the poorer the condition of adult males and females) ... Ultimately, if sea ice disappeared altogether, polar bears would become extinct.
>
> (Stirling and Derocher, 1993; Stirling, 2004)

Demise of the polar bear, an icon of the Arctic world, would be a tragic loss in itself, and would diminish the polar experience for many hundreds of Arctic visitors. Particularly affected would be those who come north mainly to meet the species in such locations as Churchill, Manitoba and Wager Bay, Nunavut.

Both marine and freshwater fish stocks, to which the lucrative Arctic sport fishing market is sensitive, would be strongly influenced by climatic changes.

Fig. 3.4. Polar bears, here wandering 50 miles from the North Pole, rely on sea ice for seasonal feeding. (Photo: B. Stonehouse.)

Adverse changes in populations, geographic distribution and seasons of fish species sought by the sport angler would have a considerable impact on Arctic economies. Conversely, reduced sea ice might improve access to fisheries and boost this form of tourism.

Wildlife habitat changes will necessitate re-consideration of wildlife management practices by Arctic jurisdictions. Land-based recreation activities such as wildlife viewing, photography and sport hunting will be subject to new management practices. Likewise, fisheries management will need to change rules and regulations regarding allowable sport fishing activities, seasons and catch limits. Necessary changes in international wildlife management cooperative agreements, such as those for polar bears, migratory birds, commercial fishing and the conservation of marine mammals, may further affect tourism activities.

Antarctic ecology is relatively simpler, though neither the extent nor the consequences of climatic changes are correspondingly easier to predict. Already substantial losses of shelf ice from the flanks of the Antarctic Peninsula and other coasts have opened channels to tourist ships that were hitherto ice-filled (Crosbie and Splettstoesser, 1997), and more new routes will become available as warming proceeds. Changes have been detected in the relative breeding success of different species of penguins, ascribable to instability of winter sea ice in the Bellingshausen Sea (Patterson *et al.*, 2003). Climatic amelioration may ease environmental constraints on all bird and mammal species, allowing breeding seasons to start earlier and end later each year, and reducing the likelihood of mid-season breeding disasters due to unseasonable bad weather. Constraints may be relaxed on flora and soil microbiota: the spread of maritime climatic conditions from Antarctic Peninsula and Scotia Arc to ice-free continental coasts might encourage a progression from lithosols and regosols to brown soils, and from desert and semi-desert cryptogamic flora to the richer vegetation of Antarctic fringe islands. For discussions of Antarctic coastal soil processes and flora, see individual papers in Beyer and Bölter (2002).

Interesting though these changes will be for scientists, they impinge only slightly on Antarctic tourism as it is currently practised, generally relieving constraints of time and locality imposed by severe climatic conditions and favouring future developments. As for the Arctic, there is little evidence that either man-made or natural changes will diminish tourism. Practically all pointers indicate opportunities for increase.

Transformation of protected areas

Climate warming, with accompanying reductions in sea and land ice and shifts in ecological zones, is rendering substantial changes in the Arctic's protected areas. National parks, wildlife refuges, wilderness areas, World Heritage Sites and marine sanctuaries were established with the explicit intention of conserving unique environmental resources. Several are now increasingly challenged to protect the elements of landscape for which they were designated, and will probably have to re-evaluate their purposes. For example, Glacier Bay National Park, the USA's largest marine park, was created for the purpose of 'providing the opportunity afforded here for the scientific study of glacial action'. Current reality is that most of its tidewater glaciers are in rapid retreat, and resource management objectives, indeed the whole reason for the park's existence, will need to be re-considered.

Staple and Wall (1996) draw attention to similar problems affecting Canada's national parks – notably Nahanni National Park – where climatic warming is transforming vegetation, modifying hydrologic cycles, lengthening water sport seasons, increasing wildfire threats, and altering the numbers, species and migratory patterns of wildlife that visitors come to see. Hunting and fishing regulations for the areas will need re-evaluation and neighbouring communities are seeking to increase economic benefits from longer seasonal use of the parks.

Visitor perceptions

Tourists' perceptions of polar environments have changed radically during the past century. The early 1900s' image was one of hostility, reinforced by contemporary accounts of the perils and hardships of polar exploration. This has largely been replaced by the belief that polar regions offer some of the safest tourist destinations in an increasingly dangerous world.

Media attention to environmental changes in the polar regions may be contributing to a growing interest in polar travel. News releases, televized documentaries, magazine articles and popular films featuring wilderness, abundant wildlife, splendid scenery on one hand, and threats from global warming on the other, either way provide polar tourism with priceless advertising and increased bookings. Obviously all ecological systems throughout the world are changing, but perceptions of dramatically rapid change in polar regions, constantly stressed by the media, may help to increase public interest.

Environmental hazards

Attention has already been drawn to man-made pollution in both polar regions. Wittingly or unwittingly, tourists may add to pollution by trampling, leaving litter or introducing alien species from plant material carried on their clothing. Climatic warming may encourage the survival of plant propagules (seeds, spores, etc.) that would previously have been destroyed by cold and aridity. A standard prophylaxis, now applied on Antarctic cruise ships, is to ensure that boots are washed (and ideally, clothes well brushed) before going ashore.

Conversely, tourists themselves are subject to greater hazards from existing pollution by encountering contaminants that were previously off-limits. Such historic sites as abandoned canneries, DEW-line stations, mining installations and World War II camps in the north, and explorers' huts, derelict whaling stations and scientific bases in the south, may contain asbestos and other currently proscribed building materials, chemicals used to process ores, caustic cleaners, machinery lubricants, pesticides and other contaminants that pose health threats, if only to visitors who defy guidelines by interfering with them. The most serious risk arising from this particular hazard may be the threat of expensive litigation between tour operators and disaffected clients.

Cultural resources

The Arctic environment is the setting for its indigenous peoples, containing the vital resources on which their livelihoods and cultures depend. Climate changes and their consequences are of critical importance to the cultural and economic well-being of Arctic peoples. Changes are happening quickly and indigenous peoples are displaying a sense of urgency to find a response. Sheila Watt Cloutier, International Chair, Inuit Circumpolar Conference, passionately articulated those concerns when she stated:

> What is at stake here is not just the extinction of animals but the extinction of Inuit as a hunting culture. Climate change in the Arctic is a human issue, a family issue, a community issue, and an issue of cultural survival.

(Pegg, 2004)

Commenting on recent changes in sea ice distribution, Alaskan Native Elder Warren Matumeak commented:

These changes I and other indigenous people see, can be perceived as positive and negative. The Inupiaq welcome the warmer temperatures but do not appreciate the lack of multiyear pack ice and the increased difficulty of whaling these temperatures bring. The subsistence way of life has to adapt to the environment. The world is changing and I and other indigenous people are bearing witness.

(Matumeak, 2004)

Fast ice permits travel along the shore by dog sled or snow machine, enabling people to circumvent mountain travel and hazardous sea routes. It provides also access to seals, polar bears, whales and fish that sustain traditional ways of life and value systems. George Porter, Inuit leader of the community of Gjoa Haven, King William Island, in the Northwest Passage, tells of a danger arising from increasing use of their local waterway:

You know for Inuit people the land and the water are the same thing – here the sea is frozen over for most of the year. So to us driving a ship through the ice is like driving a bulldozer across a field with the blade down … A few years ago a groups of hunters from Arctic Bay to the north of here were out on the ice miles from home hunting seals. Without knowing they were there, a Canadian Coast Guard icebreaker cut a lane between them and the village. They were stranded for several days until the ice closed up again. If it hadn't those men could have died.

(Bockstoce, 1990)

Any unwarranted interference with sea ice may place a community at risk. Increasing use of the Northwest Passage by tourist ships could well endanger traditional ways of life, and indeed the continuing existence of native communities in the maritime Arctic.

Does Tourism Change Polar Environments?

The presence of an already large and rapidly expanding industry in wilderness or semi-wilderness environments raises the question of how much the industry's own activities contribute to environmental changes. Tourism, particularly mass tourism involving tens of thousands of visitors, has an unenviable reputation for riding roughshod in sensitive areas: has it already left its mark on the polar regions?

There appears to be general agreement that the hundreds of thousands of tourists who have visited Antarctica in the half-century of commercial tourism have left surprisingly few traces. Stonehouse and Crosbie (1995) attribute this squarely to the benign pattern of shipboard tourism pioneered by Lars-Eric Lindblad, and since maintained by almost every cruise operator. The 'Lindblad pattern' involves landings by not more than 100 passengers at a time, who remain ashore for 2–3 h accompanied by well-informed guides. On board there is strong indoctrination in the conservation ethic, based on guidelines issued by the International Association of Antarctica Tour Operators (IAATO), which Lindblad co-founded. (For details of IAATO see Chapter 12, this volume.) In consequence visitors avoid walking on vegetation, disturbing nesting birds, leaving litter and otherwise damaging the environment. Stonehouse and Crosbie (1995: 222) wrote:

This pattern of education ... has recommended itself strongly to the kinds of tourists who have so far made up the majority in Antarctica. Many claim that they would avoid tours which did not feature similar levels of concern. As a consequence of the Lindblad pattern, in an environment that ... many regard as hypersensitive to visitor impact of any kind, there is so far very little evidence of damage from tourism ... The Lindblad pattern of tourist management has ensured high standards of behaviour among tour operators and tourists alike, and far less environmental damage than might have been expected had Antarctic tourism developed without it.

Cruise touring still accounts for more than 90% of Antarctic tourism (Chapter 17, this volume), and environmentally sensitive behaviour still prevails among this majority of tourists. In the Arctic the greater variety of tourism leads to wider possibilities for environmental damage, but it is probably no less true that those who visit the Arctic are likely to be concerned for maintaining its integrity. At neither end of the world is there evidence that tourism brings about environmental changes that match either the natural or man-made changes discussed above.

The most severe tourist-induced changes occurring in the Arctic are probably those affecting indigenous human communities. Specifically, existing climatic and cultural changes impose risks on native cultures, and also bring benefits to them. An expanding tourism industry brings similar risks and benefits, which further complicate an already complex situation. The ability of the Arctic's indigenous peoples to survive has always depended on their capacity to adapt to change. Now, as they witness the changes wrought by reduced Arctic sea ice, and the influx of tourists that is at least partly consequential upon it, they face challenges that again put their cultures and livelihoods at serious risk. Should they be forced to abandon their communities, fewer opportunities will exist for tourists to experience cultural traditions, and for tour operators to prosper. If for no other reason, it is very much in the operators' interests to regard the problems of native communities as their own problems, and take care to plan their operations accordingly.

Summary and Conclusions

Changes during the last two centuries, both man-made and natural, have radically affected and continue to influence both polar regions, in particular their wildlife and human ecology. In relation to tourism, continuing shifts in polar ecological systems are likely to result in both gains and losses of tourism attractions and amenities. Accurately discerning how tourism is affected by these processes, and how tourism itself contributes change, is essential for understanding how the industry should be managed in the future. Impacts induced by the industry itself are small compared with other changes described, ranging from near-negligible (for example, viewing scenery from cruise ships) to potentially injurious (ill-planned visits to indigenous settlements and wildlife sites). Future problems for polar tourism are more likely to be generated by growth of the industry facilitated by environmental changes, than by the changes themselves.

References

ACIA (2004) *Impacts of a Warming Arctic; Arctic Climate Impact Assessment*. Cambridge University Press, Cambridge, UK.

AMSP (2004) *The Arctic Marine Strategic Plan*. Arctic Council, Reykjavik.

Beyer, L. and Bolter, M. (eds) (2002) *Geoecology of Antarctic Ice-Free Coastal Landscapes*. Springer-Verlag, Berlin.

Bockstoce, J. (1990) Northwest Passage. *National Geographic Magazine* 178(2), 2–33.

Corell, R.W., Prestrud, P. and Weller, G. (2004) International assessment enumerates climate change impacts across the Arctic. *Witness the Arctic* 11(2), 1–3.

Crosbie, K. and Splettstoesser, J. (1997) Circumnavigation of James Ross Island, Antarctica. *Polar Record* 33(187), 341.

Feshbach, M. and Friendly, A. Jr (1992) *Ecocide in the USSR. Health and Nature Under Siege*. Basic Books/HarperCollins Publishers, New York, New York.

Goldman, M.I. (1972) *Environmental Pollution in the Soviet Union; the Spoils of Progress*. MIT Press, Cambridge, Massachusetts.

Lewis, E.L. (1982) The Arctic Ocean: water masses and energy exchanges. In: Rey, L. (ed.) *The Arctic Ocean: The Hydrographic Environment and The Fate of Pollutants*. Macmillan, London, pp. 43–68.

Matumeak, W. (2004) Local observations of recent change. In: *Arctic Forum Abstracts 2004*. Arctic Research Consortium of the US, Fairbanks, Alaska, p. 16.

PAME (2006) *Protection of the Arctic Marine Environment: Progress Report, October 2006*. PAME International Secretariat, Akureyri, Iceland.

Patterson, D.L., Easter-Pilcher, A. and Fraser, W.R. (2003) The effects of human activity and environmental variability on long-term changes in Adelie penguin populations at Palmer Station, Antarctica. In: Huiskes, A.H.L., Gieskes, W.W.C., Rozema, J., Schorno, R.M.L., van der Vies, S.M. and Wolf, W.J. (eds) *Antarctic Biology in a Global Context*. Backhuys Publishers, Leiden, The Netherlands, pp. 301–307.

Pegg, J.R. (2004) *The Arctic: Earth's Early Warning System*. AlterNet web page; available at http://www.alternet.org/story/19930/ (accessed December 2006).

Pryde, P.R. (1991) *Environmental Management in the Soviet Union*. Cambridge Soviet Paperbacks: 4. Cambridge University Press, Cambridge, UK.

Shear, J.A. (1964) The polar marine climate. *Annals of the Association of American Geographers* 54(3), 310–317.

Staple, T. and Wall, G. (1996) Climate change and recreation in Nahanni National Park Preserve. *The Canadian Geographer* 40(2), 109–120.

Stirling, I. (2004) Polar bears, seals, and climate change in Hudson Bay and the High Arctic. In: *Arctic Forum Abstracts 2004*. Arctic Research Consortium of the US, Fairbanks, Alaska, p. 21.

Stirling, I. and Derocher, A.E. (1993) Possible impacts of climatic warming on polar bears. *Arctic* 46(3), 240–245.

Stonehouse, B. (ed.) (1986) *Arctic Air Pollution*. Cambridge University Press, Cambridge, UK.

Stonehouse, B. (1989) *Polar Ecology*. Blackie, London.

Stonehouse, B. (1990) *North Pole South Pole: a Guide to the Ecology and Resources of the Arctic and Antarctic*. Prion, London.

Stonehouse, B. and Crosbie, K. (1995) Tourist impacts in the Antarctic Peninsula area. In: Hall,
    C.M. and Johnston, M.E. (eds) *Polar Tourism: Tourism in the Arctic and Antarctic
    Regions.* John Wiley & Sons, Chichester, UK, pp. 217–233.
Vinokurov, Y. (1986) The Katun River's 'cheap' kilowatts. *Pravda* 1 December [as translated in
    CDSP (1986), 38(48), 4–5].
Zelikman, L. (1989) A large scale ecological disaster is threatening from the north. *Environ-
    mental Policy Review* 3(2), 1–8.

II

Economic Roles of Polar Tourism: Introduction

JOHN M. SNYDER

Tourism is producing substantial economic impacts in both polar regions. Arctic jobs, household incomes and government revenues rely increasingly on the tourism industry. In the Antarctic, to the amazement of many and consternation of some, tourism has become the continent's largest human activity, yielding profits to tour operators and service industries in gateway ports, but as yet none to benefit the continent or its governance. The chapters in this section offer insights into ways that economic activities and development decisions are affecting both the Arctic and Antarctic.

Historically, Arctic natural resources sustained small nomadic human populations at subsistence levels. But since their discovery by European and American enterprises, the Arctic's whales, seals, fish, mineral ores and hydrocarbons have been extracted industrially with little regard for consequences. Antarctic waters have been stripped successively of whales and seals. Defence facilities throughout the Arctic and scientific stations in both polar regions have left legacies of waste materials that are only slowly being absorbed or tidied up. For readers new to polar regions, this is the setting within which polar tourism has developed.

The regions' newest industry – tourism – involves governments, local communities, foreign businesses, international gateway cities and of course the tourists themselves, all playing economic roles that incur costs and derive benefits. Judgements on whether polar tourism itself is good or bad – and for whom or what – depend on an understanding of how costs and benefits are distributed among the stakeholders, including the polar regions themselves. These issues are explored in the present section, 'Economic Roles of Polar Tourism'.

In Chapter 4 John Snyder presents the distinguishing features of the several polar tourism markets, including mass tourism by cruise ships and commercial air transport, sport fishing and hunting, nature tourism, adventure tourism, and culture and heritage tourism. Among symptoms of growing polar tourism he includes constantly improving transportation access, the opening of huge, previously restricted regions in the Russian Arctic, strong promotional efforts of Arctic governments and a growing demand among tourists for safe destinations.

Chapter 5 presents three case studies of developing tourism in rural Arctic communities, by five authors – Henry Huntingdon, Mike Freeman, Bill Lucey, Grant Spearman and Alex Whiting – all of whom have direct personal involvement. Based on their collective experience, the authors stress the key importance of willing involvement by the local people, willingness of governments, people and operators to seek compromises, and effective forward planning. They conclude that tourism in rural Alaska is there to stay, and that it can, by heeding lessons from the recent past, be managed to provide economic benefits and visitor opportunities without undue disruption.

Similar forces are operating in Arctic Canada. In Chapter 6, based on a quarter-century of work with the Nunavut people in the Canadian Arctic, Mike Robbins traces tourism development at the time when the Nunavut had recently attained self-government, again providing valuable lessons in achieving economic self-sufficiency. The author concludes that the future for tourism in Nunavut is promising, but continuing community and government recognition and support will be required to ensure that, 25 years hence, cultural ecotourism plays an integral role in sustainable community economies and Inuit cultural preservation throughout Nunavut and other northern provinces.

In Chapter 7 John Snyder describes the economic roles that growing and diversifying tourism are playing in the economies of each of the Arctic's eight sovereign nations. The author outlines each nation's separate approach to Arctic tourism development. All are to one degree or another seeking to expand tourism within their Arctic sectors, in attempts to reduce chronic south-to-north drains on national or provincial exchequers, and effectively to increase the role of tourism in the economies of their native peoples. The author illustrates both the significant economic role that tourism is playing, and the development pressures it exerts, throughout the entire Arctic region.

Chapter 8, by Esther Bertram, Shona Muir and Bernard Stonehouse, examines the roles of six southern ports – Ushuaia (Argentina), Punta Arenas (Chile), Stanley (Falkland islands), Cape Town (South Africa), Hobart (Tasmania) and Christchurch (New Zealand) – that have become involved the development of Antarctic tourism. The ports through which tourists pass on their way to and from Antarctica, provide the services that keep the ships, aircraft and passengers moving for three to five months each year – some benefiting only marginally, others flourishing as a direct result of the new industry and in consequence making every effort to promote the source of their new-found prosperity. Their economic development policies, including substantial investments to improve infrastructure and tourist services, cannot fail to contribute to the growth of Antarctic tourism.

4 The Polar Tourism Markets

JOHN M. SNYDER

Strategic Studies, Inc., 1789 E. Otero Avenue, Centennial, CO 80122, USA

Introduction

Based on more than 150 years of commercial activity, polar tourism is now a mature industry that provides diverse attractions in both polar regions. Expanding numbers of attractions, recreational activities, international destinations, visitor accommodations and convenient modes of travel are enticing an increasing clientele. And now that regularly scheduled excursion travel is provided to both regions, year-round polar tourism has become a reality.

This chapter examines five sectors of commercial tourism that at present dominate and provide variety within the polar market. Not surprisingly these have reached more advanced development in the Arctic, which tourists have visited for almost two centuries, than in the Antarctic where tourism is approaching its first half-century. The chapter concludes that the distinctions that characterize each of these markets must be accurately understood in order to create appropriate and relevant resource management and community development responses.

Polar Tourism Markets

Polar tourism cannot be characterized as a single, monolithic market. The industry is growing and expanding for an obvious reason: that it appeals to tourists who are willing to pay for the unique experiences it offers. Less obvious is the fact that its appeal is to several distinct, highly specialized market segments. The diverse attractions of polar regions appeal to a no-less-diverse clientele.

The five market segments identified in this chapter are best defined in terms of their primary attractions, an approach to classifying tourist markets that explicitly acknowledges the high expectations of the tourist and the service delivery methods used to realize those expectations:

1. The *mass market*, comprised of tourists primarily attracted to sightseeing within the pleasurable surroundings of comfortable transport and accommodations.

2. The *sport fishing and hunting market*, with participants who pursue unique fish and game species within a wilderness setting.

3. The *nature market*, consisting of tourists who seek to observe wildlife species in their natural habitats and in the solitude of natural areas.

4. The *adventure tourism market*, providing a sense of personal achievement and exhilaration from meeting the challenges and potential perils of outdoor sport activities.

5. The *culture and heritage tourism market*, a very distinct market comprised of tourists who want to experience personal interaction with the lives and traditions of native people, learn more about a historical topic that interests them, or personally experience historic places and artefacts.

Each of these markets has its own distinguishing visitor experiences and economic dimensions, involving different tourists' motivations, expectations, on-site behaviour and resource uses. Sophisticated tourist industries have evolved to provide each market with travel and support services, equipment, clothing, transport and accommodations. Each is energized by promotional campaigns and specialized publications dedicated to sustaining special interests and active involvement.

Tourists themselves are certainly not constrained by this classification: they participate freely in as many types of activity as they wish. However, distinguishing these markets provides a useful organizational framework for better understanding polar tourism in terms of economic activity, visitor behaviour and resource uses.

Mass Tourism

Mass commercial tourism involves group travel provided by tour and transport companies. Participants desire to experience beautiful sights, new territories and different cultures while travelling comfortably and safely. The key to economic success in mass marketing was expressed by Thomas Cook in the 1850s: 'The largest profits come from intensive use by the greatest number of people at the lowest cost' (Brendon, 1991).

The invention of steamships and steam locomotives heralded mass tourism throughout the world. Mass polar tourism began in the mid-1800s when cruise ships first steamed to Arctic destinations. Since then continuously modernized ships, augmented by trains and commercial jet aircraft, have enabled mass tourism to grow and expand throughout the Arctic. Luxurious cruise ships navigate Norway's coastal fjords and many other attractive destinations throughout the Scandinavian Arctic. Canadian ports serve hundreds of large and small cruise ships travelling to the North Pacific, North

Atlantic and Arctic oceans, and Canadian railroads convey tourists to numerous inland attractions. Iceland's popularity as a cruise venue is steadily increasing, and Greenland is aggressively seeking to accommodate increased numbers of cruise tourists. Russian icebreakers routinely each summer take tourists to the North Pole and the Barents and Kara seas (Armstrong, 1991; Brigham, 2000) (Fig. 4.1). Large and small cruise ships, including icebreakers, regularly visit the South American sector of Antarctica (Fig. 4.2).

Cruise ships

As the single largest provider of mass tourism in the polar regions, the cruise-ship industry has a huge economic role. In 2004, the most recent year for which complete data are available, more than 1.2 million passengers travelled to polar destinations aboard cruise ships. Based on a review of rate schedules for the major cruise lines in 2004, passengers paid between $2000 and $20,000 each for their cruises. In addition, each person spent an additional $82 per port visit (International Ecotourism Society, 2004). Shore-based travel by train and motor coach, land, air and sea excursions (Fig. 4.3), and accommodation, food and beverage expenditures further increased the economic value of this mass tourism market. The examples noted below offer compelling evidence that polar cruise operations are not only popular, but also economically significant:

Fig. 4.1. Russian nuclear icebreaker *Yamal* takes tourists to the North Pole. (Photo: B. Stonehouse.)

Fig. 4.2. Cruise ship serving Antarctic tourists. (Photo: J.M. Snyder.)

- In 2004, Alaska received 876,000 cruise passengers from May to September. According to the state Chamber of Commerce, this number represents an increase of 100,000 passengers from 2003 (Harpaz, 2005).
- Non-resident cruise-ship tourists visiting Norway in 2004 (the most complete data available as of December 2006) spent 2.383 million NOK. Expenditures for this form of Norwegian tourist travel have increased from 2.196 million NOK in 1998. These economic measures are representative of Norway's long-established, stable cruise-ship industry (Statistics Norway, 2006a).
- Greenland hosted 56 cruise ships in 2005, during a season that now extends from May to September. The cruise industry is steadily adding Greenland to its sailing itineraries, some ships making direct trips to and from Greenland, others including travel to Iceland and the Faroe Islands (Greenland Tourism, 2005).
- Since 1990, Iceland has experienced tourism growth at an annual rate of 9% or more. Of the 320,000 foreign tourists who visited Iceland in 2003, cruise-ship tourists numbered 31,200 visitors (Icelandic Tourist Board, 2005).
- The number of cruise-ship passengers travelling to Antarctica has also increased dramatically. During the last quarter of a century the number of sea-borne passengers has increased from a mere 855 passengers in the 1980/81 season (Enzenbacher, 1992) to more than 20,000 in 2005 (IAATO, 2005).

The size and configuration of the ships on which most cruise passengers travel have changed dramatically during the last 30 years. In the 1970s and early 1980s cruise ships accommodated an average of 500 to 800

Fig. 4.3. Alaskan local cruise ship *Executive Explorer.* (Photo: J.M. Snyder.)

passengers. From 1997 onwards these figures increased to between 2600 and 3800 passengers (Klein, 2003). The addition of numerous amenities and glamorous services continues to accelerate in pursuit of this extremely lucrative market. Although increasingly larger ships make regular appearances at polar destinations, the smaller 'expedition ships' continue to expand their itineraries and conduct a brisk business. Regardless of size, all represent themselves as the most appropriate way to 'explore' the polar regions. The earliest polar explorers would probably be amused to learn that large cruise-ship companies, such as Radisson Seven Seas, attempt to capture a polar tourism market with the slogan 'Luxury Goes Exploring'.

The sheer magnitude and increasing dominance of the cruise industry within the entire tourism industry are destined to create substantial impacts in the polar regions. In 1990 the entire cruise industry transported 4.5 million tourists to diverse international ports and destinations. By the year 2003 this number had more than doubled to 9.5 million, yielding staggering profits. Record cruise industry profits have been posted for the past several years: the most recent data indicate a $15.3 billion profit in 2004 (International Ecotourism Society, 2004). Again, that number is exclusively one year's profit, not total revenues. This rapid growth and profitability has made cruise-ship travel the fastest growing sector of the tourism industry (Honey, 2004).

Economic forecasts by cruise industry experts indicate that by 2010 at least 17 million passengers will travel by cruise ship. The positive outlook for cruise industry growth has been matched by an equally ambitious ship-building programme lasting more than 15 years. The economic prosperity of the 1990s well justified the cruise industry's pursuit of an aggressive ship-building programme. Substantial numbers of vessels were added to their fleet during that decade and it is evident that expansion will continue. Since 2000, shipyards have not only been building more vessels, the passenger capacity of the vessels has grown enormously. For example, between 2002 and 2006, 49 new vessels were launched at a cost of more than $12 billion and this included four immense new ships that entered service in 2006. Based on new ship orders placed by Norwegian Cruise Line, Costa, Cunard and Royal Caribbean, it is certain that more ships will be joining this fleet to transport the 17 million passengers anticipated by 2010 (Brown, 2005).

Given the rising popularity of cruise ships and the industry's commitment to growth, the economic outlook for the polar cruising market can only be for more growth. Patrick Shaw, president of Quark Expeditions (a company prominent in polar travel), has witnessed a steady 5% per year growth in polar tourism during the last several years (Nelson, 2004). Significant growth is reported from the Antarctic: visitor statistics published by the International Association of Antarctic Tour Operators (IAATO, 2005) indicate that numbers of visitors increased from 6704 in 1993 to 19,772 tourists in 2004 (see also Chapter 13, this volume).

High visitor satisfaction, personal safety considerations, perceptions of good value for the cost and the relative ease of travel in today's threatening world are cited by the cruise industry as the economic reasons for its rapidly growing popularity. These findings are validated by the strong evidence of growing bookings despite an industry-wide cost increase of 4–6% per year (Peisley, 2005).

For local Arctic economies, visits by cruise ships are a mixed blessing. The huge size of the industry suggests that it must bring substantial wealth to impoverished Arctic communities. However, the host community bears virtually all the costs of constructing, operating and maintaining the port facilities and other infrastructure needed to serve the ships and their passengers. Another economic disadvantage is that locally owned businesses must compete fiercely with the cruise ship for tourist expenditures. Ships attempt as much as possible to capture passenger souvenir expenditures on board. Money that does 'go ashore' most likely benefits transport companies and travel wholesalers located outside the region. Local work generated ashore and the income it generates is very important, but highly seasonal. Thus 'annual incomes' must be made within a few months, and it is difficult to repay home mortgages, business and personal loans. Collectively, these economic features essentially contribute to a leakage of revenue and capital. In terms of actual revenue received, the primary economic beneficiary of cruise-ship operations is not the host region.

In summary, the total economic value of cruise-ship operations in polar regions adds up to many hundreds of millions of dollars – a source of funding that the polar destinations are eager to tap. As Rudyard Kipling (1899) satirized during his own cruise-ship experiences: 'Granted that the tourist is a dog, he comes at least with a bone in his mouth, and a bone that many people pick.' But the total number of cruise-ship passengers far exceeds the number of residents at all polar destinations, and this can cause stress and disruptions. Thus while Arctic economies and governments aggressively seek the economic benefits of cruise ships, residents often lament the social impacts, cultural intrusions and economic costs they have to bear. Given the inevitable growth in cruise-ship demand supported by the cruise-ship industry's extraordinary capital and marketing investments, polar regions will face increasing pressures to reconcile those motives and concerns.

Commercial air transport

Mass tourism via air travel was made possible by the introduction of commercial jet service. Jet service substantially reduced travel times and progressively transported large numbers of passengers. In 1954 Scandinavian Airlines System (SAS) pioneered Arctic Ocean overflights (Armstrong, 1972). By the 1960s regularly scheduled air transport service was well-established in the Arctic. In 1970 the introduction of Boeing's 747 jumbo jet ushered in a new era of mass tourism via air transport. The mass tourist market now utilizes modern airport facilities to gain direct access to the Arctic. All major cities and national capitals in the Arctic are now served by regularly scheduled commercial air services. This year-round access has provided convenient tourism gateways to the Arctic. In addition, it has expanded tourist seasons and enabled visitors to spend more time at their Arctic destination rather than travelling.

Even communities located throughout remote locations in the Arctic are now served by commercial air services, provided by either charter carriers or major commercial airlines. For example, previously remote Arctic destinations such as Iqaluit (Baffin Island), Longyearbyen (Svalbard) and Petropavlovsk -Kamchatsky (Kamchatka) now serve the tourism market with regularly scheduled flights.

Modern air transport currently facilitates both mass and small-group tourist access to Antarctica. The mass market clientele flies to commercial air facilities located in Chile, Argentina, the Falkland Islands, New Zealand and South Africa (Chapter 8, this volume). From these gateway facilities tourists are transported to ports where they embark on cruise-ship voyages to and from Antarctica. This highly integrated transport system supplies a steady stream of tourists to and from Antarctica throughout the November to March period. Another aspect of Antarctic mass tourism is the overflight experience that departs from and returns to Australia (Chapter 11, this volume).

Commercial air transport of small groups of tourists to the Antarctic continent has been operating continuously since 1985. Pioneering aviation by Adventure Network International (ANI) discovered that 'blue-ice runways' could be used to land tourists on the continent. ANI established an encampment at Patriot Hills in the Heritage Range to provide a variety of land-based tourism opportunities. Other charter air companies have provided tourist transport between South Africa and Dronning Maud Land, and between Punta Arenas, Chile and King George Island (Swithinbank, 2000a,b).

Sport Fishing and Hunting

Anglers and hunters have been attracted to the Arctic for nearly two centuries. Throughout that period their economic contributions to the polar regions have continuously strengthened. Arctic wildlife possesses all the traits needed to attract and capture a large sporting market. The seasonal concentration of huge numbers of animals and fish species, trophy-size wildlife and the relatively high probability of harvesting them are enormously appealing to anglers and hunters. This appeal is further enhanced by especially attractive settings characterized by their remoteness and wilderness beauty.

Following World War II, the economic benefits of sport fishing and hunting became especially evident to Arctic communities and governments. The availability of reliable bush planes, four-wheel-drive vehicles and efficient boat engines combined with new personal wealth and the booming demand for outdoor recreation. These ingredients provided both the means and the incentives needed to fuel the sport angling and hunting market.

Arctic peoples and communities generally appreciate the economic benefits resulting from angling and hunting. From their perspective, the single greatest economic benefit of this type of tourism is the fact that expenditures remain in the community. In pursuit of their sport, anglers and hunters pay many thousands of dollars to local people and establishments. They employ local guides, pilots, charter-boat captains and crews, outfitters and suppliers. They use local transport, stay in local accommodations and eat in local establishments. An equally significant economic benefit is that many of the same anglers and hunters return year after year. The economic loyalty of this market, which translates into economic stability for Arctic businesses, is a consistently documented fact. In many Arctic locations, for example, sport fishing lodges must be booked years in advance because their repeat clientele keep them full.

It should also be noted that during the last century the environmental ethos of anglers and hunters has changed substantially. Rigorously enforced fishing and hunting regulations, 'catch-and-release angling', and selective hunting supervised by licensed guides have combined to substantially alter the personal behaviour of this tourist market. The anglers and hunters who comprise today's market have a far higher regard for environmental quality, habitat conservation and wildlife management practices than the 'hook-and-bullet' sporting crowd of the 19th century.

Arctic jurisdictions at all levels of government benefit economically from the employment, personal income and taxes that angling and hunting create, as well as the fees derived from licenses and permits. Collectively, these economic benefits enable jurisdictions to politically justify their support for fish and wildlife management programmes. The economic and political clout of the government agencies that manage the Arctic's wildlife resources is substantial. Their impact on environmental management decisions is considerable.

The US *National Survey of Fishing, Hunting, and Wildlife Associated Recreation* provides accurate information concerning the economic importance of angling and hunting in the state of Alaska. The most recent survey, conducted in 2001, indicates that total of 239,000 non-resident anglers fished in Alaska. They spent a total of 556,000 days fishing. The economic contributions of the sport fishing market to Alaskan communities are enormous. Specifically, it generated over $659 million in 2001. More than $424 million of this amount was spent on trip-related expenses such as food and lodging, transport and guide services, licensing and other items (US Fish and Wildlife Service, 2002).

In 2001 the number of non-resident hunters in Alaska exceeded 21,000 persons. The hunters sought a diversity of regulated wildlife species for a total of 193,000 days. The dollar value of this market is significant. Hunters spent nearly $217 million and the survey results indicate that the vast majority (approximately 80%) of those expenditures occurred in Alaska. For example, Alaska's direct economic benefits included hunting expenditures of nearly $27 million for food and lodging, nearly $40 million for in-state transportation, more than $38 million for equipment and $93.7 million for other trip costs (US Fish and Wildlife Service, 2002).

A strong case can be made that the world's highest costs for sport fishing are to be found in Iceland. The economic benefits that the nation and its citizens derive from this tourist market are stunning. The angling experience begins with the purchase of a fishing license that costs 200,000 Ikr *per day*. That cost does not include a guide, transportation or equipment and, of course, it does not include food, lodging and transport. For those essentials, Iceland's Tourist Bureau suggests that the cost of fly-fishing can run as high as $2000 per day. Year after year, many thousands of anglers pay these amounts for the opportunity to stalk Iceland's salmon, trout and Arctic char. But not surprisingly, celebrities and royalty comprise a noticeable part of this fly-fishing market (Harding and Bindloss, 2004; Weinman, 2004).

All the Scandinavian countries have extensive fresh- and saltwater sport fisheries. Salmon, trout, grayling, pike, perch, whitefish, break and Arctic char are the species most actively sought by anglers. The diversity of angling experiences, types of water, well-managed fisheries, and fine environmental quality of the water resources sustain both the popularity and economic value of this recreational activity.

Scandinavian sport fishing has a rich history. Since the 19th century wealthy anglers have been attracted to the region's coastal and inland fisheries. The trophy fish has traditionally been salmon, but Norway, Finland and Sweden have as many as 41 species of fish to tempt the angler. Professional resource management techniques are utilized throughout Scandinavia to conserve fish habitat. The economic benefits of these efforts include personal income, employment, government revenues in the form of licensing fees and permits, and capital investment. However, like all Arctic fisheries, the cyclical behaviour of sea-run fish such as salmon, sea trout and Arctic char are subject to the vagaries of climate and environmental change. These global events can cause economic uncertainty that radically alters sport fishing success and visitor satisfaction. For example, according to Statistics Norway, 2004 was the worst year for salmon angling in Norway since the 1997 season. A total of 404 t of salmon, sea trout and sea char were taken in Norwegian rivers, representing a decline of 176 t since 2003. That decline was immediately followed by a sharp increase of 518 t of total river catch in 2005 (Statistics Norway, 2006b). Norway's 200,000 rivers and lakes will remain critically important fish habitats and the country is steadfastly committed to the perpetuation of healthy fisheries. But it is also apparent from this example that Arctic nations economically dependent on angling are particularly vulnerable to global environmental events that are well beyond their control.

Sport fishing in Canada is a major recreational and economic activity that is carefully regulated and monitored by Canada's Department of Fisheries and Oceans. That department conducts comprehensive surveys of the uses of its fisheries and like many resource agencies there is a delay between the time the survey was conducted and the date of publication. The information contained below is based on surveys conducted in 2000 and then published in 2005. Most of the 2000 fishing activity (93.7%) took place in fresh water. Canada carefully monitors both commercial and sport fishing. Based on Canada's Department of Fisheries and Oceans *2000 Survey of Recreational Fishing in Canada*, the following information clearly summarizes the economic magnitude of this from of tourism (Canada Department of Fisheries and Oceans, 2005):

> Recreational fishing is an important economic activity in the natural resources sector. In total, anglers spent $6.7 billion in Canada in 2000. Of this amount, $4.7 billion was directly associated with recreational fishing. Anglers spent over $2.4 billion on trip expenses such as package deals, accommodation, food, transportation, fishing supplies and other services directly related to their angling activities. Investments in 2000 totaled close to $4.3 billion for such durable goods as fishing equipment, boats, motors, camping equipment, special vehicles and real estate. Anglers estimated that almost $2.3 billion of these investment expenditures were wholly attributable to recreational fishing.
>
> Nonresident anglers took over 3.5 million trips in Canada for fishing and other reasons. Visitors to Canada made over 2 million of these trips with the balance being trips by Canadians visiting other provinces and territories. Overall, nonresident anglers fished on 52% of their trips. Non-Canadians fished on 69% of their trips across the border. Of the total days spent in other provinces and territories (4.3 million), visiting Canadian anglers fished on almost 30% of these days. This compares to visiting anglers from other countries, who fished on over 64% of their days spent in Canada.

The province of British Columbia provides especially strong evidence of the economic value and impact of Canada's Arctic and sub-Arctic sport fishing. Since 1998, British Columbia has hosted over 600,000 anglers per year. In

1998, the average revenue per salmon caught by sports anglers was estimated at nearly Can$500, compared with less than Can$7 for a commercially caught fish. During the same time period, the sport fishing industry created approximately 5990 person-years of employment, compared with 2300 person-years of employment in the commercial fishery. More recently, in 2002 British Columbia's sport fishing sector revenues exceeded Can$675 million and 8900 jobs (Government of British Columbia, 2005).

Foreign access to Russia's remote Arctic regions is a recent phenomenon, but notably among the first allowable recreational uses of those regions were sport fishing and hunting. Foreign entry to the Kola Peninsula and the Russian Far East for angling was not permitted until 1991. At that time trains to Murmansk provided access to the Kola Peninsula and foreign air carriers such as Alaska Air were allowed to transport tourists to Russia's Far Eastern cities from which Aeroflot transported anglers to Siberia. More recently, the Kamchatka Peninsula, once an entirely off-limits military zone, was opened to anglers, trophy hunters in pursuit of bears and mountain goats, and backcountry enthusiasts wishing to experience the volcanoes of Kamchatka World Heritage Sites. A brief description of the angling and hunting recreational experiences and their economic impacts is provided below.

Each summer, 500 non-resident anglers are allowed to fish for king salmon and rainbow trout along the rivers of the Kamchatka Peninsula. Anglers generally fish the Ozernaya, Two Yurt, Opala, Kolpakova, Tigil and Sopochnaya rivers at a weekly cost of between $2000 to $5000 per person, plus transportation to Russia. International organizations such as the United Nations Development Programme and the Wild Salmon Center are seeking to promote catch and release fishing as an integral part of multi-million dollar ecotourism development and habitat conservation projects. Their primary objective is to demonstrate that angling can be used as an economic catalyst to promote Kamchatka's World Heritage Sites.

For anglers willing to pay $4950 to $9950 per week for a tent in the Russian Arctic wilderness, the Ponoi River on the Kola Peninsula provides the world's largest runs of Atlantic salmon. (It should be noted that the 2005 prices do not include transportation.) Since the timing of the runs coincides with the endless Arctic night, the angler need not spend much time in the tent (Rizzo, 2005).

The Kamchatka Peninsula is also the location of a thriving trophy hunting economy. The prized species is the Kamchatka brown bear. Attaining a height of nearly 10 feet and a weight of 1200 pounds, the Kamchatka bear is Eurasia's largest bear species. As of 2005, the Kamchatka Department of Wildlife Management issued 500 hunting permits for trophy hunts guided by licensed outfitters. Clients paid up to $10,000 for the opportunity to hunt the Kamchatka bear. Consequently, the economic impacts of recreational hunting are significant for this Russian region. The realization of economic gain from trophy hunting may be abhorrent to some people, but the implementation of sustainable wildlife management practices is probably preferable to the death of approximately 450 additional Kamchatka bears killed illegally by poachers in 2005 (Russell and Enns, 2003; Meier, 2004; Raygorodetsky, 2006).

Russia's recent entry into the polar tourism market deserves special attention. The Russian Arctic contains the world's largest Arctic land mass, extending across 11 time zones. As previously stated, it is only since 1991 that foreigners have been allowed to visit this part of the globe. Although Russia's evolving polar tourism industry seemingly replicates economic development processes that occurred in other polar regions, the sheer vastness of the Russian Arctic will most probably write a distinctly new chapter in the development of polar tourism. The anglers, hunters and adventurers in the Russian Arctic, like other polar destinations, are the pioneers of Russian Arctic tourism. The primary attractions of these persons are the 'undiscovered' and previously inaccessible outdoor recreation experiences located in Russia's extremely remote wilderness. The Government of Russia is attempting to lure these types of tourists with wildlife attractions such as fishing and hunting and extreme sports including kayaking and mountaineering (Whelan, 2004). The Russian Ministry of Tourism is aggressively promoting polar tourism development extending from the White and Barents seas to the Kamchatka Peninsula. The construction of additional tourist infrastructure, facilities and services throughout this vast region are intended to support this market. Russia's entry into the polar tourism market represents the single largest geographic expansion of tourism in the Arctic.

Nature Tourism

Seasonally abundant wildlife populations and immense wilderness areas attract thousands of nature tourists to the polar regions. Wildlife enthusiasts, photographers and birders perceive their visits to the polar regions as a unique opportunity to witness diverse species of wildlife in their natural habitats. The migration of large numbers of marine and land animals and birds in the polar regions enables the nature tourist to simultaneously view both large congregations of animals and numerous species.

Nature tourists are generally as impressed with the enormous size, beauty and remoteness of natural habitats as they are with the wildlife. Consequently, the environmental setting is valued as an inseparable part of the wildlife viewing and nature tourism experience. Polar regions generally enjoy a competitive advantage in this regard. The nature tourism market is especially attracted to the numerous national parks, wildlife refuges and World Heritage Sites located throughout the entire polar world. National parks in the polar regions can often be enormous. For example, the world's largest protected area is the Northeast Greenland National Park, covering nearly 1 million sq km, and Europe's largest protected area consists of Finland's Oulanka National Park and Russia's Paanajarvi National Park.

Within the vastness of the polar regions wildlife migrations can be predicted with near certainty, thus virtually ensuring quality wildlife-viewing opportunities. This in turn increases visitor satisfaction and contributes to the growing popularity of polar destinations. Given this high degree of certainty, travel arrangements can be reliably secured and the tourist's wildlife viewing expectations will, most probably, be met.

Participants in the wildlife tourist market include avid wildlife photographers, birders seeking to add to their 'life lists', inquisitive persons seeking to view wildlife in their natural habitats, and numerous societies and clubs devoted to wildlife viewing and conservation. The number of wildlife watchers and the economic size of this tourist market are substantial. For example, in 2001 Alaska's tourism industry hosted 420,000 wildlife watchers. This tourist market provides vital benefits to polar economies such as Alaska. According to the *National Survey of Fishing, Hunting, and Wildlife Associated Recreation*, the expenditures of Alaska wildlife-watchers reached nearly $499 million in 2001. Of this amount, more than $386 million were spent on trip-related costs, over $53 million on equipment, and other trip expenditures exceeded $59 million (US Fish and Wildlife Service, 2002).

The nature tourist market is also strongly represented by people seeking to relieve the stress of daily life in environmental settings that they perceive provide solitude and serenity. These nature tourists experience natural areas in a variety of ways. Their activities can range from passive viewing to active hiking and backpacking. Their extended stays may involve the use of remote camps or the use of recreational vehicles. The tourist's search for peaceful destinations has increased since the tragedy of 11 September 2001 and subsequent terrorist events. Personal safety is now a motivating factor in selecting tourist destinations. Survey research by the tourist industry reveals that the public perceives polar regions as among the world's safest tourist destinations. One prominent survey, conducted in 2002 by the travel magazine *Blue*, listed Iceland, Antarctica, Alaska, British Columbia and Canada, Patagonia and New Zealand as the world's safest destinations (*Iceland Review*, 2002).

Numerous companies now offer a diversity of wildlife and nature tours throughout all regions of the Arctic and Antarctic. All of these purveyors emphasize their professional knowledge of both the region and its wildlife. The nature companies highlight the credentials of their guides and naturalists in an effort to gain a competitive advantage. They also make sincere representations about their respect for conserving the destination's environmental integrity and local cultures. Nature travel in the polar regions costs thousands of dollars and individual prices vary tremendously by destination, trip duration, transport modes and logistical supports.

The nature tourism market is continuously fuelled by numerous wildlife societies, zoos, natural history museums, university alumni associations, clubs and special interest groups. Membership of these organizations consists of middle-to-upper income, well-educated persons who travel frequently in pursuit of nature-based vacations. The economically attractive profile of this particular tourist has not escaped the attention of the travel industry or the Arctic residents. Many international tourism businesses have targeted the nature tourism market and more are seeking to cater to this affluent travel group (Cater and Lowman, 1994; Honey, 1999). In addition, residents of the

Arctic serve as guides and naturalists offering a remarkable diversity of nature tourism experiences. These people provide tourists with both nature tours and a personal understanding of the local culture and communities. The revenues these local guides generate and their employment strengthen not only Arctic economies, but also enhance resource management and cultural preservation.

Adventure Tourism

Perhaps the most distinguishing characteristics of the adventure tourism market are the unique motivations of its participants. Personal accomplishment is the common denominator that characterizes this market group, but beyond that description there are innumerable individual motivations. Some of these include testing one's physical and mental stamina by means of recreation activities such as technical climbing, white-water kayaking, river rafting, mountain biking, scuba diving and skiing. Still other participants actively engage in a variety of recreational activities for the sake of sheer exhilaration. Since approximately 1990, a novel approach to adventure tourism in the polar regions has occurred. Some polar adventurers have been motivated to replicate the exploits of polar explorers. Variations on this theme include travelling historic polar exploration routes by means of alternative transport modes. Clearly, personal skills and experience, appropriate equipment, competent knowledge of environmental conditions, adequate preparation and identification of emergency support services all play vital roles in the pleasurable and safe accomplishment of the adventurer's goals.

Specialization is another dominant characteristic of this tourist group. The outdoor recreation industry identifies very distinct specialities within each adventure sport. For example, mountaineering specializes in the type of rock formation, ice conditions, size of the summits and climbing seasons. Each of these conditions appeals to distinct population groups in the mountaineering recreation market. The sport of kayaking is categorized according to the classification of the rapids, by type of sea kayaking or type of flat-water kayaking. And so it goes for all types of adventure sports. Highly specialized publications, organizations and personal networks vigorously supply the adventurers with new challenges, tales of recent achievements and the availability of new technologies.

Significant management challenges confront those responsible for accommodating the adventure tourist. Principally, it is nearly impossible for recreation resource managers to competently know if either the skill levels or the health and psychological preparation of the participant are sufficient to meet the rigors of their sports. Inspection of equipment can be conducted, but resource managers rarely have that opportunity, and they probably do not have spare parts and repair facilities to correct deficiencies. Maps, marine charts and tide tables can be supplied and it is hoped that adventurers will responsibly seek these essential aids to navigation. Emergency communication instructions in the form of emergency radio frequencies, protocols, directional beacons and Standard Operating Procedures for search and rescue (if available) can also be provided. Again, it is essential for adventurers to avail themselves of these in order to safely and pleasurably pursue their activities.

Culture and Heritage Tourism

The culture and heritage tourism market is comprised of persons who want to personally experience the history, art and cultural traditions that distinguish a 'unique' destination. According to a 2003 study by the Travel Industry Association and *Smithsonian Magazine*, 118 million persons sought out history and culture tourism experiences. That number represented a 13% increase from 1996. The survey also revealed that this market, in comparison with the average tourist, consistently spends substantially more money and time at their destinations (Olson, 2003).

The intimacy with which cultural and heritage attractions are experienced generally defines distinct subgroups within this tourism market. The greatest personal involvement is generally sought by tourists seeking participation in living cultures. Their tourist experiences are defined in terms of personal contact with indigenous people who allow them to participate in traditional ways of life, arts, crafts, ceremonies and other cultural activities. The type and availability of interpretive services, such as signage and exhibits, and knowledgeable guides also impact the quality of the tourist experience and resource conservation. Like all dynamic situations, the relationships between tourists and their Arctic hosts are constantly reviewed, evaluated and modified.

Native peoples throughout the Arctic are experimenting with the cultural and heritage tourism venue as a way to strengthen their economies. They are attracted to this type of tourism because it can serve as a way of preserving their traditional ways of life, their language and their culture (Milne *et al.*, 1995; Amberger, 2003). This form of tourism provides jobs and income, creates markets for its products and artistry, and, ideally, it enables native peoples to determine how their culture will be shared. Level of involvement extends from direct contact to the use of the tourist market as an outlet for art and other products. Heritage tourism is also perceived to be a method for accomplishing self-sufficiency, especially for native peoples who now have sovereignty over their lands and economic resources, such as the Nunavut of Canada who govern one-fifth of Canada's land mass, the Native Corporations of Alaska, and the Saami of Sapmi, northern Scandinavia (Walle, 1993; Alaska Native Council, 2005).

The native peoples of the Arctic have approached cultural tourism development in various ways. Some have formed corporations and councils, such as the Alaska Native Council, that actively seek to finance and promote this form of development. Others, such as the Nunavut, have exercised governmental powers to implement this form of tourism. Still others, such as the Saami, employ smaller-scale, community-based lodgings to share their traditions and culture. But small-scale accommodations should not be interpreted as small visitation. For example, although there are only 4000 Saami in Finnish Lapland, individual lodges have each year hosted 130,000 day visitors and 1200 overnighters (Rennicke, 2004). Governments throughout the Arctic assist these rural economic and cultural development efforts by means of legislation, financial subsidies and promotional campaigns in support of native arts and crafts.

But the roster of positive economic and cultural benefits can be offset by the sheer number of visitors that can overwhelm both physical infrastructure and social norms. In terms of Arctic cultures, the most significant numerical fact regarding polar tourism is that the number of tourists now greatly exceeds the number of permanent residents in all polar regions. The cultural resource management implications of this fact are equally significant. Intrusive visitor behaviour can potentially violate traditional customs. The introduction of technologies and tourist service amenities can impact local people's desires to maintain traditional life-styles. And additional people increase the competition for already scarce environmental resources. Consequently, native peoples face the daunting challenge of balancing cultural preservation with economic self-sufficiency.

An especially well-defined culture and heritage tour group consists of highly specialized tours organized by museums, societies, religious organizations and other special interest groups for the express purpose of sharing their mutual interests. Examples include museums tours to view the art and architecture of various destinations; historical societies and similar organizations that visit the huts and routes of polar explorers; and religious groups that visit Arctic sites such as monasteries in the Russian Arctic (Barr, 1980). These tours are a shared experience among like-minded people. The economic value of this type of tour to a polar destination is substantial. Per capita tourist expenditures for travel, accommodations, food and beverage are large because the entire tour is customized and dependent upon charter transportation and specialized guides. The tourists exhibit respect for the host location and are especially interested in spending time at each destination along the tour route. This respect generally translates into a favourable willingness to pay for local products and services.

A third type of heritage market is evidenced in the historic towns and sites that have been restored in the hopes of attracting tourists. Dawson City, Whitehorse and Fort Selkirk in the Yukon Territory seek tourists interested in 'reliving' the Klondike Gold Rush era. Along the same theme, Skagway (Fig. 4.4) and Juneau, Alaska have refurbished their downtowns to replicate that era. During the brief summer season, communities such as these offer numerous tourist attractions in an effort to diversify and strengthen their economies. The keys to their economic success are measured in terms of expenditures per person and duration of stay. A colourful array of techniques is employed to capture the tourist's money and time. Nostalgia is flamboyantly displayed by townspeople in period costumes, readings of Robert Service poetry and Jack London abound, honky-tonk pianos play in false-fronted saloons and re-enactments of historic events are conspicuously staged

(Jarvenpa, 1994). Recreation attractions such as helicopter tours, tramway and train rides (Fig. 4.5), wildlife viewing, salmon bakes and golf are offered in an effort to further extend the tourist's visit. All of these efforts are strongly supported by government-financed 'vacation planners' that annually promote these, and many other forms, of polar tourism (Government of Yukon, 2005; State of Alaska, 2005). A considerably different 'heritage market' venue is offered to polar tourists who visit the historic towns, museums and festivals in the Scandinavian Arctic, Greenland and Iceland.

Summary and Conclusions

The polar tourism market is a complex mixture of activities that appeal to different populations. It is growing in terms of numbers of tourists, capital expenditures by the private sector, financial and political commitments by the public sector, and geographic expansion. Greater reliance upon this economic activity is being demonstrated by all Arctic peoples, while simultaneously cultural hopes are being vested in tourism by Native Peoples who struggle to preserve their cultural heritage. The geographic expansion of polar tourism will inevitably continue because it is directly tied to improved transportation access, the opening of huge, previously restricted regions such as the Russian Arctic, the strong, promotional efforts of Arctic governments and the growing demand of tourists for safe destinations.

In summary, the visitor and resource management challenges, risks and opportunities associated with each polar tourist group are unique. Visitor

Fig. 4.4. Restored building in Skagway, Alaska. (Photo: J.M. Snyder.)

Fig. 4.5. Whitehorse Railroad, Yukon, Canada. (Photo: J.M. Snyder.)

behaviour, the ways in which natural and cultural resources are utilized, the seasons and duration of resource use, the geographic distribution of tourism activity and their inherent danger vary substantially among each segment of the polar tourism market. The adequacy and relevance of economic development strategies and resource management plans will depend upon the recognition of these distinctions.

References

Alaska Native Council (2005) *Alaska Native Journeys.* Alaska Native Council, Anchorage, Alaska.
Amberger, R.L. (2003) Living cultures – living parks in Alaska: considering the reconnection of Native Peoples to their cultural landscapes in parks and protected areas. In: Watson, A.E. and Sproull, J. (comps) *Science and Stewardship to Protect and Sustain Wilderness Values: Seventh World Wilderness Congress Symposium,* Port Elizabeth, South Africa, 2–8 November 2001. Proceedings RMRS-P-27. US Department of Agriculture Forest Service, Rocky Mountain Research Station, Ogden, Utah.
Armstrong, T. (1972) International transport routes in the Arctic. *Polar Record* 16(102), 375–382.
Armstrong, T. (1991) Tourist visits to the North Pole, 1990. *Polar Record* 27(161), 130.
Barr, W. (1980) The first tourist cruise in the Soviet Arctic. *Arctic* 33(4), 671–685.
Brendon, P. (1991) *Thomas Cook: 150 Years of Popular Tourism.* Secker & Warburg, London.
Brigham, L.W. (2000) Polar icebreakers at the end of the twentieth century. *Polar Record* 36(198), 247–249.

Brown, C.S. (2005) Special Report: Cruises. Quiet year for big ship debuts. *The Denver Post* 30 January, Travel Section, pp. 1 and 4 (also available at www.CruiseCritic.com).

Canada Department of Fisheries and Oceans Pacific Region (2005) *2000 Survey of Recreational Fishing in Canada.* Canada Department of Fisheries and Oceans, Vancouver, British Columbia; available at http://www.dfompo.gc.ca/communic/statistics/recreational/canada/index_e.htm (accessed January 2006).

Cater, E. and Lowman, G. (1994) *Ecotourism: A Sustainable Option?* John Wiley & Sons, New York.

Enzenbacher, D.J. (1992) Tourists in Antarctica: number and trends. *Polar Record* 28(164), 17–22.

Government of British Columbia (2005) *Fisheries Statistics*; available at http://www.agf.gov.bc.ca/fish_stats/statistics.htm and *Tourism Sector Monitor*; available at http://www.bcstats.gov.bc.ca/pubs/pr_tour.asp. Government of British Columbia, Vancouver, British Columbia (both accessed December 2006).

Government of Yukon (2005) *Yukon: Canada's True North. 2005 Vacation Planner.* Department of Tourism and Culture, Whitehorse, Yukon, 117 pp.

Greenland Tourism (2005) *News.* Greenland Tourism – The National Tourist Board of Greenland, Nuuk; available at http://www.greenland.com/ (accessed September 2005).

Harding, P. and Bindloss, J. (2004) *Iceland.* Lonely Planet Publications, Footscray, Victoria.

Harpaz, B.J. (2005) Travel forecast is sunny as 2005 begins. *The Denver Post* 23 January, 8T.

Honey, M. (1999) *Ecotourism and Sustainable Development: Who Owns Paradise?* Island Press, Washington, DC.

Honey, M. (2004) Greening the cruise industry: letter from TIES Executive Director. *Eco Currents* Second/Third Quarter, 1–3.

IAATO (2005) *Tourism Statistics.* International Association of Antarctica Tour Operators, Basalt, Colorado; available at http://www.iaato.org/tourism_stats.html (accessed December 2006).

Iceland Review (2002) News Release 20(2).

Icelandic Tourist Board (2005) *Your official guide to Iceland.* Icelandic Tourist Board; Reykjavik; available at http://www.icetourist.is/ (accessed December 2006).

International Ecotourism Society (2004) Special Issue: The Cruise Industry. *Eco Currents* September, 7 pp. International Ecotourism Society, Washington, DC.

Jarvenpa, R. (1994) Commoditization versus cultural integration: tourism and image building in the Klondike. *Arctic Anthropology* 31(1), 22–26.

Kipling, R. (1899) *From Sea to Sea: Letters of Travel.* Doubleday & McClure Co., New York.

Klein, R.A. (2003) *Cruising Out of Control: The Cruise Industry, The Environment, Workers, and The Maritimes.* Canadian Centre for Policy Alternatives, Halifax, Nova Scotia, available at http://www.policyalternatives.ca/index.cfm?act=news&call=21&do=article&pA=BB736455 (accessed December 2006).

Meier, A. (2004) A message in blood. *Outside Magazine* 29(12), 151–175.

Milne, S., Ward, S. and Wenzel, G. (1995) Linking tourism and art in Canada's eastern Arctic: the case of Cape Dorset. *Polar Record* 31(176), 25–36.

Nelson, A. (2004) Polar express: the Antipodes heat up as travelers head to the Arctic and Antarctic. *National Geographic Traveler* November/December, 17–18.

Olson, M.S. (2003) Heritage tourism becoming booming business in US. *The New York Times* 19 October, 8T.

Peisley, T. (2005) *Global Changes in the Cruise Industry 2004–2010.* Seatrade Communications, Ltd, Colchester, UK; available at http://www.seatrade-global.com/cruise_report (accessed December 2006).

Raygorodetsky, G. (2006) Giants under siege. *National Geographic Magazine* 209(2), 50–65.

Rennicke, J. (2004) Finnish Lapland – the culture of cool. *National Geographic Traveler* October, 72–78.

Rizzo, W. (2005) Fly right. *Outside Magazine* 30(5), 42–45.

Russell, C. and Enns, M. (2003) *Grizzly Seasons – Life with the Brown Bears of Kamchatka.* Firefly Books, Buffalo, New York.

State of Alaska (2005) *Alaska Official State Vacation Planner.* State of Alaska, Anchorage, Alaska.

Statistics Norway (2006a) *Tourism consumption in Norway by non-residents. 1998–2003. NOK million, current prices*; available at http://www.ssb.no/english/subjects/09/01/turismesat_en/arkiv/tab-2004-10-15-02-en.html and *Tourism consumption in Norway by non-residents. 2004 and 2005. NOK million, current prices*; available at http://www.ssb.no/english/subjects/09/01/turismesat_en/tab-2006-12-19-02-en.html. Statistics Norway, Oslo (both accessed December 2006).

Statistics Norway (2006b) *Increase in river catch of salmon.* Statistics Norway, Oslo; available at http://www.ssb.no/english/subjects/ 10/05/elvefiske_en/ (accessed December 2006).

Swithinbank, C. (2000a) Non-government aviation in Antarctica 1998/99. *Polar Record* 36(196), 51.

Swithinbank, C. (2000b) Non-government aviation in Antarctica 1999/2000. *Polar Record* 36(198), 249.

US Fish and Wildlife Service (2002) *2001 National Survey of Fishing, Hunting, and Wildlife Associated Recreation.* FHW/01-NAT. US Department of the Interior, Washington, DC.

Walle, A.H. (1993) Tourism and traditional people: forging equitable strategies. *Journal of Travel Research* Winter, 14–18.

Weinman, E. (2002) Casting flies. *Iceland Review* 20(2), 32–35.

Whelan, H. (2004) Walking the bear. *Outside Magazine* 29(3), 61–66.

5 Tourism in Rural Alaska

HENRY HUNTINGTON,[1] MIKE FREEMAN,[2] BILL LUCEY,[3]
GRANT SPEARMAN[4] AND ALEX WHITING[5]

[1]23834 The Clearing Drive, Eagle River, AK 99577, USA; [2]Alaska
Department of Fish and Game, PO Box 63, Yakutat, AK 99689, USA;
[3]Yakutat Salmon Board, PO Box 160, Yakutat, AK 99689, USA; [4]Simon
Paneak Memorial Museum, PO Box 21085, Anaktuvuk Pass, AK 99721,
USA; [5]Native Village of Kotzebue, PO Box 296, Kotzebue, AK 99752, USA

Introduction

Tourism, the third largest sector of Alaska's economy behind oil and fishing, reaches across the entire state (Goldsmith, 1997). Tourist activities range from cruise-ship voyages to day-long flight-seeing visits to native communities, from motor-home vacations along the road system to sport hunting adventures in remote areas. This chapter examines three examples of tourist activity and its interactions with local inhabitants in rural areas. In Anaktuvuk Pass, cultural tourism brings day-trippers to the village and its museum, while the surrounding Brooks Range attracts hikers lured by relatively easy access to wilderness. In northwestern Alaska, sport hunting for caribou, moose and bear is highly popular, as are kayak and raft trips on the Kobuk and Noatak rivers. In Yakutat, cruise ships visit tidewater glaciers in fjords that are also home to seals and other marine mammals.

Origins and Development

Tourism in Alaska began in the late 19th century, with visitors such as John Muir travelling north to experience the scenery and wildlife of Alaska (Muir, 1915). Indeed, the first hunting regulations in Alaska were a response to trophy hunting in the southern part of the state in the early 20th century (Wilson, 1903). Following World War II, increasing affluence in the USA as a whole, combined with greater access to Alaska through the construction of the Alaska Highway, increased the attractiveness of the state for tourists of all kinds. For the most part, tourists then and now are concentrated in the areas most easily accessible by road or ship. Recent trends show that the number of visitors each year continues to increase, but that the rise is confined to cruise ships. The number of tourists arriving by air or road has remained stable (Colt and Huntington, 2002).

Tourism is viewed with some ambivalence in the populated parts of the state. The influx of cash and associated employment is beneficial to local economies, but the number of tourists can overwhelm small towns and increase traffic along major highways. None the less, the main summer tourist season is a part of the annual cycle in these areas, the populations of most of which are sufficiently large to absorb the additional visitors and in many ways are organized to accommodate them. In rural areas, by contrast, sparse populations make additional visitors more noticeable and the dependence of rural residents on fish and animals and the landscapes that support them can make interactions among local residents and tourists more prominent, leading to conflicts. Despite Alaska's reputation for limitless space, access points, travel routes and key attractions often concentrate visitors and locals in a few areas. The case studies below, written by residents of each of the three areas, describe the dynamics, costs and benefits of such interactions in three rural areas of Alaska.

Anaktuvuk Pass

In many ways, the community of Anaktuvuk Pass has long been accustomed to drawing people's attention. As the last remaining enclave of inland Iñupiat, or Nunamiut as they are better known, these once-nomadic people have attracted visitors since the 1940s. Initially visitors were mostly from the scientific community, especially archaeologists and anthropologists who, in particular, have held a long and abiding interest in the history and cultural traditions of the inlanders. Along with them also came scientists from a host of different fields, including physiologists, biologists, botanists, geologists and even nuclear physicists, who likewise came to study, work with and learn from the Nunamiut. Over the past 60 years they have collectively produced a voluminous scientific literature replete with studies to which local residents have actively contributed, although often without receiving full credit.

In more recent years however, while researchers still come to study, tourism now accounts for the vast majority of visitors to the community. They number between 1000 and 1200 per year, with most coming between early April and late September. Not unsurprisingly, tourists are often drawn by some of the very same things that drew the original researchers: an interest in a fascinating group of people as well as the plants, wildlife and magnificent scenery of the central Brooks Range. Ever adaptable, the people of Anaktuvuk Pass are exploring and learning how to accommodate and benefit from this interest.

Visitors today are made up of a variety of constituencies ranging from business or governmental representatives who take the opportunity to enjoy the sights on an otherwise mission-oriented trip, to boaters intent on floating the John River, to hikers who use the village as either a destination or departure point for their adventures. But, increasingly and overwhelmingly, they are day tourists, especially those associated with commercially organized tours.

Coincidentally, and perhaps fortuitously depending on one's point of view, this increase in tourism has occurred at a time when the North Slope Borough – the regional municipal home rule government, and primary employer in the community for the past 30 years – has begun a substantial and ongoing programme of cutbacks in both jobs and services. As this former, seemingly secure, mainstay of the local economy continues to shrink, local community members, encouraged by the Borough, have begun to look to their own resources to help build a broader-based local economy. Fortunately, the community possesses several inherent advantages over many other rural villages when it comes to attracting visitors:

- Anaktuvuk Pass has long enjoyed a well-justified reputation as one of the most open, friendly and welcoming villages in the entire state of Alaska, always greeting visitors with smiles and open hospitality.
- The village is located in the heart of the north central Brooks Range and is naturally blessed with some of the most enchanting landscapes in the Far North. It is also located within Gates of the Arctic National Park and Preserve, America's premier wilderness park, which further helps draw the traveller's attention to the area. Although protective of the park's wilderness nature, the National Park Service does relatively little to actively promote the park as a visitor destination.
- The Nunamiut possess a compelling history as America's virtually last nomadic people to settle into a permanent village, a process not complete until 1960. They also benefit from the presence of the small but excellent Simon Paneak Memorial Museum, which vividly presents their story and history to the visiting public, in essence making their cultural heritage a tangible and highly visible central attraction.
- There is ready, easy and affordable access to the village, which is served by several air services providing daily flights from Fairbanks for about $300 round trip. The flight is only an hour and a half in length, and thereby permits easy access for day trips.
- Some of these same air carriers are closely affiliated with large, multinational commercial tour companies that market Anaktuvuk Pass as a destination, often in combination with other attractions in the region, providing broad and high-quality publicity that is likely beyond the means of most small communities.

One operator, Northern Alaska Tour Company (NATC) of Fairbanks, Alaska, offers village trips in affiliation with Princess Tour Company. These trips include a bus tour up the Dalton Highway to Coldfoot, then a flight to Anaktuvuk Pass before returning by air to Fairbanks. Another operator, Warbelow's Air Ventures, also based in Fairbanks, in cooperation with Holland America Cruise Lines markets a different experience, combining a village tour with guided trips in the countryside by all-terrain vehicle.

Once in Anaktuvuk Pass, the NATC village-based tour is conducted both on foot and by van, led by student guides who escort visitors around the community, including a visit to the local museum where they have the opportunity to learn more about the people they are visiting as well as to buy local craft items at the museum gift shop. The tour also includes a stop at the community hall, where they are treated to a performance by one of the local dance groups, which typically take turns so that both groups can earn money.

The Warbelow's tour involves a guided walking tour of the community and the museum, given by the designated local guide, followed by lunch at the village corporation restaurant, then a several-hour tour out in the countryside.

While the NATC approach involves a greater segment of the community, the Warbelow's approach is more entrepreneurial, contracting with a single individual who provides an all-terrain vehicle to take visitors around the village and several miles out on to the surrounding tundra, giving a better sense of place, of the Arctic environment and of the tundra itself.

For nearly all visitors, the local museum serves as a central feature of their experience. It is a drawing card for visitors and a selling point for tour companies, as well as an educational and orientation vehicle to promote understanding of the people and the place. It is noteworthy that Holland America's customer surveys show their Anaktuvuk Pass trip as the top-selling side-trip offered in the state.

The museum has been keeping detailed attendance records since it opened in 1986, and while the level of visitation has grown it appears that the community remains comfortable with the current level of visitation. As part of their effort to ensure that the community does not feel overwhelmed, tour operators always provide comprehensive orientation sessions to their arriving passengers, describing proper etiquette while in the village so as to protect people's privacy, for example by asking permission before photographing people.

One challenge for the Nunamiut is how to develop effectively their own capacity to provide tourist services and thus to benefit more directly from the influx of visitors. Achieving this goal is not necessarily simple or straightforward. For example, some years ago, an enterprising elder built a traditional caribou-skin tent and hand-fashioned an impressive array of traditional tools and implements. He set up the tent in the willows across from his house, and eagerly awaited the arrival of visitors whom he could entertain with stories of traditional life and demonstrations of the making and use of the artefacts he had on display. While the potential customer would most assuredly have had a rare experience and gained a tremendous amount of pleasure from an hour-long visit with this man, the presentation was priced at $75 per person. Predictably the elder attracted very few customers, partially because people had already spent a fair amount for their basic package tours, compounded by the fact that most day visitors travel as couples and even families. The price was simply too high for the circumstances. Unfortunately this elder was unable to set aside his preconception that all visitors are 'rich' (which every once in a while actually is true) and resisted gentle suggestions that by cutting the price to a more affordable $25 he could earn much more by attracting a higher volume of visitors. Disappointed in the negligible results of his summer endeavour, the elder chose not to re-open the following year.

This example illustrates just one of the numerous, but perhaps the most important, potential difficulties of creating new businesses centred on local tourism: not fully understanding the market or the customers.

For this particular community, ecotourism presents one of the most promising and as yet untapped opportunities for local businesses, by capitalizing on their extensive knowledge of the local environment and providing a means of generating relatively greater income from fewer visitors. For example, locals could cater to birdwatchers, as Anaktuvuk is a major flyway through the Brooks Range. Wildlife aficionados could witness the spring and autumn caribou migrations through the valley, plant enthusiasts could witness the greening of the tundra and the profusion of wildflowers that bloom each June. In spring visitors could travel and camp with guides in traditional skin tents and moss houses while they are taken around the countryside by snow machine and sled, learning about place names and hunting areas as well as engaging in the excitement of ice fishing in local rivers and lakes.

The opportunities are almost endless yet they require a substantial investment in time, money, initiative and patience. Structures and sleds must be built, new reliable machines must be purchased along with spare parts, guides will need to be bonded and insured and take advanced first-aid courses, as well as develop the personal skills of tactfully dealing with the demands of tourists who pay a lot for their trip and want things their way.

Whether such services could be provided solely locally or in cooperation with a tour operator, and the extent to which it could become a sustained business, remains to be seen, but what is certain is that local people will have to learn, as all service industry people already know, the customer is always right – even when they aren't.

Northwest Alaska

Tourism as an enterprise in northwest Alaska began during the mid-20th century with its focus on visiting an Eskimo community (Kotzebue) above the Arctic Circle. Entertainment was comprised of viewing traditional activities such as dancing, blanket tossing, reindeer herding, walking around the community and occasionally local people provided boat rides out in the Sound in front of the town. Photographing locals along the beach engaged in processing fish and marine mammals was also common, but this became a nuisance for locals, many of whom eventually relocated these activities to campsites away from town to get away from prying tourist eyes. This type of tourism was packaged and promoted by air transportation companies as a way to increase ticket sales to northwest Alaska, with little economic benefit accruing to the population at large.

Northwest Alaska Regional Corporation (NANA), founded in the 1970s under the Alaska Native Claims Settlement Act, began efforts in the late

1970s and early 1980s to promote tourism in Kotzebue. It opened the NANA Museum of the Arctic to provide cultural and natural history information, including slideshows and dance displays, and Arctic Tours to provide bus tours of the community and the local tundra. Most travellers spend relatively little time in Kotzebue (usually half a day and/or one night) or any of the villages. The major economic benefit of their presence accrues largely to the major airlines, although there are seasonal employment opportunities for museum and tour personnel and local artists selling items in the gift shops. More recently the village of Kiana, in cooperation with Tour Arctic, initiated an effort to bring tourists from Kotzebue out to their community, to provide them with a view of smaller village life and their surrounding environment.

During the 1960s Kotzebue, the regional centre, became well known for polar bear hunting, with dozens of guides working out of the town with Super Cub aircraft before sport hunting of marine mammals was outlawed in the USA in 1972. The presence of Dall sheep, caribou, moose, wolves and bears continues to attract hunters (Fig. 5.1). Fishing, especially for the large and abundant sheefish in the Kobuk River, is another draw to the area, although there are only a couple of local guides and most visitors fly out for do-it-yourself drop-off hunts and fishing. Direct competition stems from allocation of wildlife harvests and management priorities for animal populations and from limited space on the narrow river corridors, where local use patterns conflict with an increased presence of non-locals. Limited numbers of some species, particularly sheep, bear and moose, mean limited harvests (caribou at present are sufficiently abundant to make harvest limits effectively meaningless). Having to accommodate visiting hunters, and allocate to them some portion of the harvest, means that fewer animals are available to local residents.

The debate about allocation and management must be viewed in the context of a state-wide controversy concerning the preference given to rural residents for subsistence harvests. Although the Alaska Constitution gives all state residents equal status regarding fish and wildlife, the federal Alaska National Interest Lands Conservation Act of 1980 grants rural residents a priority for subsistence hunting and fishing. The result is a split in wildlife management, in which federal agencies regulate hunting on federal lands (approximately two-thirds of the state, including some two-thirds of northwest Alaska) while the state manages hunting on state and private lands. Rural residents, not surprisingly, find the federal provisions preferable. Thus, the allocation debate reflects a much larger conflict involving culture, modernization, equality and equity, and so on.

As mentioned above, the issue of space has grown to become the major overriding concern of the region's residents because the time of year (autumn) and the places (along the river corridors) where visiting hunters concentrate coincide with the largest use of these times and places by the residents in the region, causing user conflicts. There are large numbers of local and non-local transporters that make money by dropping off visiting hunters and quite a few private pilots who also visit the region for hunting. Many of them, especially

Fig. 5.1. Hunters and fishermen prepare camping equipment in Kotzebue.
(Photo: A. Whiting.)

the non-local transporters, have little regard for the conflict this causes for local people and some have no regard to the impact this has on the wildlife resources, especially in the smaller drainages which are more vulnerable to over-harvesting and space conflicts. Unlike earlier tourism efforts based on localized non-consumptive uses of both space and animals, lately the increasing number of visitors and their consumptive use of both space and animals have grown to a point where local people can no longer ignore their impacts, but there has been an inability to implement effective control mechanisms. Both federal and state land managers have also been unable to resolve this issue satisfactorily, leading to increasing disenchantment among the local populace with regard to visiting sportsmen.

Since the 1980s, the designation of the Kobuk Valley National Park, the Noatak National Preserve, Cape Krusenstern National Monument and the Selawik National Wildlife Refuge have generated interest in and attention to the region's stunning scenery, abundant wildlife and potential for relatively straightforward journeys by boat or airplane. These visitors take away no living resources, but their presence still raises issues of space, and the question of providing what visitors feel are proper activities for the area they have come to enjoy.

The US Wilderness Act defines wilderness as an area in which humans are just visitors. Tourists to northwestern Alaska fit this definition. Local residents, obviously, do not. While there are undoubtedly many wilderness travellers who are attracted by the possibility of encountering local residents living on the land, there are also many others who regard some local practices, such as the hunting of animals, as antithetical to the wilderness experience. Ironically, both groups may cite the spiritual aspects of wilderness as a major value, but in one case it implies solitude, leaving animals alone and the absence of humans, whereas in the other it implies a continuation of occupancy and activity over countless generations. The gap between these views is not so much wide as it is deep. Both groups want the landscape to remain essentially undisturbed and both claim a substantial stake in how the lands and resources are managed. But close proximity exacerbates differences, and tourism thus remains an activity imposed from afar rather than one embraced and engaged in locally.

Yakutat

Ice bits clink quietly in the rolling swells. The hunters, bundled in white, idle their skiff near a group of harbor seals, hauled out and basking in the sun. Clutching a rifle, one man peers over the bow, while the tillerman cuts the motor. The boat slips forward, shards of ice sliding gently off the hull. As the bowman raises his gun, the fragmented ice shoots sunlight back in a thousand wavering sparkles. Alerted, some of the seals slip from their ice shelf, popping up in the open leads. Several minutes pass. The rifleman sits still, waiting for the right seal to present itself. Finally, a report echoes off the glacier and the engine kicks to life. Racing to the seal, now rolling with the swell, both hunters bend to the water and hoist the large bull into the boat.

(Ray Sensemeier, *Kwa'ashk'i Kwaan* clan)

The Yakutat area's first people arrived over a thousand years ago. Arriving in successive migrations, these clans acquired their existing territories over time, with the ice-laden waters of Disenchantment Bay – formed by several glaciers including the Hubbard – ultimately falling within the territory of the *Kwa'ashk'i Kwaan* clan, the people of the humpback salmon.

Each spring the different clans sent people north, paddling the inter-coastal canoe route to Disenchantment Bay. Here the ice floes and fish-rich waters drew seals in great numbers, and these in turn drew the people. One selected man watched the seal herd until he judged it time to hunt. This annual hunt was extremely important as the original population relied heavily on both seal meat and oil, as well as curing the hides for clothing. Even today, meat and oil are still highly valued and skins are manufactured into many items for sale. Over generations seals became inseparable from the people's lives, with the annual harvest providing not only food and clothing but the spirituality and need for ritual required of any culture. In time the seal was woven into their myth and lore, buttressing an already thriving society. This harvest continued unsullied into the 20th century. Photos taken during the Harriman Expedition of 1899 (Burroughs *et al.*, 1995) showed three seal camps located around the bay, and it is estimated that as many as 1500 animals were taken annually.

Things, however, changed. In 1950 the territory of Alaska initiated a seal bounty, and local hunters – both native and non-native – would cross Yakutat

Bay to Disenchantment, shooting animals off the ice ribbons flowing out with the tide. The harbour seal population plummeted across the state. In part to curb this slide, the Marine Mammal Protection Act was passed in 1972, though it preserved indigenous hunting rights. Yakutat's original people, now mostly Tlingit, continued their traditional hunts. However, harvest numbers have fallen by 30 to 60% over the last 10 years and it is obvious from Harriman expedition reports that the historical abundance has declined. The recent decline may be due to reduced population or changing distribution patterns which make the animals difficult to locate.

A similar decline has been seen at Johns Hopkins Inlet in Glacier Bay National Park, though hunting is not presently allowed. While many variables are involved, one common factor in the two areas is increased boat traffic. Large cruise ships are banned from Johns Hopkins Inlet itself, but they approach the mouth. Small tour vessels, pleasure boats and kayaks enter the inlet only after the end of pupping season in July. Though some researchers believe this suggests vessel disturbance has nothing to do with the decline, traditional non-Western views see animals as fellow inhabitants of the planet with memory, communication and spiritual presence. They ask, 'How can you study the seals that aren't there?' Around Yakutat, other factors may be inhibiting the population's recovery, or even driving seals out of the bay altogether. Many local residents believe the bay's exploding cruise-ship traffic, in part a result of Royal Caribbean's expulsion from Glacier Bay for dumping violations and subsequent shift in itineraries to Yakutat Bay, has become the seals' governing menace.

This thought began circulating around 1993 among local tribal leaders and hunters. Cruise ships during that time numbered 15. Today, 160–170 vessels plough up to Disenchantment Bay's glaciers, a rise which has shaken the community so much that blockades were threatened in 1999. During the Alaska Native Harbor Seal Commission's winter meetings, Ray Sensmeier, a local hunter and *Kwa'ashk'i Kwaan* member, spread the word of fellow hunters. Like him, they were having difficulty harvesting enough seals to keep stocks of smoked meat and oil year-round, as is customary. While ships approaching the ice floes may disturb resting seals, the more alarming factor may be the stress they put on pups and mothers. It was thought this might be driving seals north to Icy Bay, an area receiving little boat traffic. Overwhelmingly concerned, and working within the political framework, the hunters and other tribal members pushed for an investigation.

As an international, multi-billion dollar industry, the cruise-ship interests have substantial clout politically. It takes a great deal of time, resources and research to force a policy change, but the City and Borough of Yakutat decided to try. Working off a monitoring plan drafted by local Forest Service biologists and the Tribal Environmental Director at the time, the city proposed a tax on passing ships. A portion of that tax would subsequently fund research exploring whether or not the cruise industry adversely impacts the Disenchantment seal population. Ambient factors such as water and air quality would be monitored, as the cruise ships, large and powerful, produce a great deal of smoke as well as other wastes known as black and grey water.

While these wastes may affect seals, they are undoubtedly an overall ecological concern, and would be addressed in a comprehensive monitoring programme regardless. The overriding factor concerning seals, however, was and remains disturbance on the ice floes. The idea of a tax was foregone during negotiations in favour of a direct transfer of monitoring dollars from the industry to the Yakutat Tlingit Tribe. The city received additional annual funds to support tourism and recreation infrastructure and to cover costs for emergency services utilized by the industry. The monitoring funds have been used to design a disturbance-monitoring programme with the Seattle-based National Marine Mammal Laboratory, under the National Marine Fisheries Service (NMFS). The first phase of this study was implemented in 2002, with the assistance of tribal personnel.

The study was completed and released in March 2006. Basic findings suggest that cruise ships are indeed encountering large numbers of seals during their journey to the calving Hubbard. On-board NMFS and tribal observers recorded multiple interactions, and calculations showed seals consistently abandoning the icebergs when ships approached within 500 m. More importantly, mothers and pups left at a higher rate when a ship was present, supporting hunters' concerns that seals receive increased disturbance during the critical pupping season. In addition seals tended to form larger groups as traffic increased, which, according to biologists of the National Oceanic and Atmospheric Administration, is a sign of stress. However, by far the most interesting finding was that both Disenchantment Bay and Icy Bay seals begin the pupping season in May with similar numbers of animals. After pupping and into moulting in August, the Icy Bay population more than doubles while Disenchantment Bay seals remain stable or decline slightly.

These preliminary results pose several questions. What is the overall effect on the population? Is disturbance enough to interrupt feeding and nursing behaviour? Are large numbers of seals abandoning their traditional pupping grounds in favour of Icy Bay, 40 miles to the north? To address these questions the tribe has hired a wildlife biologist, with settlement funds, to cooperate with NMFS researchers. Additionally, the Northwest Cruise Ship Association has agreed to five additional years of payments to the tribe, a portion of which will go towards the biologist and the hope of answers. There may be ways, for instance, for the industry to alter its travel corridor to avoid high seal concentrations, or to perhaps simply limit the number of vessels. As traditional life-styles and the modern economy continue their attempts to mesh, it is hoped by all parties that the future holds not only a business opportunity for an international market seeking Yakutat's beauty, but also a promise of sanctuary for the seals and the people who depend on them.

Conclusions

Tourism can create economic opportunity, can be immensely rewarding for the tourists themselves and can generate understanding about different ways of life and different environments. At the same time, the presence of visitors, even in relatively small numbers, can disrupt local patterns and overwhelm local services in remote areas. Furthermore, tourists may have different views from locals about appropriate ways to use lands and resources, views that may have clout through the greater political power or access to management agencies that tourists possess through affluence or sheer numbers. Even in the smallest community, it is unlikely that the costs and benefits of tourism will be evenly distributed. Instead, what is opportunity for one is burden for another, what is invasion to one is income to his or her neighbour.

In Anaktuvuk Pass, tourism supports local cultural resources such as the museum and the dance groups, but most of the economic benefits go to the tour operators. Developing local value-added services such as photo safaris or tours of a traditional tent have promise, but face pitfalls such as setting prices and getting liability insurance. Using culture as a marketing tool is apparently effective in this case, but runs the risk of implying that the whole village is on display. Currently, tour operators are careful to instruct their clients in proper etiquette, but it is not clear what will be the impacts of further development of tourism in either numbers or types of activities. In short though, and due to many of the particular features of Anaktuvuk Pass, its setting and its residents, tourism here offers an important component of the local economy.

In northwestern Alaska, tourism contributes to the economy as well, but the benefits do not appear to be either as extensive or as unambiguous. In the context of state-wide conflicts over land and resource use and priorities, tourist traffic on the land can feed into pre-existing controversies and can heighten the tension that exists about who determines what appropriate human activity is. This situation may be the result of reaching a certain level of saturation in terms of the number of visitors the region can absorb. A handful of visitors may be merely a curiosity, whereas a steady stream of hunters or even trickle of boaters on a river keeps user conflicts over space in the foreground.

In Yakutat, the stunning scenery draws visitors while seals and seal hunters suffer the incidental impacts of the cruise ships. Benefits to residents are less clear, but cooperation has increased with industry donations to the local school system, the town's emergency services and the initiation of a tribal on-board interpreter programme. The potential conflict with the hunters is more difficult as tourists and local residents have different interests that conflict. Tourists want to see the glaciers and to get as close as they can, but are likely to be largely indifferent to the disturbance of seals or the success of seal hunters. The hunters want to protect the seal population and seal habitat, and gain little from the presence or satisfaction of the tourists. As is the case elsewhere, the volume of tourism is a crucial factor. Short of either a ban on Hubbard Glacier cruise traffic or seal hunting itself, neither the cruise industry nor the Yakutat residents will get all of what they want. Compromise, though, however frustrating, should be obtainable, and money, as always, is certainly an option.

From these experiences, we draw three lessons that we believe are relevant to tourism operation in rural Alaska and perhaps beyond.

Local involvement is essential. Tour companies and the communities should work together from the outset, determining what scenic and cultural resources the community has to offer, and how willing and able local residents are to show and interpret those resources to best advantage. Communities in this context need to feel that they are partners in the enterprise, realizing financial benefits, rather than just being observers. Hospitable community members can greatly enhance a visitor's experience, whereas unfriendly receptions may make a destination unpopular. At the same time, tour companies and communities need to have realistic expectations of one another, for example with regard to the frequency with which local cultural programmes can be offered without disrupting other priorities of local residents.

Compromise is likely to be necessary. Not all potential tourism can be accommodated without undue impact on the locals. The sport/hunting conflicts in northwestern Alaska or the cruise ships/harbour seals issues in Yakutat cannot easily be turned into win–win solutions. Instead, tour operators have to accept that some activities will have negative impacts and will either have to be stopped or will continue to arouse the ire of the locals, unless some compromise is reached. In the Yakutat case, cruise lines benefiting from the Hubbard could continue to pay a negotiable fee to the City and Borough of Yakutat and the Yakutat Tlingit Tribe each year, thereby augmenting the town's ability to improve itself. In addition, the research completed to date suggests that a sea-lane could be designated that would significantly reduce interactions with hauled-out seals, as would a cap on the number of vessels entering the bay. For their part, hunters might assist in monitoring population trends with the tribal biologist and continue their participation in the annual harvest survey. This information could be used to fine-tune a management plan. As in all compromise, of course, communication and a willingness to bargain will be crucial on both sides if an accord of some kind is to be reached.

Good communication and good planning can help a great deal. Local involvement and compromise clearly require effective communication, a key component of which is planning. Tour companies and communities can and should establish common goals, identify potential points of conflict, and create mechanisms to address problems that arise. Communities should designate a specific community organization (or consortium of local organizations) to work with tour companies. In Anaktuvuk Pass, the NATC had a representative live in the village for the summer to get a sense of community interests, help organize local activities such as dance programmes and craft tables, and generally keep relations smooth. Communities can also share their experiences with one another, for example by allowing community leaders to go on tours offered elsewhere, or by arranging for residents of different communities to talk with and learn from one another.

Tourism in Alaska is undoubtedly here to stay. How it can be managed in rural areas so as to provide economic benefits and visitor opportunities without undue disruption remains a challenge, but one that can perhaps be met by heeding the lessons from past experience.

References

Burroughs, P., Grinnell, G.B. and Wyatt, V. (1995) *Alaska 1899, Essays from the Harriman Expedition*. University of Washington Press, Seattle, Washington.

Colt, S. and Huntington, H.P. (2002) Oceans, watersheds, and humans: facts, myths, and realities. In: *The Status of Alaska's Oceans and Watersheds 2002*. Exxon Valdez Oil Spill Trustee Council, Anchorage, Alaska, pp. 109–117.

Goldsmith, S. (1997) *Structural Analysis of the Alaska Economy: A Perspective from 1997*. Institute of Social and Economic Research, Anchorage, Alaska.

Muir, J. (1915) *Travels in Alaska*. Houghton-Mifflin, Boston, Massachusetts.

Wilson, J. (1903) *Regulations for the Protection of Game in Alaska*. Biological Survey Circular No. 39. US Department of Agriculture, Washington, DC.

6 Development of Tourism in Arctic Canada

MIKE ROBBINS

The Tourism Company, 146 Laird Drive, Suite 201, Toronto, Ontario, Canada, M4G 3V7

Introduction

This chapter outlines some of the author's experiences in helping Inuit communities of the Canadian Arctic to develop carefully controlled ecotourism during their early years of self-government, when catering for tourists indicated a possible route towards both economic self-sufficiency and cultural preservation.

In early May 2003 I found myself travelling 'on the land' in northern Labrador, with Innu from Sheshatshiu and Natuashish (formerly Davis Inlet), some of Canada's northern people. Historically known as the Naskapi-Montagnais Indians, these people have for centuries occupied their homeland, Nitassinan, covering the eastern portion of the Quebec/Labrador peninsula. I was working on a contract for the Tshikapisk Foundation, a cultural organization, preparing a business plan for a proposed cultural ecolodge being built at Kamestastin Lake near the Quebec border. Their concept was to build a log wilderness lodge and cabins on the shores of the lake where their ancestors used to camp when harvesting caribou crossing the narrows. The ecolodge as envisioned would cater to international travellers interested in indigenous culture combined with adventure, wildlife viewing and archaeology.

We travelled through the day and well into the night (14 h) before we finally reached Kamestastin Lake, having traversed close to 100 miles up a river valley from the coast, over the tundra plateau and then back down a river valley to the lake. That night we sat around the stove in a small temporary cabin built at the ecolodge site, talking about the plans for the lodge and what it meant to the Innu. My companions expressed their hopes of what this facility would bring to the communities and particularly the youth.

They also expressed frustration about their inability to secure much-needed funding support from the government. The Canadian federal government agency responsible for economic development in eastern Canada was not willing to provide capital support for the project. The government representatives could not understand the concept of a cultural ecolodge, instead deferring to their experience with southern-owned sport fishing camps in remote northern locations. And yet this project offered the potential to create a positive direction in these two Innu communities being torn apart by high rates of alcoholism, drug use, solvent abuse and youth suicide. As I lay on my bed in my arctic sleeping bag that night my mind began drifting back to my experiences 25 years earlier with the Inuit in Canada's eastern Arctic.

Early Tourism Ventures

Nunavut (Canada's newest territory) and the Inuit had their first experiences with tourism in a planned manner through the efforts of Parks Canada, which began operating Auyuittuq National Park Reserve near Pangnirtung back in the 1970s. This was followed by Ellesmere Island National Park Reserve (now Quttinirpaaq National Park), formally established in 1988.

My own involvement began in the early 1980s, when the then Northwest Territorial Government initiated the first pilot project for community-based tourism planning and development in Pangnirtung (Pang), on Baffin Island. I was part of the planning team assigned to work with the community in developing the plan. At the time most of the Inuit in the community did not differentiate between actual tourists and federal or territorial government officials (non-Inuit) who had been coming in and out of the community on business trips for years. Their perception was that all Qallunaat (non-Inuit) were anxious to complete their business and leave the community as soon as possible. There were, in fact, few tourists coming to Pang or any of the Baffin communities. The tourism industry that did exist in the Northwest Territories (NWT) consisted primarily of remote fishing camps owned and operated by southerners (resulting in little local economic benefit), a few independent adventurers and the odd big-game hunter.

The perceived success of the Pangnirtung community-based tourism initiative led to the completion of community-based tourism plans for each of the other Inuit communities in the Baffin, Keewatin (now Kivalliq) and Kitikmeot regions through the 1980s. These initiatives were all government-funded, but in the Keewatin and Kitikmeot regions the regional Chambers of Commerce, rather than government, were the proponents. Once again I was part of the team that completed much of this work. Community-based tourism in this context referred to adoption of the following principles:

- The community makes decisions to pursue tourism (or not) based on knowledge of the pros and cons of tourism.
- Extensive community involvement and consultation is completed.
- The community develops a strategy (with outside expertise and assistance)

which defines how much control the community should have, and the events, activities and places that can and cannot be shared with tourists.

- The community makes strategic investments to be a catalyst for private sector investment.
- The community continues to monitor and evaluate tourism initiatives and development.

From the mid-1980s, when most of this community-based planning was completed, much progress was made on implementing the recommendations. Through the late 1980s and into the early 1990s the federal government invested in tourism infrastructure in many of the communities. The focus on tourism as an important form of community economic development resulted in significant growth in licensed outfitters and new Inuit-owned and operated businesses. In the summer of 1998 the author travelled as a tourist to Nunavut with his son to directly experience the progress that had been made.

Post-millennium Developments

In 2001 the Conference Board of Canada singled out tourism as one of the growth pillars for the Nunavut economy, along with mining and fisheries development. Aboriginal Tourism Canada (ATC), a partnership between business and government with a mission to develop tourism policies and programmes to benefit Aboriginal people in Canada, identified the following benefits from Aboriginal tourism:

- Helping cultural revival within a community.
- Fostering a sense of pride.
- Teaching young people about their history and heritage.
- Helping employees to develop front-line and management skills.
- Helping to dispel the stereotypical image of Aboriginal people.
- Helping employees to gain transferable skills.
- Allowing new Aboriginal partnerships with neighbours and businesses.
- Sharing Aboriginal culture and heritage with the rest of the world.

A National Study on Aboriginal Tourism was completed in 2003 and published by ATC (Bearing Point LP, Goss Gilroy Inc. & Associates, 2003a,b). This research helped to quantify the existing Aboriginal tourism sector in Canada. The following information is excerpted from a speech made by the Chair of ATC in April 2003:

- The economic activity generated by Aboriginal tourism businesses is significantly higher than previous estimates.
- In 2001 the *total* economic activity generated by Aboriginal tourism establishments, inclusive of casinos, was Can$4.9 billion, of which Can$2.9 billion resulted from tourism expenditures (direct, indirect and induced Gross Domestic Product, GDP). Thus tourism expenditures represented 59% of the total output.

- The *direct* contribution of all Aboriginal tourism businesses to GDP was Can$596 million, about Can$290 million when casinos were excluded.
- Using survey findings, direct employment by Aboriginal tourism businesses including casinos was estimated to be 13,000 jobs (full-time equivalents). Of these, 9500 were a result of tourist spending.

These figures may be compared with the total tourism expenditures in Canada in 2004 of Can$55.5 billion and total employment across the industry of 578,000 full-time and part-time positions in 2004. The Aboriginal tourism sector in Canada is a very small part of the overall tourism industry; yet there is substantial latent demand and huge potential for growth.

The study reaffirmed the continuing need for support to the industry, in order to address issues relating to:

- Building community capacity and tourism awareness.
- Increasing the number of market-ready products.
- Accessing markets/financing.
- Training to raise the level of business, hospitality, product packaging and marketing skills.

These issues were very relevant to Nunavut tourism 25 years ago, and are still relevant today.

Planning for Pangnirtung

My interest in tourism in Arctic regions began in the mid-1970s when I worked for a summer on a building construction crew in Yellowknife, to pay for university. I was intrigued by the allure of communities further north and actually applied, unsuccessfully, for an economic development position in a Far North community. My opportunity came about 5 years later, while working in the Environmental and Tourism Planning Department of Marshall Macklin Monaghan, a major Canadian multidisciplinary engineering and planning firm. My educational background was in environmental planning but my true interest lay in the tourism industry. We won a contract to conduct a pilot project on community-based tourism in the community of Pangnirtung.

In 1981 a colleague, Harry French, and I left for our first of many trips into Pangnirtung, or Pang as it is called locally. We carried with us a small portable video machine with a presentation that we had translated into Inuktitut. The presentation provided a very basic introduction to the pros and cons of community tourism, with images and examples from down south. Prior to our visit we had hired a local sub-consultant, one of whose first tasks was to circulate a notice in Inuktitut, to every house in the community, of our impending visit. Throughout our work in Pang our local sub-consultant worked as our assistant to facilitate knowledge transfer and to interpret. Accommodation had been arranged in the Anglican mission house, and we rented a small building for our office down by the shoreline (Fig. 6.1).

Fig. 6.1. Marshall Macklin Monaghan Office in Pang. (Photo: M. Robbins.)

Pang (Fig. 6.2) is a community situated on the edge of a very scenic fjord that leads from Cumberland Sound all the way down to Auyuittuq National Park. The airstrip divides the community, and provides for only two landing approaches up against the steep fjord wall. The surrounding area is rich in pre-historic and historic resources, ranging from Thule sites to old whaling stations, in addition to spectacular natural features and sites.

This, our first contract in the NWT (now Nunavut), was with the Territorial Government, Department of Economic Development. Fortunately we had sufficient budget to spend several weeks in the community in each of the four seasons. This enabled us to participate in such community activities as bingo, and weekend fishing trips and seal hunting trips in Cumberland Sound. From the outset we knew we had to differentiate ourselves from typical government employees, who would fly into the community and leave as soon as possible after their work was done. The Community Economic Development Officer at the time was Katherine Trumper, a southerner who had spent a number of years in the community learning the language, the customs and culture. We benefited immensely from her wise counsel on how to conduct our business.

The government realized that tourism represented one of the few opportunities to create a local economy, which at the time was largely dependent on government employment. A fish plant had opened around 1980, but the local economy was being devastated by the seal fur ban which began in the 1970s, and there was substantial unemployment.

Over time we became accepted in the community; people began to open up and share stories and information with us. We conducted a wide range of consultation techniques including:

● Community meetings.
● Local radio talkback shows.

Fig. 6.2. Baffin Island, showing the position of Pangnirtung (arrow). (Source: Nunavut Parks website, http://www.nunavutparks.com/on_the_land/map.cfm.)

- Drop-in sessions in our office.
- Individual group meetings with the many community groups, such as Hunters and Trappers, Drug and Alcohol Committee, Education Committee and Women's Committee.
- Individual meetings.

We also made a concerted effort to learn some Inuktitut, at least some of the basic words, so the local Inuit could see we were trying to understand and work within their unique culture.

At the time Pang did not have much to offer visitors other than a trip down the fjord (with just one outfitter) to Auyuittuq National Park Reserve, including a visit to the Park headquarters, with limited interpretive displays or a visit to the local print shop and weaving centre. The hotel, owned and run by a southerner, was run-down and overly expensive, and focused on southern fishermen and contractors. The few tourists who were coming to the community at this point were typically self-sufficient independent adventure travellers, mostly hikers and climbers travelling into Auyuittuq, or fishermen staying at Peyton's Lodge. There was very little benefit to the local Inuit, other than those few who were working for Parks Canada and/or the lodge.

We came to realize some of the unique characteristics of the Inuit:

- Local people did not understand tourism or what tourists might be interested in seeing or doing.
- Individuals often sat on numerous groups and committees and would show up at separate meetings, asking the same questions repeatedly for the benefit of the different audiences.
- Many people were more comfortable speaking anonymously to us over the radio than in person, even though others knew who they were by their voice.
- Their greatest love was to be out 'on the land' – most residents go out on the land for extended periods during the summer months.
- A major concern was to ensure younger generations retained the culture. This was being threatened as they were under strong influence, through school and the media (such as the Atlanta Super Station), to move towards a southern type life-style.
- Many of the elders in the community did not speak any English.

The most rewarding experiences came for us both with opportunities to visit the seniors' room in the back of the church (Fig. 6.3), where elders would meet to work on crafts, talk and eat rotted seal meat by the *qulliq* or seal-oil lamp. With our interpreter we could listen in on some of the stories, like the one old woman who told us of her first experience of seeing an airplane, which she initially believed to be a giant mosquito.

One of the most effective communication tools we used was large-scale topographic maps of the area around Pang, on which we inventoried sites of potential interest for tourists with coloured pins. The Inuit proved to be very adept at reading maps and we soon had two pin-covered maps – 'Resources of the Land' and 'Resources of the People', based on the input of the locals (Fig. 6.4). Educating the Inuit in what visitors might be interested to see and do in and around the community was a major part of the process. Valuing places less for scenic beauty but more for usage (fishing, hunting, berry picking, way-finding, camping, etc.), they initially ignored many potential tourist sites as not particularly interesting. This process of mapping cultural values as a means of preserving and interpreting cultural heritage, and helping to establish traditional territories, has since become increasingly commonplace with First Nations in the southern provinces like Ontario and Manitoba.

Fig. 6.3. Pang seniors' room, 1982. (Photo: M. Robbins.)

We visited these sites in all four seasons. One boat trip by 'Moosehead' canoe into Cumberland Sound, to visit an old whaling station, ended up being part hunting trip and part tourist trip. Our guide with the ever-present rifle could not resist the opportunity to fire at a seal that popped up some 100 m from the boat. The next thing we knew, following a high-speed burst in the boat, the seal carcass was being hauled bleeding across the gunwhales. Another time, travelling back from Auyuittuq National Park with the licensed

Fig. 6.4. The author with Pang Tourism Values maps. (Photo: M. Robbins.)

outfitter, we ran out of gas and had to drift until a passing boat stopped to check on us and tow us back to the community. He had forgotten to top up the tanks that morning. A common feature of any trip out on the land, whether in summer or winter, was the tea stop. In summer, on the water, the tea stop would often be accomplished by running the canoe up on to an ice pan (Fig. 6.5).

At the culmination of our year's work and consultation in Pang, we developed a community-based tourism plan with strong community support and involvement, and a process that could be replicated in other Inuit communities (Marshall Macklin Monaghan Ltd, 1982). The plan focused on tourism that the community could control through tour operators and organized groups, rather than independent visitors. A major emphasis was placed on conservation and preservation of natural and cultural heritage resources. The plan was entirely consistent with today's principles of ecotourism (focusing on education, conservation/preservation and community benefits), a term that had not yet been coined. A tourism committee was formed to maintain community involvement and control. Projects such as a visitors' centre with an elders' room, boat tours to the Kekerten Island whaling station, hiking trails accessible from the community, spring snowmobile trips and a new lodge were included in the plan (Fig. 6.6).

What has been achieved in the 25 years since the Community-Based Tourism Plan was launched in Pang? The government focus on tourism as an important form of economic development continued through the 1980s. In Pang the government invested in building an interpretive centre with a museum and elders' room (Fig. 6.3), building a new community-owned lodge,

Fig. 6.5. Tea time in Cumberland Sound. (Photo: M. Robbins.)

Fig. 6.6. Conceptual drawing of visitors centre, 1982. (Source: Pangnirtung Community-Based Tourism Plan.)

developing a territorial historic park in Cumberland Sound (the Kekerton Island whaling site) and funding the restoration of the Hudson's Bay Company Blubber Station for interpretation. Today the community has a much wider range of tourism facilities, attractions and services as follows (most are Inuit- or community-owned):

- Lodge with 25 rooms and a dining room more closely matching today's tourist standards.
- Community campground.
- One bed-and-breakfast accommodating up to four visitors.
- Fast food restaurant.
- Four outfitting companies.
- A visitor interpretive centre with an elders' facility and community library.
- Parks Canada interpretive centre.
- Hudson's Bay Company historic attraction – Blubber Station.
- Local taxi service.
- Three arts and crafts shops.
- Travel agency.
- Kekerten Territorial Park (a restored whaling station accessible only by boat with a local guide).
- Several nearby hiking trails.

By 1999 it was estimated the tourism sector in Pang accounted for 35 permanent and 14 seasonal jobs, in addition to providing income for over 100 home-based artists (10% of local jobs) (Budke, 1999).

Significant progress has been achieved but there are still many issues to be resolved and worked on. The seasonality of the tourism industry is one major hurdle. The prime tourism season of July and August could be expanded to include the spring snow season in April and May and the pleasant weather during the transition months of June and September. There is also potential for winter activities during the long winter months from November to March. As with Aboriginal tourism in most parts of Canada there is still need to work on improving the local community capacity and awareness for tourism, providing enhanced training for tourism employees and business operators, and upgrading the market-readiness of products and creating more packaged experiences to sell.

Other Developments in Nunavut

Pangnirtung, the first community-based tourism plan, is a useful case to view in terms of progress achieved over the past 25 years. However, further significant progress has been made across Nunavut. In the 1980s, when we worked in each of the communities, many did not even have hotels or lodges, let alone tourism programmes and attractions. The many exciting adventures and experiences we had in other communities we visited are perhaps worthy of a book in their own right. The Inuit love to be on the land, and we were paying them to take us out there. In the community of Igloolik, a very traditional community with many working dog teams, I had announced over the radio that I would be interested in going out with a team over the weekend. The old man who offered to take me out could speak no English so he communicated only by example. We travelled across frozen sea ice and barren ground for many hours. When we returned to the community his son came out to help un-harness the dogs. He took one look at me in my oversized southern parka with a long fox-fur-trimmed hood and told me my face was frozen. How could that be when I was able to turn my face from the wind and hide in the comfort of my huge hood, when my Inuk driver was barefaced to the wind and yet he showed no signs of frostbite? But my face was badly frozen – luckily the face heals quickly.

Another adventure took me to Coats and Bencas islands south of Coral Harbour travelling in a longliner fishing boat with an Inuit family. We crossed this part of northern Hudson Bay in eight hours. At Bencas Island we landed and started walking downwind from a walrus herd. Someone spotted what appeared to be two remnant snowdrifts on the ridge. Walking towards the 'drifts' we soon realized they were in fact sleeping polar bears that now began to sense our presence and rose to move their heads back and forth and sniff the air for our scent. Our guide picked up a piece of driftwood and began to whack at rocks like a baseball batter, making a loud cracking sound which eventually scared the bears and they hightailed it into the sea. Figure 6.7 was taken from the boat as we followed them out in the water to get a closer look – not necessarily the best approach for a tourist guide, but it made for great adventure for the consultants and the Inuit kids.

Fig. 6.7. Polar bears off Coats Island. (Photo: M. Robbins.)

Nunavut Tourism

The diversity of Inuit-owned and operated tourism products and experiences across Nunavut today is extraordinary, ranging from boat trips to view old whaling stations, or wintering sites of old whaling ships, or staying at a remote ecolodge like Bathurst Inlet in the Kitikmeot Region or Sila Lodge in Ukkusiksalik National Park (closed for the past two seasons but hoped to open again soon), to travelling the Arctic with an Inuit-owned expedition cruising company like Cruise North (launched July 2005). All of the 25 Nunavut communities have at least basic tourist lodge or hotel accommodations, and some have alternative accommodation such as bed-and-breakfasts. And there are many trained professional outfitters and guides throughout Nunavut.

To help place tourism in Nunavut on the international map, Nunavut Tourism was established in 1996/7 as an industry organization (funded by membership fees and government contributions) with the responsibility for marketing, product development and training within the tourism sector across Nunavut. The efforts of Nunavut Tourism are guided by a Board of Directors comprised of 14 people with ten private sector representatives and four government appointees. This is similar to the joint private/public sector model applied in many other Canadian provinces. Nunavut Tourism currently manages a considerable amount of marketing and promotion with a relatively tiny budget of Can$2.3 million per year (compared with certain city-destination marketing organizations in southern Canada with ten times this budget). Nunavut Tourism is financially constrained from fully implementing its mandate for product development and training across the vast geography of Nunavut. The Government of Nunavut still retains responsibility for capital planning, Parks, licensing, regulations and enforcement through two divisions – Parks and Conservation, and Tourism.

National and Territorial Parks

Today there are 13 Nunavut (Territorial) Parks (under historic, campground, destination and wildlife sanctuary categories) and plans for additional parks in four other communities as well as four National Parks (Auyuittuq, Sirmilik and Quttinirpaaq and Ukkusiksalik; Fig. 6.8), one proposed National Park (North Bathurst Island), six National Bird Sanctuaries, two established and one proposed National Wildlife Sanctuaries, three National Heritage Rivers and 11 National Historic Sites. The value of parks and protected areas to the tourism industry as both attractions in themselves and to provide credibility to a destination is well established in the tourism research literature. These are all strong additions to the Nunavut experience and they provide significant benefits to the local Inuit.

The Nunavut government and its NWT government predecessor have also established ten visitor and interpretive centres housing a range of

Fig. 6.8. Nunavut National Parks of Canada. (Source: Parks Canada, Nunavut Field Unit.)

orientation, visitor information, historic and archaeological displays in addition to providing community uses such as elders' rooms and community libraries. These centres have proved to be critical infrastructure in support of cultural preservation. In addition, most of the 25 communities now have hotel and/or lodge accommodation designed to meet the modern visitor's needs and expectations. Inns North, a division of Arctic Co-operatives (service organization owned and controlled by 35 communities), owns and operates hotels in 18 Nunavut communities as well as several in the NWT. The tourism infrastructure across Nunavut has expanded significantly over the past 30 years; however this was in large part the result of capital availability through a Canada–NWT Economic Development Agreement that was in place during the 1980s, but no longer exists.

Strategic planning

From a capital investment perspective the 1990s appears to have been a decade of stagnation for the tourism industry in the NWT, and the creation of Nunavut in April 1999 did little to change this situation. The Nunavut Land Claims Agreement in 1993 did however result in a financial settlement that provided capital to fund the eventual ownership in Canadian North Airlines (50% held by Nunasi Corporation) and in Top of the World 2000, a Nunavut-wide travel agency. Despite this lack of investment in the communities, tourism visitation to Nunavut increased throughout the 1990s. A Strategic Plan for Tourism Development in Nunavut (Blackstone Corporation, 2001) completed for Nunavut Tourism and the Department of Sustainable Development in 2001 identified the following visitor numbers for the year based on discussions with airlines, tour operators and outfitters (there is no quantitative visitor research conducted in Nunavut so these figures are estimates):

- 250 hunters.
- 500 fishers.
- 1500–2000 ecotourists.
- 100 extreme adventurers.
- 500–800 campers and hikers in the parks.
- 1000–1725 cruise-ship passengers.
- 13,000+ business (including government) travellers.

Most of these tourists are still being packaged by southerners, suggesting a significant future opportunity for the Inuit. The same report identifies that approximately 500 people across Nunavut are employed directly by the tourism industry, making tourism the largest sector employer outside government. The arts and crafts sector is an important part of the tourism industry. There are an estimated 3000 people involved in this sector.

There are many good models of Inuit-owned and operated tourism businesses throughout Nunavut. Both Bathurst Inlet Lodge and Sila Lodge are owned and operated as partnerships between Inuit families or communities and non-Inuit living in the north. Bathurst Inlet Lodge has operated successfully since 1969. Sila Lodge was built in 1986 and operated for many seasons but has been closed for the past couple of seasons due to financial sustainability issues caused in part by difficult access and high operational costs in a remote location. Bathurst Inlet Lodge hires locally and buys needed goods and services locally to the fullest extent possible. These and other examples clearly illustrate the benefits of tourism in Nunavut:

- The strong interest in Inuit culture from visitors creates pride for local people in their culture and home (Fig. 6.9).
- Provides steady seasonal employment in communities with high unemployment rates and little opportunity other than government.
- Tourism business assets can also benefit the local people – for example a bus used to transport guests is often chartered by the community in the off-season for conferences or government meetings.
- Stimulates preservation of artefacts, customs and traditional knowledge.
- Education through tourism training helps to encourage an interest in knowledge and learning.
- Encourages realization of the value and sustainable use of cultural assets like archaeological sites.
- Visitors learn about, and begin to understand, the Inuit culture.

Parks and protected areas have also proved to be beneficial tourism attractions. In 1994 an Economic Impact Assessment was completed for Katannilik Territorial Park located between Iqaluit and Kimmirut (Downie and Monteith, 1994). The area encompassed by the park was recognized in our original community-based tourism planning for these two communities as a potential visitor attraction for conservation-related activities. Today the Park covers 1500 sq km along the Soper River valley. In 1989/90 the Department of Economic Development completed a feasibility study which recommended the creation of a territorial park and the designation of the river as a Canadian Heritage River. The park master planning and development process took place through the early to mid-1990s. One of the key objectives in creating the park was to make a contribution to the expansion of the economic base in the community of Lake Harbour (now Kimmirut) and the region.

In 1993, when the economic impact work was being done, community employment in Kimmirut stood at 66 with only 12 positions in the private sector – the Co-op and Northern stores. The others were all government, local, territorial or federal. In 1993, the first year of operations, the park catered to 115 visitors, most of whom were in tour groups. These new visitors spent Can$49,395 in total in the community, an average of Can$430 per person/trip, and Can$109,905 in the region. This compared against the total park operation costs that year of Can$183,047, including capital costs for development of the park infrastructure. The most important conclusion was that the park is resulting in a small but significant new injection into the local economy. Prior to the development of the park, tourism to the community was negligible, with an estimated 50 visitors over the previous 5 years.

Cruise tourism

Another important and growing sector for Nunavut tourism and the communities is cruise tourism. Expedition cruise ships generally are smaller in size with capacities of 60–160 passengers. The cruise industry has been both benefiting and negatively impacting the communities in Nunavut. On the positive side cruise ships typically visit communities for several hours only, and do not require expensive tourism infrastructure such as accommodation and food services. They can result in significant new money in the community if planned properly with involvement of the community. A number of communities have developed visitor or heritage centres as a focal attraction for cruise visitors. On the negative side the communities often lack control over when cruise ships visit and what they do when they are in the community. Cruise ships have also been known to visit sensitive and inappropriate locations without permission and supervision – such as the 40 million year-old petrified forests on Axel Heiberg Island or Hebron and Nachvak Fjord in northern Labrador.

With the help of Nunavut Tourism the communities are working to mitigate the negatives by working proactively with the cruise companies, developing programmes for cruise visitors in the communities, and consideration is being given to requiring walk-on local guides while in Nunavut waters. In northern Quebec the Inuit have begun to take charge of these issues by starting their own expedition cruise line, Cruise North, with monies from their land claims settlement. Arctic cruises typically travel the Labrador coast en route to northern Quebec and Nunavut. In Labrador the Inuit communities, strung out along the northern coast, signed their land claims agreement in 2004. They were the last Canadian Inuit group to finalize a land claims agreement. These communities are playing an active role in Destination Labrador, the industry advocacy and marketing organization for Labrador, but from a tourism infrastructure perspective these communities are at an even earlier stage of development.

Summary and Conclusions

In summary, the tourism sector in Nunavut is still at a very early stage of development. The industry is small but has a significant impact in the communities by bringing in 'new money' and providing stimulus to preserve the local culture and traditions. The new government in Nunavut is unfortunately not placing high priority on tourism as a form of economic development. Priorities instead seem to lie with other industrial sectors such as mining and fisheries as well as the domestic needs of schooling and health care. This fact was even reflected in the Inuit land claims agreement in Nunavut, back in 1993, which does not address tourism in a direct way. The agreement does, however, directly address mining, petroleum and resource development.

The government division responsible for tourism has been placed under the Department of Economic Development and Transportation, separate from Parks, and the Division has been decentralized to Pangnirtung. The three Regional Inuit Associations that have been created to assist the communities with business development and training are for the most part not focused on tourism. The Inuit Corporations are strongly profit-oriented and have not yet played a major role in community-based tourism development. All of this leaves Nunavut Tourism working with limited funds to develop and market the Nunavut tourism sector, a sector with huge growth potential but a lack of catalyst. Nunavut Tourism does not have designated Inuit status, but rather is considered a government entity with non-Inuit members. As a result the organization does not have access to funds beyond their existing sources. Parks Canada continues to be a major player in the development of the tourism sector in Nunavut. It is hoped that the Nunavut government will begin to realize that tourism offers significant opportunity for sustainable economic development in many of the 25 communities.

The Inuit in Nunavut, as in the beginning some 25 years ago, are interested in tourism as a form of economic development and employment, but they are also still concerned with community control to minimize the intrusive nature of tourism. There is a need, and a very significant opportunity, for the Inuit to be further involved in the tourism supply chain, as most tourists coming into the territory are being packaged by southern companies. The keys to future growth of tourism in Nunavut may well continue to lie in more and enhanced training opportunities for those working in the sector, better community and political awareness of the benefits of tourism, access to capital, development of more export-ready products and experiences, and involvement in all stages of the visitor booking and travel process to allow for more control of tourism in Inuit hands.

These are common issues across the Aboriginal tourism sector in Canada. In an effort to provide the Aboriginal tourism sector with some competitive advantage and at the same time develop product to higher standards and maintain cultural integrity, perhaps there should be consideration given to an Aboriginal certification or branding programme, certifying products as authentic and as reaching certain minimum service standards. This approach has proved successful with Inuit art. The Australian Nature and Ecotourism Accreditation Program (NEAP) provides a good example of the effectiveness and benefits of such a programme.

The future for tourism in Nunavut is positive but will require both community and government recognition and support to reach its potential. Much has been accomplished over the past 25 years; the hope is that, 25 years hence, cultural ecotourism will play an integral role in sustainable community economies and Inuit cultural preservation throughout Nunavut, and in other northern provinces such as Labrador.

Fig. 6.9. Baker Lake traditional camp, 1985. (Photo: M. Robbins.)

References

Bearing Point LP, Goss Gilroy Inc. & Associates (2003a) *Aboriginal Tourism in Canada, Part I: Economic Impact Analysis*. Prepared for Aboriginal Tourism Team Canada, Ottawa.

Bearing Point LP, Goss Gilroy Inc. & Associates (2003b) *Aboriginal Tourism in Canada, Part II: Trends, Issues, Constraints & Opportunities*. Prepared for Aboriginal Tourism Team Canada, Ottawa.

Blackstone Corporation (2001) *A Strategic Plan for Tourism Development in Nunavut*. Prepared for Nunavut Tourism and Department of Sustainable Development, Toronto, Ontario.

Budke, I. (1999) Community-based tourism development in Nunavut, involving people and protecting places. In: Budke, I. and Williams, P.W. (eds) *On Route to Sustainability, Best Practices in Canadian Tourism*. Canadian Tourism Commission and The Centre for Tourism Policy and Research, Simon Fraser University, Ottawa, pp. 39–47.

Downie, B.K. and Monteith, D. (1994) *Economic Impacts of Park Development and Operation: Katannalik Territorial Park, Lake Harbour, NWT*. Prepared for Workshop on Benefits of National Parks and Protected Areas in the North, University of Ottawa, 19–20 April.

Marshall Macklin Monaghan Ltd (1982) *Pilot Project for a Community-based Tourism Plan for Pangnirtung*. Marshall Macklin Monaghan Ltd, Toronto, Ontario.

7 The Economic Role of Arctic Tourism

JOHN M. SNYDER

Strategic Studies, Inc., 1789 E. Otero Avenue, Centennial, CO 80122, USA

Introduction

Arctic tourism plays an increasingly significant role in the local and national economies of the eight nations – Canada, Finland, Greenland, Iceland, Norway, Russia, Sweden and the USA – that ring the Arctic region. Contributing jobs, personal incomes, business revenue, capital expenditures and government taxes, tourism provides a measure of stability for historically volatile Arctic economies, and opportunities towards economic self-sufficiency for recently enfranchised indigenous peoples. There is a growing recognition that sound tourism can represent a benign use of natural resources: its successful development is a goal shared by both private and public sectors. Tourism is developing in different ways among the eight national economies, several of which have recently experienced radical political transformations. This chapter outlines each nation's approach to Arctic tourism development, illustrating both the significant economic role that tourism is playing and the development pressures it exerts throughout the entire Arctic region.

Canada

Tourism has contributed to Canada's northern economies for two centuries. Canadian innovators led in the creation of the modern tourism industry, for example in using railroads to promote mass tourism and developing the world's first National Park System. Canadian ports have long served international cruising, and Canadian bush pilots helped to make possible and popularize backcountry fly-out fishing and hunting. Collaboration between Canada's indigenous peoples has created cultural tourism experiences, for example in Nunavut, distinguished by their cultural integrity (Chapter 6, this volume).

Between 1993 and 2003 Canada hosted between 17 and 20 million tourists annually (Canadian Tourism Commission, 2005; UNWTO, 2005a), mostly from the USA, with a consistent second-largest group from the UK. Tourism spending in 2003 reached Can$52.1 billion, about one-third of which derived from foreign tourists. In 2004 tourism employment provided 578,000 full-time and part-time jobs, distributed among more than 400 communities throughout Canada (Parks Canada, 2004). Tax revenues from tourism provided some Can$7.7 billion to the federal government, Can$7.0 billion to provincial coffers.

The polar component of Canadian tourism is small but significant. The Canadian Arctic contains one of the world's largest expanses of wilderness, the seasonal habitat of migrating wildlife, which is attractive to tourists but relatively remote and inaccessible. The human population, estimated in 2004 at 92,985, includes many indigenous people who rely on local resources for subsistence and the preservation of their cultural heritage. Their way of life is itself a tourist attraction: to them issues of natural resource sustainability are daily realities. They recognize the importance of tourism, but express caution regarding its impacts.

Limited access, high costs of hospitality infrastructure development, shortage of labour, scarce support services and supplies, limited access to capital, severity of the climate and short seasons all serve to restrict tourism to the Arctic. However, in 2004 tourism provided 13,000 full-time equivalent jobs to indigenous people – ample evidence of its economic significance for this population. The tourism industry created jointly by the government and northern peoples now plays a vital role in the economic stability of the region.

Anglers, hunters and 'official tourists' travelling aboard Hudson's Bay Company vessels were the first to enjoy the Canadian Arctic. Following World War II, bush planes provided improved access, but Canadian Arctic tourism remained small. The Division of Tourism of the Government of Northwest Territories (NWT) reported only 600 tourists in 1959, using four tourist establishments, and tourist expenditures totalling Can$350,000. By 1970 the NWT reported 20,000 visitors spending Can$5.2 million dollars (Government of the Northwest Territories, 1972). From the mid-1980s to the mid-1990s the numbers rose to approximately 190,000 tourists. A precedent-setting development was the advent of the first cruise ship to the Canadian Arctic in 1984 (Snyder and Shackleton, 2001).

Eleven of Canada's 40 National Parks, and five of its 146 National Historic Sites, located within Yukon, Nunavut and NWT, between 1996 and 2003 attracted an average of 128,742 tourists annually: the largest number (158,078) visited in 1996, the least (107,693) in 2002. During the same period Newfoundland and Labrador's two National Parks and nine National Historic Sites attracted more than 455,000 annual visits. In either case visits to Parks Canada sites alone consistently exceeded the number of local inhabitants.

An analysis by Stanley and Perron (1994) revealed that:

Park Canada's programs in the north created 515 person years of employment in 1992–93 and generated $12.86 million in labor income for northern residents. Spending by park visitors and by Parks Canada also increased the GDP of the territories by $15.2 million.

Tourism clearly serves a vital role in the Canadian Arctic economy by bringing stability, much-needed jobs and personal income to the region. Further development will be especially important for the recently created province of Nunavut, which comprises the largest part of the Canadian Arctic. It will be no less important to the Yukon, a more mature tourism market that is growing by means of product diversification and the expansion of its tourist season. The primary economic objective of both private and public sectors is to increase the duration of the average tourist visit. To this end tourism products and services are being developed and diversified by creating or expanding heritage, recreational and cultural attractions as well as hospitality services.

Vancouver, British Columbia, a modern port in an attractive setting, is proving a gateway for Arctic tourism by providing cruise-ship and air-transport terminals. Cruises along Canada's Coastal Range and Alaska's Inside Passage continually attract increasing numbers of tourists. The Pacific Northwest region also anticipates increased tourism benefits from the selection of the Whistler Blackcomb Ski Resort as the site for the 2010 Winter Olympics.

Since the mid-1980s tourist access to the Canadian Arctic and the Arctic Ocean has been transformed by the progressive opening of the Northwest Passage to ship travel, attributable mainly to reduced sea ice, lengthening of the sailing season, improved charts and aids to navigation, and the increased availability of icebreakers and ice-strengthened cruise ships. Between 1984 – when the *MS Lindblad Explorer* initiated cruise-ship travel through the passage – and 2004, a total of 23 commercial cruise ships and 15 recreational yachts accomplished transits of the Northwest Passage (Headland, 2005). Remarkably, these 38 voyages constitute more than half of the 70 passage transits during that time period. These facts present a rather compelling case that the Canadian Arctic, and this point of entry to the Arctic Ocean, is becoming more accessible and attractive to tourists and recreational sailors.

Finland

Finland's distinctive culture, hospitality and natural features have attracted generations of international tourists. The Saami people of Finland have sustained a traditional way of life and cultural heritage that they are willing to share with tourists. A survey by the Finnish Tourist Board in 2002 (MEK, 2004) showed that vacationers chose Finland for its nature, culture and way of life:

> The most fascinating sights in Finnish nature seem to be the lakes and the beautiful landscape. As far as culture is concerned sauna is the phenomenon

that attracts the most. Other impressive issues are architecture, design, Nordic way of life and Lappish culture.

Despite an abundance of natural resources – particularly minerals and timber – and history, Finland is a relatively new entrant to international tourism. After World War II it diversified its economy by technological innovations and introducing new industries. Tourism is now emerging to play a more prominent role. One notable development is the recent creation of national parks and wilderness reserves: most of Finland's 35 national parks were established since 1982, and they now provide 8150 sq km of recreational land use. Oulanka National Park adjoins Russia's Paanajärvi National Park to create Europe's single largest protected area (Woodard, 2005). All 12 of Finland's recently established wilderness reserves are located in the country's arctic region (Metsähallitus, 2005).

This recent dedication of natural resources to recreational uses has been accompanied by institutional efforts to both improve and promote tourism. The Finnish Tourist Board, MEK, was established in 1973 under the Ministry of Trade and Industry in order to diversify tourism and make Finland's hospitality products more competitive. National measures of Finland's tourist industry are available from Tourism Satellite Accounts (TSA) that have recorded a variety of economic information since 1999. Additional information is available from the Nordic Model that was created in 1980 to evaluate the income and employment impacts of Scandinavian tourism. These economic models provide a valuable indication of the overall economic dimensions of tourism to Finland's economy, but they should not be interpreted as specific measures of polar tourism (Vuoristo and Arajarvi, 1990).

TSA-based information for 2001–2004 indicates tourism's significant contributions to the Finnish economy. For 2001, the total demand of tourism (i.e. domestic tourism, inbound tourism and the part of the costs of outbound tourism that stays in Finland) was €8.015 billion and the value-added generated by tourism contributed a total of €2.79 billion to Finland's Gross Domestic Product (GDP). In 2001, tourism provided 57,000 persons with employment. In 2004, the total demand of tourism in Finland was around €9 billion. The value added by tourism was around €3.117 billion, comprising 2.4% of the Finnish GDP. Tourism provided employment for around 60,000 persons (MEK, 2006).

Foreign tourists in Finland increased steadily from approximately 3.8 million visitors in 1998 to 4.5 million in 2005 (MEK, 2006). More than 90% of the foreign tourists are citizens of Europe, particularly Sweden and Russia. According to surveys of foreign tourists, approximately 40% spent one day in Helsinki and the remainder spent on average 4 days visiting the western and central regions.

The rapid expansion of the national park system and wilderness reserves created many new recreation attractions, especially in the country's polar region. Roads, rail, marine ferry and air networks provide excellent transport. However, the Finnish language makes access to information difficult (MEK stresses the need for multilingual signs and guidebooks), and high service costs provide a further challenge to future tourism development.

Throughout the Cold War both tourist and business travellers relied particularly on Finland as a gateway to the Soviet Union. Since the creation of the Russian Federation 1.5 million Russians leave their own country annually through the same gateway (MEK, 2004), indicating at least a potential for further tourism from this source. Simultaneously, Russia's increased accessibility and tourism development efforts represent direct competition to Finland's transportation and tourism industries. For example, Arctic tourists attracted to the Kola Peninsula region or to the Saami residing in the high Arctic are no longer dependent upon Finnish travel companies and tourist services to access those regions. From this perspective, new Arctic transportation patterns and the recent entry of a direct, regional competitor will, most probably, cause Finland to re-evaluate its future role in the delivery of Arctic tourism experiences.

Between 2003 and 2004 Finland implemented an extensive study of its image in the major tourism markets and re-evaluated its strategic position. One result of that effort was the decision to focus on tourism product development (MEK, 2003). A national study that contributed to the Finnish Tourist Board's Operating Strategy for 2004–2007 concluded that:

> there are especially versatile possibilities in Finland to develop new activity products for tourists. The problem remains that marketing is focused on the routes instead of the products. At present, mostly hiking-type products are offered to company groups. In addition, products are individual and do not cover the whole country. As far as product development and marketing efforts are concerned, more attention should be paid to Finland's varied natural conditions: hardwood forests in the south, the Lakes, archipelago, wilderness and northern tree-covered hill landscape. Active product development should be started immediately and with the help of true experts.
>
> (MEK, 2003)

Finland's intention to utilize its natural resources for tourism products will add numerous attractions to a growing polar tourism market. The prospects associated with successfully implementing Finland's Operating Strategy will be the inevitable expansion of the polar tourism market.

Greenland

Greenland's population in 2005 was 56,989 persons, of whom 47,000 (over 80%) live in towns, the rest in small settlements along the west and east coasts (Greenland Homerule, 2005). Between west and east lies the world's second-biggest ice sheet. Not surprisingly, roads are limited and there is no railway system. The country's sparse population prevents economies of scale and inhibits access to financial markets – circumstances which, in combination with severe climate and geographic remoteness, constitute major economic development challenges. Since its achievement of Home Rule in 1979

Greenland has depended on Danish subsidies associated with the exploitation of its fisheries, wildlife, minerals, oil and gas resources, and to a lesser degree on US support for the use of military and meteorological installations. Economic independence demands the development of new strategies, of which tourism appears to be the most promising.

The Home Rule and Danish governments cooperated with community leaders and the Inuit people to determine that Greenland's primary tourist attraction was its Arctic wilderness, which would appeal to all segments of the polar tourism market including mass, nature, cultural and adventure tourism. The cruise ship and air transport mass markets, as well as cultural events, wildlife viewing, kayaking, sport fishing, mountaineering, mountain biking, heli skiing, adventure racing and dog sledding, were identified as primary attractions, for which key locations were designated. In addition to existing towns and Inuit settlements that are receptive to tourism, designated sites include Ilulissaat Fjords UNESCO World Heritage Site, the Melville, Lyngmarken, Parasdalen, Quinnguadalen and Akilia Protected Areas, and the North-East Greenland National Park, the world's largest national park.

Greenland's tourism development strategy was built on an extensive but sporadic tourism history. The first 'excursionists' arrived in 1869 when artist William Bradford and polar explorer Dr Isaac Hayes brought a party of sightseers and artists aboard the steamship *Panther* (Chapter 2, this volume). In 1902 Mylius-Erichsen unsuccessfully sought permits to bring a ship with 100 tourists twice per summer to Greenland from England. In the 1930s ships from the USA and France carried tourists along Greenland's coast. However, it was 1959 before the Danish government sanctioned a flight from Copenhagen and one-day tourist flights from Iceland. From then until the late 1980s Greenland hosted only a few thousand tourists a year: tourism investment amounts were correspondingly small, businesses were highly fragmented, and the country lacked an effective marketing programme.

The Home Rule Government, faced with economic dependence on a commercial fishery based almost entirely on a single species of deep-sea shrimp, quickly recognized the potential of tourism, in 1991 making it one of four objectives in a commercial development strategy. A master plan, quickly implemented, was to increase tourism to an annual intake of 35,000 by 2005. The goal for expenditures per tourist was set at 15,000 DKr (approximately $1800). The plan sought to create 2200–2500 full-time jobs in tourism, plus 1000–1500 in associated enterprises, and to generate annual income of 500 million DKr in Greenland, all by 2005. It established four tourism development zones: Kangerlussuaq (Sondre Stromfjord), Narsarsuaq (South Greenland), Kulusuk/Ammassalik (East Coast) and Nuuk, the capital city. Most importantly, the government stipulated that tourism development had to be conducted in an environmentally and culturally responsible manner (Kangerlussuaq Tourism, 2005).

In 1991 also the Greenland Tourism and Business Council was created, with objectives of encouraging forms of tourism that accurately identify Greenland's unique features and to maintain the country's cultural values. Greenland Tourism, which is entirely government-owned, concentrates on five prime areas: consultancy, marketing, education, documentation and services towards a wide range of customers. Its main office, located in Nuuk, handles product and structure development, consultancy, analysis, documentation, internal marketing and international cooperation. The Greenland Tourism Information Office in Copenhagen, Denmark, takes care of external marketing, distribution and information flow to the press, travel agencies and private individuals. All these marketing endeavours benefit from the fact that Greenland was the second country in the world to have a fully digitized service network (Greenland Tourism and Business Council, 2006).

Tourism development objectives were soon matched with substantial investments by the Greenland and Danish governments, as well as community commitment. By example, in 1992 the level of tourism investment rose from approximately $300,000 to well over $2 million. Human resource development programmes and community participation programmes were also implemented to support the tourism development strategy. In 1992 visitor numbers remained low (3500), but commitments to the tourism development plan were firmly in place. By 1997 numbers increased to 17,000, and an economic impact assessment for that year reported an increased income of 130 million DKr, with 220 full-time jobs created. The report also noted that tourism was heavily subsidized by the government (Lycke, 1998). In 2000 Greenland hosted 31,351 arrivals, and numbers have climbed steadily since then (Statistics Greenland, 2005). Most tourists to Greenland come from Denmark (79%), an additional 8% from other Scandinavian countries, and the remaining 13% from other countries (Statistics Greenland, 2005). On average tourists stay for 15 days. The dominant age group is 50 years and older.

Visits from foreign cruise ships have risen steadily since 1994. In 2003, 14 ships made 164 port calls carrying 10,152 passengers; in 2004, 29 ships made 232 port calls with 13,420 passengers; and in 2005, 52 ships were involved (Greenland Tourism Board, 2005). Expectations for future growth will be strengthened by Greenland's desire to expand port operations and thus strongly promote this form of tourism.

Designation of the majestic Ilulissat Icefjord as a UNESCO World Heritage Site in July 2004 is predictably accelerating the popularity of the region, helped by publication of a well-timed book by the Geological Survey of Denmark and Greenland (Bennike, 2004). Approximately 10,000 tourists visit the region every year. Ilulissat, with a population of 5000, is Greenland's third-largest town. The Greenland Tourism Board's 'absolutely essential requirement that tourism be developed in harmony with, and not at the expense of, the landscape and culture' (Greenland Tourism Board, 2005) will be severely tested at Ilulissat.

Greenland is seeking to attract adventure tourism and a significant share of the extreme sport tourism market. Specific attractions cited include:

- Arctic Team Challenge: a 5-day adventure race in East Greenland covering over 200 km in mountain biking, trekking and canoeing.
- Arctic Marathon: occurring in August and in the vicinity of Nuuk.

- Greenland Adventure Race: a 5-day event in late August over a distance of 200 km around Narsaq, South Greenland, involving kayaking, mountain biking and long-distance running.
- Arctic Circle Race: rightly called the world's toughest skiing race, it is a 3-day event around Sisimiut (west coast) with overnight camping in tents.

Focusing tourism development on high-quality services and sustainability principles, Greenland Tourism is obtaining professional advice, implementing methods for collecting and assessing information, pursuing advanced marketing techniques, and creating educational programmes to achieve its objectives (Kaae, 2002).

Iceland

Located on the edge of the Arctic Circle, and benefiting from excellent commercial air and sea transport services, Iceland has increasingly attracted hundreds of thousands of visitors since the 1980s. Between 1990 and 1999 foreign arrivals nearly doubled from 142,000 to 263,000. By 2000 they had reached 303,000, for the first time outnumbering the local population of approximately 283,000. In 2003 the Icelandic Tourist Board reported that the number of visitors had grown steadily at 9% over the past decade, and in 2004 over 320,000 foreign tourists were hosted – a trend that is expected to continue for another 20 years (Iceland Statistics, 2005).

Iceland's tourism industry has grown and diversified since the 1980s to become a major contributor to the country's economy, and remains one of the fastest-growing sectors. In 2001 tourism became the nation's second-largest foreign currency earner, providing 13% of the country's foreign income and accounting for about 4.5% of the Gross National Product. Since then it has contributed more than 35 billion ISK annually to the national economy, a figure surpassed only by the fishing industry. In 2003 tourism provided an estimated 5400 jobs, a number that continues to grow. Based on natural attractions, outdoor recreation opportunities, high-quality visitor services and a well-integrated hospitality industry, tourism is clearly an industry on which the country now relies for its economic well-being (Icelandic Tourist Board, 2005).

The tourism businesses that provide hospitality services and tourism experiences are predominantly privately and family-owned (Icelandic Tourist Board, 2005). Surprisingly, rapid growth has not been dampened by the fact that Iceland has the highest prices in Europe. Costs for excursions, sport licenses, guides, hire transport, accommodations and associated hospitality services rank among the most expensive in the world, and they too are rising at an extraordinary rate. Between 1998 and 2001, prices for tourist services rose by 25% per annum (Cornwallis and Swaney, 2001). Iceland has captured an upscale polar tourism market, and one which continues to expand.

Its secret is a unique collection of dynamic environmental attractions, notably a wide variety of volcanoes and volcanic activity: there is an eruption on the island every 4 to 5 years, a frequency that has been increasing for the past 30 years. In contrast approximately 11% of the land is covered by both inland and tidewater glaciers, all of which are very sensitive to climatic change resulting from global warming (Gunnarsson and Gunnarsson, 2001). Abundant fresh water provides spectacular settings for tourism. Unlike many other polar environments, Iceland has no large diversity of wildlife or vegetation; beyond Arctic foxes, reindeer, mink and mice, the most popular animal attraction is the Icelandic horse that provides tourists with a unique way of experiencing the landscape.

Outdoor recreational attractions include sport fishing, boating, kayaking, mountaineering, horse riding, snowmobiles, hiking and backpacking. Since most of Iceland's land use is rural, recreational venues can be pursued in a diversity of natural settings. A 1996 survey to determine tourist attractions and how they were experienced revealed that the greatest single attraction – nature – drew 80% of summer and 60% of winter visitors. Culture and history were also identified by as significant attractions, but notably, the same percentage of tourists indicated they were attracted to Iceland by price offers (Icelandic Travel Industry Association, 2006). Responses further indicated that foreign visitors were mainly interested in recreational activities connected with nature. Recreation diversification has accompanied tourism growth and there has been a dramatic increase in recent years in the range of activities offered. The survey itself was designed to enumerate 15 distinct recreational activities. As of 2004, Iceland had at least 73 operators offering horse riding tours, 27 offering various kinds of boat tours, 27 offering hiking tours, 26 offering jeep and glacier tours, and ten offering snowmobile expeditions. About 25 operators offer a wide range of day excursions (Icelandic Tourist Board, 2005).

Iceland's economy will continue to depend on utilization of its rich natural resource base. Commercial marine fisheries, energy-intensive industries dependent on abundant hydropower, and the tourism industry's environmental attractions and beauty will all compete with one another for the use of Iceland's resources. Although the Icelandic government is interested in further promoting the development of ecotourism, it recognizes the need to reconcile these potentially conflicting resource uses. One example of this dilemma occurs in the central highlands where the desire to preserve the natural beauty of region to expand the tourism industry directly competes with demands to utilize the tremendous hydro- and geothermal power potential for the economic development of rural communities located along the coast. The government is evaluating the impacts of alternative resource uses and attempting to create development programmes that establish compatibility among those uses – an enormous challenge.

Iceland also acknowledges the issues of potential social and infrastructure impacts caused by large numbers of tourists. As the least populated country in Europe, and seventh least-populated country in the world, Iceland will experience an unprecedented collection of social impacts and demands upon its infrastructure. The nation currently has both the community support and wealth needed to cope with near-term tourism increases. If at some time in the future Iceland decides to directly influence the number of foreign tourists, it will have many economic, governmental, natural resource management and transport access controls at its disposal.

Norway

Polar tourism is frequently said to have started with excursion travel to Norway during the early 1800s. By the mid-1800s steamships enabled tourism entrepreneurs to create a mass market in the 'land of the midnight sun'. Throughout the century, Svalbard, the North Cape and Norwegian fjords became popular international tourist destinations offered by Norwegian, British and German tour companies. By 1900, a year-round, highly diversified Norwegian tourism market was offering comfortable cruise-ship tours, trophy hunting for polar bears in Svalbard, sport fishing throughout Arctic waters, and from 1891 ski vacations for a mass tourism market.

Norway's polar tourism continued to expand until World War I, then enjoyed increasing popularity throughout the 1930s (Viken and Jorgensen, 1998). From the 1960s to the present, improved transport technologies, infrastructure development and private investment have enabled Norway to capture a valuable share of the polar tourism market. Examples of significant tourism investments include the use of Norway's coastal transport system, the Hurtigruten, from the 1960s, and construction of a commercial jet airport at Longyearbyen, Svalbard in 1975. These and similar Norwegian investments greatly enhanced the convenience, geographic scope and seasonal use of Norway's Arctic tourist attractions.

After two centuries of commercial tourism Norway fully recognizes the industry's vital contributions to the national economy. In order to accurately track tourism's economic role, Norway implemented a detailed accounting system, TSA (Tourism Satellite Accounts), which catalogues numbers of tourism industry jobs, tourist expenditures by type of industry, tourism's contribution to national income, and levels of capital expenditures (investments). Between 1996 and 2003 most economic indicators of the Norwegian tour industry demonstrated stability. Total full-time equivalent jobs among ten distinct types of tourism businesses ranged between approximately 125,000 and 132,000, i.e. between 6.4 and 7% of the nation's total employment during this period (Statistics Norway, 2006).

In 2003, resident and foreign tourists spent NOK 76 billion on a wide diversity of tourist services and products. Spending by Norwegian households amounted to NOK 37 billion and represents almost 50% of the total tourism consumption. Correspondingly, non-residents spent NOK 22 billion and Norwegian trade and industries' spending on business travel was estimated at NOK 17 billion, measured in constant 2001 prices. Since 2000 the only increase has been in tourism consumption by Norwegian households. Since the largest single decline is in transport services, part of the consumption decrease may be attributable to the effects of the 11 September 2001 terrorist events that had a negative impact on the entire international tourism industry.

For the 8-year period between 1996 and 2003, the tourist industry contributed an average of NOK 125.1 billion per year. Revenue generated during that period ranged from a low of NOK 115.9 billion to a high of NOK 130.5 billion (measured in constant 2001 prices). When tourism industry output is compared with national economic output, the tourism industry contributed 5% of Norway's total output in 2003.

Norwegian accommodations are currently experiencing increased numbers of guests whose sole purpose is holiday and recreation. According to Statistics Norway, hotels and similar establishments booked more than 1.71 million guest-nights during the January to March 2004 period and during the same time in 2005 the number increased to more than 1.78 million. The 2004 average price per room was calculated to be NOK 717. Recreational guest-nights at camping sites also showed an increase from more than 420,000 from January to March 2004, to more than 540,000 in the January to March 2005 period. These substantial recreational expenditures for accommodation are economically significant, but they are very general measures. It is not possible to determine how many of the guests were resident versus foreign tourists, and it is certainly not appropriate to attribute the entire amount to polar tourism. They do, however, provide one indication of the economic value of Norwegian recreational travel (Statistics Norway, 2006).

A key indicator of economic commitment to the tourism industry is the annual level of capital investment, which from 1996 to 2001 amounted to over NOK 84 million for tourism industries, or a mean annual investment of over NOK 14 million per year, measured in terms of constant 2001 prices. Regrettably specific tourism counts are not published (Statistics Norway Information Centre, personal communication).

The economic stability of Norway's tourism industry is primarily the result of its longevity: the country's tourism products and services are well known and respected in the international market. Its range of tourist attractions, outdoor recreational activities, geographic destinations and seasonal uses represent a collection of tourist products and services that have, relative to other polar destinations, been optimally diversified, accompanied by the application of increased environmental and cultural knowledge to the nation's resource management and community development programmes. Norway's 21 national parks, its Nordic and Saami cultures, and its wildlife, water and forest resources have all witnessed a long history of tourism management practices. In summary, Norway's two centuries of tourism development experiences provides a valuable case study of the evolution of the polar tourism industry, from birth to maturation.

Russia

The land mass of the Russian Federation from the Kola Peninsula in the east to the Chukotka Peninsula in the west encircles one half of the Arctic, encompassing 11 time zones. Economically the Federation is the most recent major competitor in the polar tourism market: for the first time in modern history foreign access to the Russian Arctic is now permitted. Russia possesses the world's largest fleet of icebreakers and has expressed keen interest in the development of the Northern Sea Route for a variety of commercial purposes. Given the size of the Russian Arctic and the Federation's substantial commitments to pursue its economic development, polar tourism is poised to expand. Any who summarily dismiss this prospect may wish to consider how incredible Antarctic tourism once seemed.

Intourist, the Soviet Union's state-controlled tourism sector, provided travel for selected citizens of the Soviet Union, its Eastern European allies and foreign visitors, but a developed tourism infrastructure was not deemed a national investment priority throughout the Soviet era. Except for limited entry to Murmansk and the Archangel Region, the Soviet Union prohibited foreign access to its Arctic domain. The market-based economic system created by the newly established Russian Federation included the creation of the Russian Federal Tourism Agency, to enable Russian tourism to compete favourably with other international destinations and obtain the economic benefits from foreign currency exchange, job creation, personal income and business revenues that tourism can produce.

Undaunted by historical realities and infrastructure challenges, the Federation now aggressively promotes tourism, including polar tourism, through cruise-ship voyages, sport fishing, trophy hunting, nature tourism, cultural venues and adventure, all ambitiously promoted. As Vladimir Strzhalkovskiy, Deputy Minister of Economic Development and Trade of the Russian Federation and Chairman of the UNWTO Executive Council, stated in a 2005 press release:

> Russia today is much more than a prime cultural destination. The country boasts numerous opportunities for all-year-round ecological, adventure, ethnographic and other specialist types of tourism. The Federation is dotted with new destinations and activities, still little known to the international traveler, but which are developing rapidly.
>
> (UNWTO, 2005b)

Examples of polar tourism advertised by the Federal Tourism Agency (2005) include:

> ecotourism in the Kamchatka Peninsula in the Far East, renowned for its geysers, volcanoes and hot-water rivers; white-water rafting on the Katun River; horse-riding at the foot of the Caucasus mountain; hunting in Siberia; and tours to the Arctic.

Russia's efforts to promote polar tourism are not without precedent. The tourist attractions currently being advertised build upon an existing, although exceedingly limited pre-Soviet polar tourism history, established in the White and Barents seas, Arkhangelst, Vologda, Murmansk and Karelian regions. This small market included highly regulated pilgrimages to ancient Russian Orthodox monasteries originally constructed in the 16th century as fortresses to protect the Russian Empire. Later, trophy hunting and fishing were permissible recreation activities for select members of the Russian nobility, later still for the Soviet elite. A limited number of tourist cruises in the Barents and White seas were allowed in the 1930s. Russia's intention to reinvigorate these venues is apparent in the current popularity of the Sovolki Archipelago region and the re-consecration of its monastery in 1992 (Armstrong, 1972; Wilson, 1978; Barr, 1980).

Contemporary tourism development by the Russian Federation represents far more than publicity and the enumeration of competitive attractions. The Tourism Department of the Ministry of Economic Development and Trade is striving to ease visa policies, following a course charted by President Vladimir Putin aimed at introducing visa-free travel between Russia and the Schengen agreement countries. In addition, international tourism contacts have expanded significantly: as of 2005, 48 countries have signed inter-governmental agreements on tourism cooperation (Federal Tourism Agency, 2005). The Russian tourism industry is growing as a result of the Federation's pro-development efforts. Between 2001 and 2002, the number of foreign tourists increased from 7.4 million to 8 million visitors, a creditable accomplishment considering the decline in tourism elsewhere in the world during that period. Tourist expenditures in 2001 were estimated to be $374 million (excluding airfares). The average stay of the foreign visitor was 6.5 days (World Trade Organization, 2003).

Russia's Arctic population in 2004 was 1,999,711, estimated by the Federation and the Arctic Council, representing slightly more than 1% of the Federation's total population, scattered sparsely throughout an enormous region. This population includes at least 44 distinct indigenous ethnic groups who have subsisted for millennia by traditional methods (Oakes and Riewe, 1998). These are people who have sustained international trading relationships, cultural practices and art forms, language and traditional ways of life with extraordinary integrity, most importantly maintaining a strong relationship and intimate knowledge of their land and marine environments. They have organized themselves for a variety of development purposes and have excellent telecommunication access and resources in order to strengthen their political presence. Their demonstrated capacity for collaboration also provides several economic development opportunities such as collective marketing, improved access to capital, regional planning and provision of collateral for financing. They are likely to be capable of creating a viable cultural tourism economy.

Currently faced, like other Arctic indigenes, with diverse pressures to develop their region's natural resources, Russia's northern people have serious concerns about cultural intrusion, loss of heritage and damage to the resources upon which their traditional ways of life depend. Given the extremely rapid growth of the international cultural tourism market and the unprecedented opportunity to gain access to Russia's indigenous communities, polar tourism may experience considerable expansion. In preparation for such a possibility, organizations such as the Russian Association of Indigenous Peoples of the North are determining how their cultural values and natural resources can best be shared with others (Russian Association of Indigenous Peoples of the North, 2005).

In summary, the entry of the Russian Federation into international tourism is a new phenomenon for both the Russian people and the international economy. A history of prohibited access to Russia's polar lands and seas is now replaced with active campaigns for its economic development. As a consequence, fully one-half of the Arctic Region is now positioned to compete for a share of the international polar tourism market. The Russian Arctic's heritage sites, cultural traditions, wildlife populations, sport fisheries, trophy hunting, mountains, rivers and World Heritage Sites are now being promoted to foreigners for the first time in modern history. It will obviously take time for tourism in the Russian Arctic to evolve, but its entry as a competitor in this market will ultimately affect the future of polar tourism.

Sweden

Sweden's ancient custom of *allemansratt* provides a useful insight for understanding the nation's attitude towards outdoor recreation and tourism. Since medieval times Sweden has sustained a tradition of access to all lands and waters of the country regardless of ownership, allowing everyone to hike, camp, canoe, moor a boat and gather edible plants on private lands and along waterways, so long as they do not cause damage and they respect the rights of others. As transportation developments progressively enhanced public access to its natural attractions, pioneering natural resource protection laws were introduced to safeguard outdoor recreation resources in perpetuity.

During the 19th century Sweden constructed a network of railroads and interior roads that expedited the transport of iron ores, timber and other inland resources to ports near the Gulf of Bothnia and the Norwegian Sea. By the early 20th century important east–west railway routes were constructed between Lulea and Sundsvall in Sweden to Narvik and Trondheim in Norway, and a north–south rail route, the Inlandsbahan, connected the entire nation from the northern iron ore railway network to the Stockholm rail system in the south. Simultaneously an ambitious road construction programme provided direct access between the Bothnian coast and the mountains.

Public transportation systems that traverse all regions of the country, and human settlement patterns that consciously attempt to retain a close relationship with rural areas, provide unparalleled opportunities for recreation throughout the country, not least in the arctic north. The Swedish Touring Association was an international pioneer in the use of retailing and advertising techniques to promote national tourism development. Equally significant, in

1909 Sweden created Europe's first national park system, with the
establishment of nine National Parks.

Again, Sweden was an international leader in the developing an elaborate
national trail system, involving over 10,000 km of trekking and bicycling
paths that enable hikers, cyclists and skiers to access literally all regions of
Sweden (Bain and Cornwallis, 2003). This internationally famous trail system
includes routes that are especially prominent polar tourism venues. The most
popular trail, the Kungsleden, extends from Salen to Terriksroset, with its
most popular segment in Lapland. The Arctic Trail, 800 km long, is a joint
development of Sweden, Norway and Finland connecting their arctic
domains.

Sweden currently has 28 National Parks, of which six lie within the Arctic
Circle. Abisko, Stora Sjofallet and Sarek were established in 1909,
Vadvetjakko in 1920, Maddus in 1942, and Padjelanta in 1962. Just south of
the Arctic Circle is Peljekaise National Park, also among the first to be
established in 1909 (Swedish National Park Service, 2005). Tourism
opportunities afforded by the parks are augmented by 12 UNESCO World
Heritage Sites and 2200 Nature Reserves, and four of Sweden's largest rivers
are accorded National Heritage River status to protect them from
hydroelectric power development. Perhaps the most distinguishing feature of
these tourist attractions is their accessibility. Those seeking hiking, biking and
skiing are rarely more than 25 km from a road, rail link or serviced trail, and
a variety of excellent accommodation, from luxurious to rustic, is available
throughout the entire country. A major exception is the remote Abisko–Sarek
National Park region in northern Sweden (National Atlas of Sweden, 1993).

A strong partnership exists between the Swedish government and private
businesses for the development and promotion of tourism. National policy
documents, investment strategies and marketing campaigns are designed and
implemented to advance the competitive position of Sweden's tourism
industry. According to a report published by the Swedish Tourist Authority
and the Swedish Travel and Tourism Council (Swedish Tourist Authority,
2005):

> tourism is very much a small-business industry but a sector that is vital for
> employment in many regions. The tourist industry in Sweden has an annual
> turnover of approximately 163.5 SEK billion, which represents about 2.63 per
> cent of Sweden's GDP. The number of people employed in the Swedish tourist
> industry is equivalent to approximately 126 000 full-year positions.

The Swedes and Saami of Norrland, the northernmost region of the
country, view prospects for polar tourism with both positive expectations and
caution. Comprising 60% of Sweden's entire land mass, including the 15% of
the country that is beyond the Arctic Circle, Norrland contains recreational
and cultural resources that provide quality visitor experiences, and can play an
important economic role for the 260,000-strong permanent population of
the region (Fjelheim, 2005). Following centuries of mining and other forms of
exploitation, tourism is perceived by local people to be a more acceptable way
of using Norrland's resources as a source of jobs and personal income, and a

more effective and equitable way of returning wealth to the region (Heininen and Tuija, 1993). Local Saami people are especially keen to optimize both their control of local resources and the wealth that those resources create.

Among the several tourist attractions located in Norrland, the Ice Hotel located in the Saami village of Jukkasjarvi has gained international recognition. Constructed annually from 10,000 t of clear ice cut from the Torne River, with an additional 30,000 t of natural and artificial snow, the hotel operates during a December to April tourist season, providing for 130 guests at a time to experience −5°C ambient temperatures in complete comfort. Support facilities include an Ice Cinema, Ice Church, Ice Sculptures and Japanese Wintergarden − one of the world's most imaginative polar tourism products. Another polar tourist attraction is the Hoga Kusten (High Coast) UNESCO World Heritage Site. Since the end of the last ice age, the Swedish land mass has risen at a prodigious rate: here the cliffs have lifted 285 m and are still rising (UNESCO, 2006). Former coastal towns are now far inland and former seabed has become land − a dramatic and perhaps instructive example of ways in which warming may affect polar tourism in the future.

United States (Alaska)

At the time of Alaska's purchase from Russia in 1867, popular opinion perceived no economic use for it, especially since most of the nation's land below the 49th parallel remained to be settled and developed. The federal government assigned one revenue cutter to patrol the entire expanse of the North Pacific and Arctic oceans, and garrisoned a small fort in Sitka to symbolize territorial possession. The discovery of gold in the 1880s, realization of salmon's huge commercial value, and publicity for the territory's natural wonders transformed its economy. There followed a succession of boom and bust cycles − booms generally following the discovery and exploitation of resource wealth, and declines often caused by subsequent resource depletion. Alaska's tourism industry mirrored those economic events.

A travel boom generated by the discovery of Glacier Bay and the Gold Rush was marked by a fleet of tourist steamships exploring Alaskan waters. Then, just as suddenly, an earthquake in 1899 blocked entry to Glacier Bay, and the Gold Rush collapsed. A tourism revival in the early 1900s arose from the construction of railroads, the economic resiliency of the steamship companies, renewed accessibility to Glacier Bay, the establishment of national parks and monuments, and the perennial attraction of sport fishing and hunting (Territorial Governor of Alaska, 1883–1956). From 1914 Alaskan tourism was again hit by world events, notably World War I and the Depression. Territorial status, and the fact that Alaskan islands were occupied by the enemy in World War II, contributed little to Alaska's image as safe for investment.

In July 1946 came revival, symbolized by a 71-page pamphlet, 'Alaska's Recreational Riches', that served as the foundation for a territorial tourism development programme (Alaska Development Board, 1946). The report recognized that the highest investment priorities were transportation infrastructure and hospitality facilities:

> In common with other parts of the nation and the world, transportation is the major unsolved problem for tourist travel to Alaska in the summer of 1946 ... Ever since the airplane became practical, Alaska has made extensive use of it. In this land of vast distances, few roads and almost no railroads, the airplane has played an increasingly important role ... Now that the war is over, the airplane doubtless will play a new role in Alaska as a tourist carrier. This will call for a change in Alaska's tourist economy – an economy which in pre-war days was based on keeping the tourist moving; that is he ate and slept aboard the steamer that carried him north and south. Increasing number of tourists in the future will come north by plane, utilizing new Alaska hotel and resort accommodations. Such stopover facilities are being planned.

Implied in this were increased use of the recently completed Alaska Highway and the creation of a marine ferry system from Prince Rupert to ports throughout Southeast Alaska. Between 1946 and 1952 numbers of tourists visiting Alaska exclusively for recreational purposes increased from zero to 20,252. In 1952 tourist personal expenditures exceeded $7 million annually: 'the recreation resources of Alaska have demonstrated their attractiveness (as of 1952) to the point that they already represent a business worth of nearly $19 million annually to the Territory' (Stanton, 1953). Statehood achieved in 1959, the implementation of resource conservation policies, the development of human resource skills, entrepreneurial business developments and private investment all contributed to Alaska's resurgent tourism (Rogers, 1962).

The industry continued to grow: between 1967 and 1970 the total number of visitors increased from 86,700 to 129,000. It had reached 570,660 by 1980, 774,980 by 1985 and 914,500 by 1999. Since then the rate of growth has increased further: as of 2004 the total number of vacation visitors numbered 1,076,500 – a remarkable increase of 10% over the previous season (Alaska Division of Tourism, 2005). The annual number of tourists now exceeds double the entire population of the state (estimated in 2004 at 481,054), a ratio that places inordinate demands on local infrastructure and seasonal uses of local resources.

In terms of positive contributions, the revenues and jobs derived from tourism have helped to stabilize Alaska's boom-and-bust economy. Tourism's contribution in 2002 was $1.6 billion, in 2003 $1.8 billion. Visitors in that year accounted for approximately 30,700 jobs, a rise from 9 to 12% of private sector employment since the previous year, and personal income obtained from tourism employment exceeded $600 million. Angling, hunting and wildlife watching continue to predominate among tourist attractions. A survey of wildlife tourists in 2001, produced by the US Fish and Wildlife Service, reveals that 239,000 visiting anglers fished for one million days and spent over $323 million. More than 21,000 visiting hunters hunted for 193,000 days and spent over $103 million. The single largest group were wildlife viewers, of whom 292,000 spent more than $358 million (US Department of the Interior, 2002).

Alaska residents too place great personal value on wildlife watching, angling and hunting as an important part of their life-style, and esteem jobs associated with guiding or wildlife management. Currently the Alaska Occupational Licensing Board authorizes thousands of local people to provide licensed sport fishing and hunting guide services – a number much increased since 1946, when only 60 guides were licensed. Clearly, Alaska's wildlife-related tourism provides not only economic benefits, but also supports many of the social values maintained by Alaskans themselves (Alaska Department of Community and Economic Development, 2005).

A 2004 study of the industry (Alaska Division of Tourism, 2005) concludes that travel and tourism have been engines of growth in the Alaskan economy, with an economic contribution expanding by 38% from 1998 to 2002. In summary, while other sectors of the Alaska economy that depend on natural resources continue to experience substantial fluctuations, the tourism industry steadily and increasingly provides much-needed economic benefits.

Conclusions

This survey of the development of tourism in eight Arctic countries concludes that, in each case, tourism has grown from a novelty to a significant economic contributor; that Arctic economies are increasingly reliant upon it; and that its geographic scope is expanding enormously. In sharp contrast to the extractive industries that previously dominated Arctic economies, tourism currently provides a more consistently profitable, more stable and more locally beneficial alternative use for natural resources, in particular to the benefit of local people who previously gained little from their exploitation. Vastly expanding tourist access resulting from allowable entry to the Russian Arctic and diminishing sea ice will inevitably increase its economic stature. In all eight instances tourism appears to be expanding rapidly, providing both benefits and challenges for those charged with its governance.

References

Alaska Department of Community and Economic Development (2005) *Professional Licensing Data Retrieval*. Office of Occupational Licensing, Juneau, Alaska; available at http://www.dced.state.ak.us/occ/apps/ODStart.cfm (accessed December 2005).

Alaska Development Board (1946) *Alaska's Recreational Riches*. Alaska Development Board, Juneau, Alaska.

Alaska Division of Tourism (2005) *Visitation Statistics*. Alaska Department of Community and Economic Development, Juneau, Alaska.

Armstrong, T.E. (1972) The northern sea route, 1971. *Polar Record* 16(102), 421–422.

Bain, C. and Cornwallis, G. (2003) *Sweden*. Lonely Planet Publications, Melbourne, Victoria.

Barr, W. (1980) The first tourist cruise in the Soviet Arctic. *Arctic* 33(4), 671–685.

Bennike, O., Mikkelson, N., Pedersen, H. K. and Weidick, A. (2004) *Ilusissat Icefjord A World Heritage Site*. Geological Survey of Denmark and Greenland (GEUS), Ministry of the Environment, Copenhagen.

Canadian Tourism Commission (2005) *Tourism Snapshot – 2005 Facts & Figures Year Review*. Canadian Tourism Commission, Vancouver, British Columbia; available at http://www.canadatourism.com (accessed June 2006).

Cornwallis, G. and Swaney, D. (2001) *Iceland, Greenland, and the Faroe Islands*. Lonely Planet Publications, Melbourne, Victoria.

Federal Agency for Tourism of Russian Federation (2005) *News*. Federal Agency for Tourism of Russian Federation, Moscow; available at http://www.russiatourism.ru/eng (accessed October 2005).

Fjelheim, R. (2005) Arctic indigenous peoples facing climate change – a Saami perspective. In: *Arctic Forum 2005*. Arctic Research Consortium of the US, Fairbanks, Alaska, p. 13.

Government of the Northwest Territories, Division of Tourism (1972) Tourism in Canada's Northwest Territories. *Polar Record* 16(102), 424–426.

Greenland Homerule, Greenland Tourism and *Atuagagdliutit* Editorial Board (2005) *Greenland Export Magazine 2005*. Greenland Homerule, Nuuk.

Greenland Tourism Board (2005) *Tourism Economic Development Strategy*. Greenland Tourism Board, Nuuk; available at http://greenland.com (accessed January 2005).

Greenland Tourism & Business Council (2006) *Press Release*. Greenland Tourism & Business Council, Copenhagen; available at http://www.gt.gl/Front_Page.php (accessed December 2006).

Gunnarsson, B. and Gunnarsson, M. (2001) Iceland's central highlands: nature conservation, ecotourism, and energy resource utilization. In: Watson, A.E., Alessa, L. and Sproul, J. (eds) *Wilderness in the Circumpolar North: Searching for Compatibility in Ecological, Traditional, and Ecotourism Values*. US Department of Agriculture Forest Service, Ogden, Utah, pp. 54–63.

Headland, R. (2005) Appendix F: Transits of the Northwest Passage. In: Brigham, L. and Ellis, B. (eds) *Arctic Marine Transport Workshop, 28–30 September 2004*. Institute of the North, Anchorage, Alaska, pp. 20–25.

Heininen, L. and Tuija, K. (eds) (1993) *Regionalism in the North*. Arctic Centre Reports No. 8. Arctic Centre, Rovaniemi, Finland.

Iceland Statistics (2005) *Tourism, Transport and Information Technology*. Iceland Statistics, Reykjavík; available at www.statice.is (accessed December 2006).

Icelandic Tourist Board (2005) *Tourism in Iceland in Figures*. Icelandic Tourist Board, Reykjavík; available at http://icetourist.is (accessed October 2005).

Icelandic Travel Industry Association (2006) *Travel Industry Statistics*. Icelandic Travel Industry Association, Reykjavík; available at http://www.saf.is (accessed December 2006).

Kaae, B.C. (2002) Nature and tourism in Greenland. In: Watson, A.E., Alessa, L. and Sproul, J. (eds) *Wilderness in the Circumpolar North: Searching for Compatibility in Ecological, Traditional, and Ecotourism Values*. US Department of Agriculture Forest Service, Ogden, Utah, pp. 43–53.

Kangerlussuaq Tourism (2005) *Kangerlussuaq – The Gateway to Greenland*. Kangerlussuaq Tourism A/S, Kangerlussuaq, Greenland.

Lycke, L. (ed.) (1998) *Turismestrategi og-udvikling i Gronland* [*Tourism Strategy and Development in Greenland*]. Nordic Press, Copenhagen.

MEK (2003) *Finnish Tourist Board Operating Strategy for 2004–2007*. Finnish Tourist Board (MEK), Helsinki.

MEK (2004) *Border Interview Survey*. Statistics Finland and Finnish Tourist Board (MEK), Helsinki.

MEK (2006) *Basic Facts and Figures on Tourism in Finland, Tourism Satellite Accounting*. Finnish Tourist Board (MEK), Helsinki; available at http://www.visitfinland.com/W5/index.nsf/(Pages)/Finland_Facts (accessed May 2006).

Metsähallitus (2005) *Nature Protection and Recreational Use*. Metsähallitus, Helsinki; available at http://www.metsa.fi/natural/nationalparks/index.htm (accessed October 2005).

National Atlas of Sweden (1993) *Cultural Life, Recreation and Tourism*. Swedish National Atlas Publishing, Stockholm.

Oakes, J. and Riewe, R. (1998) *Spirit of Siberia; Traditional Native Life, Clothing, and Footwear*. Smithsonian Institution Press, Washington, DC.

Parks Canada (2004) *Visitor Statistics, 1997 and 2003*. Parks Canada, Ottawa.

Rogers, G.W. (1962) *The Future of Alaska; Economic Consequences of Statehood. Resources for the Future*. Johns Hopkins Press, Baltimore, Maryland.

Russian Association of Indigenous Peoples of the North (2005) *What is RAIPON?* and *History Populations*. Russian Association of Indigenous Peoples of the North, Moscow; available at http://www.raipon.org (accessed November 2005).

Snyder, J. and Shackleton, K. (2001) *Ship in the Wilderness; Voyages of the MS Explorer Through the Last Wild Places on Earth*. Gaia Books Ltd, London.

Stanley, D. and Perron, L. (1994) The economic impact of northern national parks (reserves) and historic sites. *Northern Perspectives* 22(2–3), 3–6.

Stanton, W.J. (1953) Economic aspects of recreation in Alaska. 1. Economics – analysis of Alaska travel. 2. A recreation program plan for Alaska. In: *Alaska Recreation Survey*. US Department of the Interior, National Park Service, Washington, DC, pp. 3–84, 93–175.

Statistics Greenland (2005) *Recent Publications*. Statistics Greenland, Nuuk; available at http://www.statgreen.gl/english/ (accessed April 2005).

Statistics Norway (2006) *Tourism satellite accounts, final figures for 2004 and preliminary figures for 2005*. Statistics Norway, Oslo; available at http://www.ssb.no/english/subjects/09/01/turismesat_en/ (accessed December 2006).

Swedish National Park Service (2005) *Sweden's National Parks*. Swedish Environmental Protection Agency, Stockholm; available at http://www.internat.naturvardsverket.se/index.php3?main=/documents/nature/engpark/ (accessed December 2005).

Swedish Tourist Authority (2005) *Links/Travel and Tourism – Information about Travel and Tourism*. Swedish Tourist Authority, Stockholm; available at http://www.sweden.se/templates/cs (accessed December 2005).

Territorial Governor of Alaska (1883–1956) *Annual Reports of the Governor of Alaska to the Secretary of the Interior*. US Government Printing Office, Washington, DC.

UNESCO (2006) *The List – Sweden*. United Nations Educational, Scientific, and Cultural Organization, Paris; available at http://whc.unesco.org (accessed June 2006).

UNWTO (2005a) *Study of the Canadian Tourism Satellite Account*. United Nations World Tourism Organization, Madrid; available at http://www.unwto.org/statistics/committee/7th_meeting/phaseII.pdf (accessed December 2005).

UNWTO (2005b) *International Tourism in the Russian Federation – Presentation in Moscow, Russia*. United Nations World Tourism Organization, Madrid; available at http://www.world-tourism.org/ (accessed December 2005).

US Department of the Interior, Fish and Wildlife Service and US Department of Commerce, US Census Bureau (2002) *2001 National Survey of Fishing, Hunting, and Wildlife Associated Recreation*. US Department of the Interior, Washington, DC.

Viken, A. and Jorgensen, F. (1998) Tourism on Svalbard. *Polar Record* 34(189), 123–128.

Vuoristo, K. and Arajarvi, T. (1990) Methodological problems of studying local incomes and employment effects of tourism. *Fienna* 168(2), 153–177.

Wilson, E.A. (1978) *Soviet Passenger Ships 1917–1977*. World Ship Society, London.

Woodard, C. (2005) Iron Curtain: minefield to greenbelt. *Christian Science Monitor* 28 April, 15.

World Trade Organization (WTO) (2003) *Russia makes its mark as a tourism-friendly country. News* 3rd Quarter, Issue 3, 13; also available at http://www.russiatourism.ru/eng/section.asp@id=22 (accessed December 2005).

8 Gateway Ports in the Development of Antarctic Tourism

ESTHER BERTRAM,[1] SHONA MUIR[2] AND BERNARD STONEHOUSE[3]

[1]Royal Holloway, University of London, Egham, Surrey, UK; [2]Institute of Antarctic and Southern Ocean Studies, University of Tasmania, Private Bag 77, Hobart 7001, Tasmania, Australia; [3]Scott Polar Research Institute, University of Cambridge, Lensfield Road, Cambridge CB2 1ER, UK

Introduction

From tentative mid-20th century beginnings, Antarctic tourism has developed in recent years into a major industry. Ship-borne tourism currently involves over 30 ships annually, most of which make several voyages to Antarctica each summer, employing several thousand on-board crew and staff. Air-borne tourism involves both overflying aircraft and ground teams supporting land-based operations. In comparison with tourism elsewhere in the world, Antarctic tourism remains small: in 2004/5 some 30,000 recreational visitors came to Antarctica and its environs – fewer than might be expected in a US or European national park in a single week. This number represents, however, a substantial increase over the past 10 years (Table 8.1), and the

Table 8.1. Antarctic tourism, 1994/5 and 2004/5. (Source: Landau and Splettstoesser, 2007.)

Type of tourism	1994/5	2004/5
Shipborne with landings	8,098	22,297
Shipborne, no landing	0	5,027
Airborne with landings	104	876
Airborne, overflights	3,301	2,030
Total visitors	11,503	30,230

industry continues to grow and diversify (Chapters 9 and 12, this volume) to the benefit of its own entrepreneurs and of others who provide and cater for it.

Of the latter, many are concentrated in the gateway ports – the six ports in continents that are peripheral to Antarctica, through which pass the industry's ships, aircraft and clients. The gateway ports both benefit from and contribute to the development of Antarctic tourism. It is and always has been in their interests to encourage and foster the industry, and their roles increase as the industry continues to grow. This chapter reviews the histories of the six ports, their separate responses to the challenges and opportunities presented by Antarctic tourism, and their current involvements in the industry.

Gateway Countries and Ports

Following Bertram (2005: 148) we define an Antarctic gateway port as a coastal or island port, able by its proximity to the Antarctic to benefit from, and control access to, Antarctic and Southern Ocean resources, including fishing, tourism and scientific support. Minimal characteristics of such a port include: (i) managers who maintain political and scientific interests in Antarctica; (ii) good deep-water facilities for refuelling and re-provisioning ships; (iii) an international airport close by; and (iv) local infrastructure developed to facilitate exchanges of commodities and people.

The gateway ports here considered are Ushuaia (Argentina), Punta Arenas (Chile), Stanley (Falkland Islands), Cape Town (South Africa), Hobart (Australia) and Christchuch/Lyttelton (New Zealand). Christchurch and Lyttelton appear conjoined – Lyttelton is the sea port of the much larger city of Christchurch, linked to it by a short road tunnel through the Port Hills. Figure 8.1 shows the geographical distribution of the ports in relation to Antarctica; Table 8.2 shows their coordinates and distances from the nearest points in Antarctica. This list is not exhaustive: passengers for Antarctica have from time to time embarked in Invercargill, the Bluff or Wellington (New Zealand), Freemantle (Australia), or from several of the minor ports of southern South America.

These six ports currently divide between them practically all the tourist trade into the southern oceanic area, with the first three unequivocally taking the lion's share (Fig. 8.2). All but one belong to states that Dodds (1997: xi–xii) defines as Southern Ocean Rim States (SORS), explaining that the term specifically identifies:

> the states that are geographically proximate to the Antarctic and the Southern Ocean rather than southern hemispheric states en masse ... The designation is intended, therefore, to be explicitly geographical in the sense of location and physical proximity.

The exception, Stanley, is the capital of a British Dependent Territory, closely associated with that sector of Antarctica which Britain defines as British Antarctic Territory.

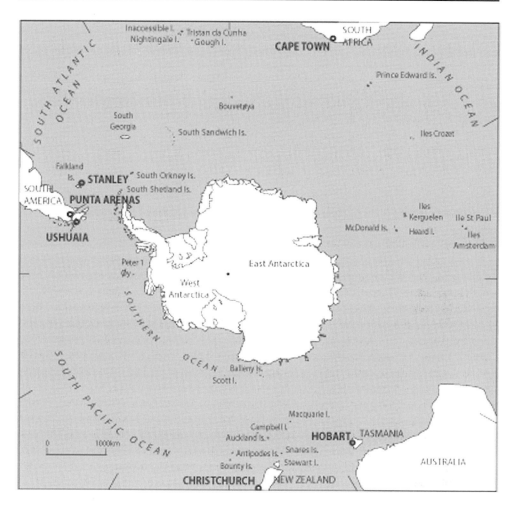

Fig. 8.1. Positions of the six Antarctic gateway ports. (Source: Bertram, 2005.)

Through these ports pass visitors, including tourists and national programme personnel, mainly from the northern hemisphere, to gain access to the Antarctic. To this extent northern hemisphere countries depend on the compliance of SORS to ensure access to the southern region, and SORS benefit by drawing revenues from visitors. This cooperation could be undermined if controlling bodies were tempted either to over-charge for their services or to increase their regulatory requirements for political purposes (see 'Port state jurisdiction' below).

From Ushuaia, Punta Arenas and Stanley, all in the South American sector, the nearest points to the continent or offlying islands (Antarctic Peninsula and the South Shetland Islands) can normally be reached in two to three days, and are reliably free of pack ice for three or four summer months each year. Voyages from South Africa and Australasia take two to three times as long in good weather, even longer in bad weather, with a stronger

Table 8.2. The six gateway ports: positions and distances to the nearest point on Antarctica.

				Distance	
Port	Latitude	Longitude	Closest point in Antarctica	km	Nautical miles
Ushuaia	54°47'S	68°20'W	Hope Bay, Trinity Peninsula	1131	610
Stanley	51°42'S	57°51'W	Hope Bay, Trinity Peninsula	1283	693
Punta Arenas	53°10'S	70°56'W	Hope Bay, Trinity Peninsula	1371	740
Hobart	42°50'S	147°20'E	Dumont d'Urville Station, Adélie Coast	2609	1409
Christchurch	44°33'S	172°40'E	Leningradskaya Station, Oates Coast	2852	1540
Cape Town	35°55'S	18°22'E	SANAE Station, Kronprincesse Märtha Kyst	3811	2057

likelihood of encountering ice close to the continent. Not surprisingly, both national expeditions and tour operations favour the shortest possible voyages, relatively free of floating ice, that will achieve their purposes. Hence the relative popularity of the South American sector for both scientific stations and tourism, and the concentration of traffic through South American gateway ports.

Origins and Development

Each of the six Antarctic gateway ports was founded during the 17th to late 19th centuries, for reasons connected with the European colonization of the southern hemisphere (Table 8.3). The earliest, Cape Town, developed as a

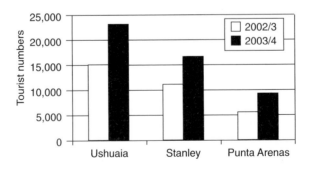

Fig. 8.2. Numbers of tourists visiting Ushuaia, Stanley and Punta Arenas, the three most popular gateway ports, in two recent seasons. Antarctic-bound passengers may visit two or all three of the ports in a single voyage. (Sources: IN.FUE.TUR, 2002/3, 2004/5; Falkland Islands Customs and Immigration, 2003, 2004; M. Bisso, Submanager, Austral Port Authority, Punta Arenas, Chile, personal communication, 2005.)

stopping-off centre for east–west shipping, and Cape Colony's administrative centre and major trading port. Hobart became Tasmania's administrative centre and a focus for South Pacific sealing and whaling activities. Stanley and Punta Arenas became naval bases and important repair and re-supply centres for ships rounding southern South America. Ushuaia developed slowly from mission centre to penal settlement and naval facility.

Apart from sealing and whaling, these ports first became involved in Antarctic traffic through the visits of 18th and early 19th century explorers, who called in to report progress and repair and restock their vessels. Almost all 18th century explorers of the southern oceans called at Cape Town; those of the 19th and early 20th centuries from Bellingshausen to Byrd used Cape Town, Hobart, Lyttelton, Ushuaia or Punta Arenas, and a very few other ports (e.g. Dunedin or Invercargill in southern New Zealand) as points of departure and first return. Explorers heading south towards the Antarctic Peninsula often called at Buenos Aires or Montevideo for social contacts, but relied on Punta Arenas (more rarely Port Stanley, which was more expensive) for essential last-minute supplies. Early 20th century whaling ships operating in the South American sector also tended to call at Punta Arenas, or sail directly to South Georgia, which quickly established its own stocks of coal and other essentials. Pelagic fleets operating in the South Atlantic Ocean used Cape Town, those in the South Pacific restocked in Australasian ports, notably Hobart and Dunedin.

The new wave of scientific exploration that began after World War II and intensified during and after the International Geophysical Year (1957/8) reaffirmed the roles of the gateway ports for Antarctic traffic. Cape Town, Hobart and Christchurch/Lyttelton, Stanley, Punta Arenas and Ushuaia, supported airstrips that made them marginally more accessible.

During the late 1940s and 1950s Chile and Argentina each used its southernmost naval stations to re-supply their growing numbers of Antarctic research stations. Chile used Punta Arenas plus Puerto Williams, a smaller port in Beagle Channel, and Argentina used Ushuaia. Similarly Britain used Stanley as 'rear base' for the Falkland Islands Dependencies Survey (later British Antarctic Survey). New Zealand and Australia re-supplied their Antarctic stations from several of their southern ports before concentrating respectively on Christchurch/Lyttelton and Hobart. Post-war US expeditions operating in the Ross Sea sector also used several southern Australasian ports before settling in Christchurch.

High costs of air-borne operations, notably the expense, difficulties and uncertainties of maintaining air facilities on or about the continent, continue to make it necessary for most goods and passengers to reach Antarctica by sea. A notable exception has been the summer-long air shuttle by the US national expedition between Christchurch and McMurdo Station. Air transport seems likely to be used increasingly by other national expeditions that are concerned to reduce time spent unprofitably at sea. This will not diminish the roles of the gateway ports, all of which now have air facilities of international standard.

Table 8.3. Founding dates, current populations and histories of development of the six Antarctic gateway ports.

Port	Year of founding	Current population[a]	Reason for founding and development, and major traffic
Cape Town, South Africa	1652	2,733,000	Port-of-call for Dutch East India Company bound for Indian Ocean and Far East territories; later used by all east–west shipping in the South Atlantic, becoming South Africa's main port
Hobart, Tasmania, Australia	1803	189,400	First port developed in the colonization of Van Dieman's Land, for export of farm produce and timber. Used by sealers and whalers, becoming Tasmania's capital and main port
(Port) Stanley, Falkland Islands	1842	2,400	Port on East Falkland, developed during the mid-19th century settlement of the Falkland Islands as a naval station, for coaling, ship repairs and export of wool
Punta Arenas, Chile	1849	122,000	Port on Strait of Magellan, developed during the settlement and Chilean colonization of eastern Tierra del Fuego. Ship repairs, bunkering, main outlet for locally produced wool, timber, coal
Christchurch/Lyttelton, New Zealand	1850	340,000	Developed during the settlement of Canterbury, Christchurch becoming the provincial capital, with Lyttelton as its neighbouring deep-water port for export of local produce
Ushuaia, Argentina	1870	60,000	Fishing village and mission centre on Beagle Channel, developed as a naval base and penal settlement during the Argentine colonization of southern Tierra del Fuego, later (1991) designated provincial capital

[a]Population figures are taken from recent websites:
http://www.capetown.gov.za/home/demographics.asp (Cape Town);
http://en.wikipedia.org/wiki/Hobart (Hobart); http://www.citypopulation.de/Falklands.html (Stanley); http://www.greatestcities.com/South_America/Chile/Punta_Arenas_city.html (Punta Arenas); http://en.wikipedia.org/wiki/Christchurch (Christchurch);
http://www.tiscali.co.uk/reference/encyclopaedia/hutchinson/m0014768.htm (Ushuaia) (all accessed 16 August 2005).

Gateway Ports and Early Tourism

The first tourist flights and ship-borne excursions to Antarctica, originating from Chile and Argentina, made use of existing military and scientific facilities. In promoting both scientific expeditions and tourism, each nation was reaffirming its claims to the southern extension of its own country. Herr (1996: 210) comments:

> The two principal gateway states – Chile and Argentina – occupied a critical position in the early years, as their ... supportive attitudes allowed commercial tourism to establish a niche for itself in Antarctica. The national and sub-national laws of other gateway states, particularly Australia, New Zealand and South Africa similarly became significant influences favouring the growth ... of tourism in subsequent years.

Despite this early start, neither Chile nor Argentina developed Antarctic tourism operations of their own, presumably for want of development capital and lack of interest among domestic consumers. The market was seized by US (and later European) operators, which have since effectively controlled it. Of the 32 Antarctic tour operators that are currently full members of the International Association of Antarctic Tour Operators (IAATO, 2005), 15 are based in North America. In the 2002/3 season, a total of 13,571 ship-borne tourists visited Antarctica of whom 5343 were North American; only 47 Argentine and 13 Chilean citizens visited (IAATO, 2003). Visitors to Antarctica are predominantly North American and European, passing annually through Chilean and Argentine gateways which the citizens of those countries do not choose to use, but from which they benefit considerably.

The economic significance of Antarctica to modern gateway ports is measured in terms of expenditure by both scientific expeditions and tourist operations, including expenditure on aviation, shipping, supplies, equipment, services, repairs, staffing expenses, port and airport dues, etc. A large cruise ship requires a host of servicing agents from television repairmen to piano-tuners, sometimes local food produce, and occasional bulk purchases of stationery, novelties and other essentials. Further revenues come from scientists, technicians and tourists travelling to Antarctica. Gateway ports also attract tourists with strong Antarctic interests, who visit but do not proceed to Antarctica. Christchurch, for example, provides a museum with permanent display of Antarctic artefacts from the 'heroic' era of exploration, and an Antarctic Centre that features such Antarctic experiences as Hagluund rides and a simulated storm, as well as halls of educational material, audio-visual presentations, a gift shop, a café and conference facilities.

The significance of the gateways ports will increase, providing both economic and political benefits to their inhabitants, so long as Antarctic tourism continues to expand. Expansion may translate into greater regulatory burdens for those maintaining control, as well as for those who organize the visiting.

Gateways in the South American Sector

Ushuaia, Punta Arenas and Stanley are currently the busiest gateways providing both for Antarctic tourism and for scientific expeditions. Each developed as a 19th century commercial and naval port; each from the late

1940s provided facilities for Antarctic research operations; and each has subsequently become involved in Antarctic tourism. Figure 8.2 compares numbers of tourists visiting the three ports in two recent successive seasons.

Punta Arenas at the end of the 19th century was already a flourishing city, strategically placed on the Strait of Magellan, with wealth based on local produce including coal and wool. In 1906/7 the Chilean–Norwegian Sociedad Ballenera de Magallanes, which had previously operated off Tierra del Fuego, began whaling from Deception Island, South Shetland Islands, using Punta Arenas as its rear base (Headland, 1989: 237). Ushuaia at the time remained a frontier town, recently enlarged from a native fishing village to accommodate a penal settlement, with a limited range of port facilities and little to trade but poor-quality timber. Argentina's main interest in Antarctica at this time was centred on Orcades, the meteorological station on Laurie Island, South Orkney Islands, which the government had inherited from the Scottish National Antarctic Expedition in 1904. In the same year the Argentine–Norwegian Compania Argentina de Pesca established a whaling station at Grytviken, South Georgia; for an account of its fortunes see Hart (2001). Its operations were serviced mainly from Buenos Aires rather than Ushuaia.

Port Stanley at the turn of the century was a small port servicing a predominantly agricultural community and naval station. Nominally the capital of the Falkland Islands Dependencies, which included the Antarctic waters immediately to the south, its government initiated no whaling but drew small annual revenues from both Argentine whaling operations on South Georgia and Chilean, Norwegian and British operations along the South Shetland Islands and Antarctic Peninsula.

The fortunes of both Punta Arenas and Port Stanley declined steeply when, in 1914, the opening of the Panama Canal diverted shipping from the South Atlantic. Punta Arenas recovered slowly with the subsequent development of local industries including fishing, sheep and cattle ranching, meat packing, wool production and (after 1945) petroleum. Port Stanley remained relatively undeveloped until well after World War II.

From the early 1940s both Chile and Argentina sent expeditions to establish bases within their Antarctic territories. Involving considerable stores and equipment, these tended at first to operate from northern ports, rather than using the limited facilities of Punta Arenas and Ushuaia. Similarly Britain's establishment of bases in the Falkland Islands Dependencies from 1943 made use of Stanley only as a final port-of-call. As more Antarctic stations were established, both Argentina and Chile re-emphasized their respective claims to the South American sector of Antarctica (Child, 1988: 115; Dodds, 1997: 50–58, 112–119), and Punta Arenas and Ushuaia became more important as gateways. Modern competition between Chile and Argentina for ownership of the most popular gateway port could be interpreted as a continuation of this long-running competition over the boundary of the two national territories and access to the Beagle Channel (Child, 1988: 115–117). Meanwhile the British Government, while in no way relinquishing its claim to the sector, saw the Falkland Islands and South Georgia as little more than declining assets.

Though all three states signed the Antarctic Treaty in 1961, none ceased to regard land south of 60°S as other than an integral part of its territories. The almost simultaneous development of tourism brought new significance to the gateway concept, tempered by slow development in early years, and only recently manifest in substantially increased traffic. As numbers of ship-borne tourists increased, especially in the South American sector, so did the importance of the three ports that became most concerned with their passage.

Ushuaia

Sandwiched between the Beagle Channel and the Martial Range of Tierra del Fuego, Ushuaia currently handles more Antarctic-bound ships and passengers than any other gateway port. This is due partly to its closest proximity to the Antarctic, but also to a massive injection of government funding from the early 1980s onwards. Development accelerated after 1991 when the city became the administrative capital of the newly defined province of 'Tierra del Fuego, Antarctica and the South Atlantic Islands', an investment that proved both timely and highly beneficial to Ushuaia's role as a gateway to Antarctica (Fig. 8.3).

Fig. 8.3. Recently extended quays of Ushuaia, Argentina. (Photo: B. Stonehouse.)

The length and breadth of the mooring quay were increased significantly between 1997 and 1999 as part of a $10 million extension programme (Johnson, 1997), allowing several large container ships, tankers and cruisers to moor alongside and receive services simultaneously. Tax incentives attracted population and light industry to the area, resulting in a rapid increase in the size of the city, with new shops, restaurants and hotels in the centre, and factories and accommodation for immigrants on the outskirts. The airport was expanded in 1995, with an extended runway accommodating larger passenger aircraft and more frequent services. These investments, during a period when Argentina's economy was greatly stretched, indicate the importance to the government of developing Ushuaia as the business and administrative centre of its remote southern province.

In 1992 the government-funded Instituto Fueguino de Turismo (IN.FUE.TUR) was established in Ushuaia to encourage tourism generally in the area, but with an additional aim of enhancing 'the participation of the Province of Tierra del Fuego in Antarctic affairs, particularly in the tourism field given that Ushuaia is an internationally acknowledged gateway to Antarctica' (Galimberti, 1996: 101). Ushuaia is a winter ski-resort and a summer centre for backpacking into the local national parks and other attractions. The Institute's quayside centre demonstrates Ushuaia's proximity to Antarctica in relation to other ports, describes it as the 'most active gateway', provides information on Argentina in Antarctica, and monitors the numbers of Antarctic tourism vessels and passengers passing through the port. Its annual reports provide data on the expansion of Antarctic tourism, clearly demonstrating the steady growth that has resulted in Ushuaia's premier gateway role.

Marisol Vereda (2004: 12–13), a resident of Ushuaia and lecturer at the national university, states that the development of Antarctic tourism results directly from the adoption of a suite of public policies in fields including education and the creation of a suitable commercial environment, rather than the simple availability of natural resources associated with the Antarctic. One example is the prioritization of tourism vessels and those non-tourist vessels bound for the Antarctic over other types of vessels at the port (Table 8.4). Tourist ships contributed 33.9% of port revenues in 2000/1 and 33.7% in 2001/2; Antarctic tourist ships made up 38% of total tourist vessels in the first period and 52% in the second.

Ushuaia's major advantage over its rivals is closer proximity to Antarctica: cruise ships take a day longer to reach the Antarctic Peninsula from either Punta Arenas or Stanley. This is an important consideration in the short-term cruises to the Peninsula that have become the norm for small ships, for example allowing four days rather than two in Antarctic waters during an eight-day cruise. Flights in support of scientific research operate between Ushuaia and an unpaved runway at Marambio Station on Antarctic Peninsula.

Table 8.4. Benefits (in Argentine Pesos) to Ushuaia port from tourist vessels and Antarctic tourist vessels, 2000–2002. (Source: Vereda, 2004: 3.)

	Use of port	Tourist charges	Other charges	Total charges
Non-Antarctic ships				
2000/1	148,765	260,193	348,031	756,989
2001/2	125,651	317,265	312,565	755,481
Antarctic ships				
2000/1	60,195	61,583	168,251	290,029
2001/2	74,790	106,123	211,730	392,643

Punta Arenas

Punta Arenas's fortunes revived towards the end of the 20th century with the development of a nearby offshore oil industry, for which the port became the main servicing centre. During early days of Antarctic tourism, its age, stability and better port and airport facilities made it a more attractive centre for tourists in transit than Ushuaia. However, it gradually lost ground to Ushuaia's vigorous growth. Its city dock still offers facilities for small passenger ships within easy walking distance of the city centre; there are museums, historic buildings and day-long excursions to divert both local tourists and ship-borne visitors (Fig. 8.4). The recent development of an out-of-town container port and deep-water harbour provides facilities for bigger ships, including ocean liners in transit.

Fig. 8.4. Punta Arenas: old dock area and Strait of Magellan. (Photo: J.M. Snyder.)

Visitors currently spend on average 1.5 days in Punta Arenas, compared with 2.5 days in Ushuaia (Valencia, personal communication, 2002). Chilean tour agents are considering options for encouraging their clients to spend longer in their southernmost city. However, as an Antarctic gateway, Punta Arenas's basic problem is its greater distance from Antarctica compared with Ushuaia. While it remains the port of choice (as in the whaling days) for ships heading eastwards to South Georgia, ships heading south to the South Shetland Islands and Antarctic Peninsula (currently by far the majority) must take into account an extra day's sea-time and pilotage (Captain E. Lampey, personal communication, 2002). In 2004/5 only one cruise ship made repeated voyages to Antarctica from Punta Arenas.

Two air companies operate over the Antarctic from Punta Arenas. The Chilean Aerovias (DAP) offers flights within Tierra del Fuego, as well as to the Falkland Islands and Antarctica. This was the first company to carry out regular commercial flights to the Antarctic in 1987 and in 1990 was the first to restart regular flights to the Falkland Islands. ANI/ALE, a US-based company that provides adventure tourism in Antarctica, also operates from Punta Arenas.

Puerto Williams, a former fishing settlement on Navarino Island, on the southern shore of Beagle Channel, was used in the early 1990s as an exchange point for Antarctic tourists, providing a shorter route to Cape Horn and Antarctica than the day-long run from Punta Arenas (Guzman, personal communication, 2003). This was discontinued after an accident in which a passenger-carrying aircraft crashed off the end of the runway, killing a number of Antarctic-bound tourists. However, Puerto Williams has continued to develop slowly as a naval facility, possibly in response to Argentina's intermittent hostility over ownership of Navarino, Picton and Nueva islands.

Stanley

Formerly Port Stanley, this small port on the eastern flank of East Falkland is the capital of the Falkland Islands, and currently the second most-popular gateway for Antarctic tourists (Fig. 8.5). Possession of the Falkland Islands remains a contentious issue between the UK and Argentina: in committing the UK to the islands' defence in June 1982, the British Prime Minister Margaret Thatcher pointed out that the Falklands had strategic value as 'the entrance to the Antarctic' (quoted in Beck, 1988: 191). Since the Argentine invasion and defeat of that year, the UK has invested heavily in the island's economic infrastructure, developing both the port facilities and a military airport.

Though neither was provided specifically to benefit tourism, both have offered considerable opportunities for tourism, especially Antarctic tourism, to develop, with considerable economic benefits (Table 8.5). The new quay and servicing installations in the inner harbour allow small and medium-sized cruise ships to transfer passengers and replenish stores and bunkering directly. Unlike Ushuaia (see above), Stanley does not prioritize Antarctic

Fig. 8.5. Stanley waterfront, Falkland Islands. (Photo: B. Stonehouse.)

vessels at the port. Liners remain in the outer harbour and transfer goods and passengers by tender, sometimes impeded or altogether prohibited by strong winds. Stanley has occasionally had to find overnight accommodation for 100 or more passengers stranded ashore by a gale in the harbour. From Mount Pleasant airport the Royal Air Force provides flights to the UK (RAF Brize Norton, Oxfordshire). LanChile from time to time makes regular scheduled flights between Santiago and the Falklands, but these may be interrupted arbitrarily by Argentina's withdrawal of permission to fly over its territory.

Stanley itself is village-sized and can be traversed in a brisk half-hour walk, with little more than a small cathedral, a museum, a general store, a few gift shops and a recent history of warfare to intrigue visitors. In a town with a population of 2400, the presence of 500–1000 passengers from a large liner is noticeable. Local agencies provide day-long and half-day excursions to

Table 8.5. Economic benefits (in £) to Stanley from Antarctic ship-borne tourism, 2000–2005. (Source: Falkland Islands Customs and Immigration, 2005.)

Season	Total ships	Exit fees	Entry fees	Passenger taxes	Customs service	Harbour dues	Total charge
2000/1	41	3,030	3,870	29,750	7,593	38,577	53,070
2001/2	46	3,300	4,740	39,670	9,394	47,594	65,029
2002/3	49	3,036	6,336	18,980	8,457	45,920	63,749
2003/4	61	5,821	6,981	25,370	12,014	57,910	82,726
2004/5	59	5,880	7,800	114,328	10,832	57,480	81,992

penguin colonies, beaches and battlegrounds. Smaller cruise ships can visit several of the smaller islands, for walks and a farmhouse tea.

South African and Australasian Gateways

Cape Town, Hobart and Christchurch developed as provincial capitals under colonial rule, primarily as gateways to their own hinterlands. Only relatively recently, as ports already well established for commercial traffic, have they taken on new responsibilities as Antarctic gateways. As British colonies, South Africa, Australia and New Zealand showed no interest in Antarctica and assumed no gateway roles except as occasional hosts to passing Antarctic expeditions. Later as dominions they assumed more responsibilities, though remaining very much under British influence in foreign affairs. During the 1920s and 1930s, when Britain, Norway and France were asserting or re-asserting claims to sectors of East Antarctica, Britain first defined a sector bordering the Ross Sea, which in 1923 was passed to New Zealand for administration under the title of the Ross Dependency. Following the British, Australian and New Zealand Antarctic Research Expedition of 1929–1931, it defined a further sector which in 1933 became Australian Antarctic Territory. So began the direct involvement of these two dominions in Antarctic affairs.

Because of the greater distances involved, fewer than 5% of tourists visiting Antarctica pass through Cape Town, Hobart or Christchurch. All three are large cities with well-established commercial ports, in which the impacts of Antarctic science and tourism are less apparent than in Ushuaia, Punta Arenas or Stanley. However, business organizations in Hobart and Christchurch especially are forming consortia that are designed to foster and if possible increase Antarctic trade and connections.

Cape Town

A relative late-comer in Antarctic affairs, South Africa took no part in the 1921 meetings in which Britain, Australia and New Zealand annexed sectors of Antarctica (Dodds, 1997: 186). Throughout the early 20th century whaling period Cape Town provided harbourage for pelagic whaling fleets of several nations that operated in the pack ice of the Antarctic sector immediately to the south. Towards the end of World War II South Africa maintained a meteorological station on Tristan da Cunha (Crawford, 1982) and later a scientific station on neighbouring Gough Island. In 1948 it took over from Britain responsibility for the Prince Edward Islands, to avoid their use by Soviet ships en route to Antarctica; at the same time Australia took responsibility for Heard Island, doubly strengthening the Commonwealth's role in the southern region (Dodds, 1997: 193–194).

South Africa participated actively in the International Geophysical Year, thereby qualifying as one of the 12 states that originally signed the Antarctic Treaty. Since then it has retained research stations on Gough Island, Marion

Island and mainland Antarctica, and played an active role in Treaty affairs. Its involvement during the 1980s while supporting apartheid caused much criticism from opponents of the Treaty System (Dodds, 1997: 204–205), a cause for concern that disappeared with the ending of apartheid in 1994.

The 'Cape route' remains busy as many large vessels that cannot fit through the Suez Canal continue to be routed round it. Cape Town provides re-provisioning facilities for contemporary scientific and tourist vessels – high-class port facilities alongside a popular city with an efficient international airport. Since the 1980s it has been the main refuelling and re-provisioning gateway for South African Antarctic national expeditions and for increasing numbers of research ships from seven other national programmes – those of Norway, Sweden, Finland, Russia, the UK, Japan and India – that currently use it. Together they have formed a network of cooperation to aid logistical operations to and from the Antarctic, the 'Dronnland Community' (Jacobs, personal communication, 2003). Direct flights to the stations are operated by a private company, Antarctic Logistics Centre International (ALCI), from Cape Town.

Although of great interest to scientists, the ice-bound coast of Antarctica immediately south of South Africa is unlikely ever to attract small Antarctic cruise ships. However, Cape Town caters for small numbers of larger passenger vessels, mostly large liners, heading to or from the South American sector of Antarctica. In the words of a Cape Ministry representative (Valentine, personal communication, 2003): 'We are the boring part of Antarctica … Dronning Maud Land is just a vast expanse of ice.'

However, liners on round-the-world cruises that include the Antarctic Peninsula in their itinerary call at Cape Town for re-provisioning and passenger exchange, to the benefit of local enterprises.

Hobart

Tasmania gained strategic importance from the early 1800s for the re-provisioning of whaling and sealing fleets. In the 19th century Australians showed little interest in the emerging southern continent, though Hobart was occasionally used by passing Antarctic expeditions. In the early 20th century Douglas Mawson's Australasian Antarctic Expedition (1911–1914) sailed from Hobart to explore the continental mainland and establish an outpost on Macquarie Island. Later the British, Australian and New Zealand Antarctic Research Expedition (1929–1931) further explored the coastline, also from Hobart.

The Australia National Antarctic Research Expedition (ANARE), based originally in Melbourne, has long operated from Hobart, as has the French national expedition Institut Français pour la Récherche et la Technologie Polaires (IFRTP) and national expedition ships from Japan, the USA, Russia and China. Hobart also houses the headquarters of the Commission on the Conservation of Marine Living Resources (CCAMLR), the Council of Managers of National Antarctic Programs (COMNAP) and the Agreement on

Conservation of Albatrosses and Petrels (ACAP), and is the home of the Australian Bureau of Meteorology and the Commonwealth Science and Industrial Research Organisation's (CSIRO) Division of Marine Sciences. These latter institutions are partners in the Antarctic Climate and Ecosystems Cooperative Research Centre (ACE-CRC). The University of Tasmania in Hobart includes the Institute of Antarctic and Southern Ocean Studies (IASOS) and provides an undergraduate degree course in Antarctic studies.

An independent quarterly magazine, *Ice Breaker*, keeps the city's business community informed on Antarctic and Southern Ocean affairs. Since 1998, when Dick Smith and Giles Kershaw made a pioneering flight from Hobart to Casey Station, Antarctica (Bauer, 2001: 98), Hobart has sought to establish a regular air link between the two, and appears at last to have succeeded. A note by the General Manager of Antarctic Tasmania (part of the Tasmanian Government Department of Economic Development) in the June 2005 issue of *Ice Breaker* congratulated the Australian Antarctic Division on securing funding of Aus$46.3 million for the link:

> Concerted efforts, over many years by the State Government, the Tasmanian Polar Network and others in the Tasmanian Antarctic building community, have contributed to the allocation of this substantial Commonwealth Government funding. Together, we combined our strengths and: (1) actively lobbied the Commonwealth Government to secure Hobart as the departure point for the Australian Antarctic Airlink; and (2) built the case for Commonwealth funding to be allocated to this pivotal project. The Tasmanian Antarctic community has much to be optimistic about at this juncture.

It would be hard to find a clearer exposition of the positive role available to a dedicated business community in an Antarctic gateway port. The publication *Tasmania's Antarctic, Sub Antarctic and Southern Ocean Policy* (Antarctic Tasmania, 2004) provides further evidence of the community's coordinated efforts. This development echoes Australia's long-standing interest in combining tourism with other Antarctic activities (Government of Australia, 1989).

The Tasmanian Government's Wildlife and Heritage Department currently controls Macquarie Island as a nature reserve, with exemplary management provisions for tourists who visit on their way to or from Antarctica (Selkirk *et al.*, 1990).

In establishing 200-nautical-mile Economic Exclusion Zones around the mainland, Macquarie, Heard and the McDonald islands, and Australian Antarctic Territory, Australia significantly extended the area of southern oceans under its control (Dodds, 2000: 241). Its increasingly militant stance on illegal fishing is demonstrated by funding a non-government organization (NGO), the International Southern Ocean Longline Fishing Information Clearing House (ISOLFICH), to monitor the industry. Greenpeace, Ecofleet and the Australian Foundation, all NGOs, have all actively campaigned in Hobart against illegal southern ocean fishing and Antarctic environmental issues. The Antarctic and Southern Ocean Coalition (ASOC) has a presence at CCAMLR meetings and publishes *ECO*, an environmental view of the convention (Kriwoken and Williamson, 1993: 101).

As a gateway for Antarctic tourism Hobart shares with Christchurch the disadvantage of distance from the continent (Table 8.2). However, its excellent port and airport facilities are used for passenger exchanges and re-supply during circumnavigation cruises, and as a starting point for cruises to the Ross Dependency, bringing hundreds of visitors who contribute to the city's growing tourist industry.

Christchurch/Lyttelton

New Zealand's proximity to Antarctica has resulted in close links with the continent since the 1840s. Christchurch/Lyttelton was the departure point of Scott's *Discovery* Expedition (1901–1904) and *Terra Nova* Expedition (1910–1913), and Shackleton's *Nimrod* Expedition (1907–1909). The Ross Dependency, which New Zealand acquired on British initiative in 1923, is half as large again as New Zealand itself, and was at the time considered a significant asset. Yet it quickly transpired that New Zealand's chief responsibility – dealing with Norwegian whaling in the Ross Sea region – would yield little profit (Logan; cited in Dodds, 1997: 112) and there was no capital available for further research or development. Though whaling fleets heading towards the Ross Sea used New Zealand ports en route, New Zealand took little official interest in its dependency from 1925 until 1955 (Auburn, 1972: 157).

The New Zealand Antarctic Society was established in 1933 to encourage national interest in Antarctica (Peat, 1983). This small but influential society provided much of the momentum required to involve New Zealand in the Trans Antarctic Expedition of 1955–1958 and the International Geophysical Year of 1957/8, which jointly founded Scott Base, New Zealand's permanent Antarctic research station. Logistics for these two expeditions operated mainly through Christchurch/Lyttelton, launching them into their joint modern role of gateway port.

During the same period New Zealand encouraged the USA to use Christchurch airport and Lyttelton for its access to Antarctica, an arrangement which has proved highly beneficial to both parties. Cooperation was enhanced in 1974 when the two countries signed the Science and Technological Cooperation Agreement, helped by alliances formed through Antarctic operations (Prior, 1997: 18). Italy also uses Christchurch/Lyttelton as its gateway port. Close contact between these expeditions has led to many examples of cooperation in the field, and understanding and development of common policies on environmental issues. Dodds (1997: 183) comments:

> Old discourses about territorial sovereignty and resource control appear to have been replaced by environmental stewardship/security and exploitation of commercial tourism originating from Christchurch.

A number of Christchurch businesses, organizations and institutions have strong Antarctic associations, specializing in the publication of Antarctic books, the collection of Antarctic-related postage stamps and Antarctic photography. Antarctic-oriented international committee meetings and conferences are frequently held there. Gateway Antarctica, a research centre administered through the University of Canterbury, is dedicated to multidisciplinary research of Antarctica and the Southern Ocean, benefiting Christchurch by promoting it in the national and international community. The centre coordinates the internationally renowned Graduate Certificate in Antarctic Studies (Henzell, 2003). The permanent Canterbury Museum Antarctic display at Christchurch Art Gallery is one of the city's most popular attractions. An Antarctic Heritage Trail has been established which is also a popular tourist activity.

While Lyttelton provides for Antarctic-bound shipping, Christchurch markets itself as an 'aerial gateway' to Antarctica. In 1992 the International Antarctic Centre was established at Christchurch International Airport. In addition to departure terminals, cargo handling areas and aircraft maintenance hangars, the Centre accommodates the offices of Antarctica New Zealand (New Zealand's national Antarctic programme) and of the US Antarctic Program (USAP) with related US naval, army and air force support units, the liaison office for the Italian Antarctic Programme (ENEA), an International Antarctic Visitors Centre and the offices of the Antarctic Heritage Trust.

Antarctica New Zealand, previously a division of the Department of Scientific and Industrial Research, was established in its current form in 1996. Its programme, centred on Scott Base, receives approximately NZ$7 million in funding per year, much of which is spent in Christchurch. That Christchurch/Lyttelton also serves major Antarctic programmes both of the USA and Italy enormously extends its role as a gateway port and airport, and also brings annual revenues to the city conservatively estimated by Muir (2004) at over NZ$25 million (Table 8.6). Prior (1997: 16) provides even higher estimates of NZ$40–50 million (£14–18 million) annually.

International Antarctic committees and conferences are frequently held in Christchurch. The Antarctic connection is present in and encouraged through academic institutions. Gateway Antarctica, a research centre of the University of Canterbury, is dedicated to multidisciplinary research of Antarctica and the Southern Ocean. The centre coordinates the Graduate Certificate in Antarctic Studies, an internationally renowned course, international conferences, projects, research and presentations.

The International Antarctic Visitors Centre is a popular interactive feature providing information and 'virtual experience' of Antarctica. Widely advertised around the city, it provides a strong visual indication of Antarctic associations and an Antarctic 'presence' in Christchurch. The Antarctic Heritage Trust, a small voluntary organization that cares for the historic huts in the Ross Dependency, is also accommodated in the International Antarctic Visitors Centre.

Place promotion is an important component of New Zealand government regional policy (Hall, 2000; Cuzens, 2003). Within the international Antarctic community, Christchurch's marketing of its aerial capacity brings it into fierce

Table 8.6. Estimated contributions (in NZ$) to the local economy of Christchurch/Lyttelton from New Zealand, US and Italian national expeditions. (Source: Muir, 2004.)

	NZ programme (2002/3)	USAP (1999/2000)	ENEA (mean, 2002/3)	Total
Aviation support	1,763,000	2,500,000	510,000	4,773,000
Shipping support		566,535	500,000	1,066,535
Supplies and equipment	513,000	1,800,000	356,290	2,669,290
Contractor expenditure		9,000,000		9,000,000
Services	788,000			788,000
Capital expenditure	3,254,000			3,254,000
Christchurch staffing expenses	1,687,000	2,000,000		3,687,000
Total	8,005,000	15,866,535	1,366,290	25,237,825

USAP, US Antarctic Program; ENEA, Italian Antarctic Programme.

competition with Hobart (see above) which is in many ways similarly placed (Hall, 2000). The Canterbury Development Corporation recognizes the value of this marketing (Pickering, personal communication, 2004), as does Christchurch and Canterbury Marketing (Hill, personal communication, 2004) and the Christchurch City Council (Hay, personal communication, 2004).

Efforts to recognize the value of Antarctic associations are actively being pursued on national and regional levels. This is reflected by the formation in 2000 of the national New Zealand Antarctic Association and the regional Antarctic Link Canterbury (ALC). A partnership between the Christchurch City Council and a number of local organizations, ALC aims to promote economic benefits to the region from Antarctic activities (Antarctic Link Canterbury, 2006). Founder members include Antarctica New Zealand, Gateway Antarctica, the International Antarctic Visitors Centre, Christchurch and Canterbury Marketing, Canterbury Museum, Banks Peninsula District Council and the Antarctic Heritage Trust (Stubenvol, personal communication, 2004). The University of Canterbury, for long associated with Antarctic research and teaching, currently provides degree, higher degree and diploma courses in Antarctic studies.

Many young New Zealanders reach Antarctica as participants in the national expedition, but relatively few become tourists. Christchurch/Lyttelton is the point of departure for a small number of Antarctic cruise ships each summer, carrying mainly American passengers. Numbers are increasing slowly, though these are far less popular than cruises in the South American sector, for reasons listed by Bauer (2001: 224). Overflights of the continent, pioneered from Christchurch in the 1970s, terminated in a disaster in McMurdo Sound in November 1979 and have not since been resumed, though similar overflights from Melbourne were resumed in 1994. New

Zealand itself is a growing tourist attraction, and day-long flights over Antarctica from Christchurch could again become an attraction.

Gateways for Air-borne Tourism

Tourists can now fly over Antarctica from Australia or Chile, and flights with landings would be possible from all the gateway ports. Will tourist flights with landings increase? Swithinbank (1993: 108–109) predicts only a modest increase, as there are only three permanent, sealed runways available in West Antarctica, none of which is suitable for aircraft larger than C-130 Hercules. While wheeled or ski-wheeled aircraft on intercontinental flights can land also on consolidated snow or on naturally occurring blue-ice runways that require no consolidation or maintenance, scheduled tourist flights would need modern approach aids and much-improved runways, at costs that would currently be regarded as prohibitive.

The tour operator ANI/ALE currently uses blue-ice runways, of which there are over 100,000 sq km in Antarctica, for small-scale operations; Swithinbank (1993: 109) comments that it would be technically feasible for a Boeing 747 to land at a blue-ice runway and bring in 300 tourists at a time. However, there would be very limited facilities waiting to receive them, and costs would be high. ANI/ALE's role as support for emergencies in the region places it in a strong position to extend its activities; even national programmes are at times dependent on the operator's capabilities and expertise.

Economies of scale dictate that transporters, whether ships or aircraft, become progressively larger with time, to carry more people more cost-effectively. It appears inevitable that tourists en masse will eventually be flown to Antarctica and accommodated in hotels, either land-based or floating, in significantly larger operations than any at present. Ship-borne tourism has already made the transition to liners carrying 1000-3000 passengers, and there is currently nothing to stop a similar trend in air-borne tourism. It will be interesting to see if airports associated with the current gateway ports, or those in a circle more remote from Antarctica, will provide the necessary facilities.

Port State Jurisdiction

Many operators of ship-borne tours – estimated in a British working paper for the 21st ATCM (Antarctic Treaty Consultative Meeting), April 1997 at about 40%, and now likely to be higher – use ships that are registered with states which, not being parties to the Antarctic Treaty, are not bound by its terms or those of its instruments. The international lawyer Orrego Vicuña (2000: 48–55) and the environmental pressure-group ASOC (2002: 3–4) have both expressed concern that tourist vessels (and similarly fishing vessels) operating under flags of convenience, and subject only to inspections under flag state jurisdiction, may not provide the safeguards needed for vessels operating in Antarctic waters.

Within the Treaty and Protocol there are currently limited requirements for Antarctic vessel inspections, mainly applicable to scientific vessels and rarely used for tour ships. As necessary adjuncts to Antarctic ship-borne tourism, the gateway ports have been cited as possible points of control through which the Antarctic environment could gain increased protection from possible rogue operators. Both Vicuña and ASOC point out that port state jurisdiction could be applied to provide rigorous inspection of fishing and tourist vessels within the gateway ports. In a draft memorandum to the 25th ATCM of 2002 (IP 63, Agenda Item 7), ASOC (2002) proposed that, to enhance the Protocol, all vessels bound for the Antarctic and calling at gateway ports should be subject to inspections which – as all gateways are owned by Treaty parties – could be designed to conform to requirements laid down by the Treaty parties.

This has precipitated discussion as to how exactly port state jurisdiction could be incorporated into the Treaty System. ASOC suggests that the Paris Memorandum of Understanding for Port State Control, signed in January 1982 and taking effect in July of that year, would be relevant to the Antarctic situation, in providing port authorities with powers to inspect all ships regardless of flag. The UK, in a working paper to the 26th ATCM (2003), supported such a regime, noting that Consultative Parties already have certain responsibilities under Article VII (5) of the Antarctic Treaty for expeditions, which include tourist vessels that are 'organised in, or depart to Antarctica from, their territories'. The UK therefore urged that parties take responsibility for expeditions to the Antarctic, ensuring that they follow the highest standards determined by the Environmental Protocol. Several other states appear to support the scheme, including Norway and Chile (Hemmings, personal communication, 2003). For Argentina and the UK, both would need to be very sure that such a system will not undermine their claimant status. The UK appears to see the scheme being beneficial; they are however sensitive to the terms 'gateway state' or 'gateway port jurisdiction' and instead use the term 'Departure State Jurisdiction' in their working paper. This is because the UK Government feels that such terms tend to exclude the Falklands, due to the UK's geographical distance from it (Richardson, personal communication, 2004). On the other hand, these terms also re-emphasize the issue of sovereignty that continues to trouble Argentina.

Considering that five of the six gateway countries that would be involved in such a development are claimants, they might interpret port state jurisdiction (or be thought by other states so to interpret it) as a tool for furthering their status in the region. This would undermine the aims of such a system and could lead to disagreements over biased decisions for inspection. For example, British ships visiting Ushuaia, or Argentine ships visiting Stanley, may be treated differently from ships visiting those ports from other countries. Such concerns are commented on by Richardson of the British Foreign and Commonwealth Office (personal communication, 2004):

I suspect that Argentina would formally counter our right to undertake such port inspections, and would raise objections if information from such inspections was to be forwarded to the ATCM or Treaty Secretariat. The sovereignty dispute over the Falklands/South Georgia and the South Sandwich Islands is probably the most significant impediment to Port State jurisdiction in respect of Antarctica proceedings.

Unless there were complete agreement on standards between the port authorities, and rigorous maintenance of standards between them, those wishing to attract ships could be tempted to apply more lenient criteria – a consideration that might well apply to the three gateway ports in the South American sector.

Port state jurisdiction would thus have serious implications for all the Antarctic gateway ports. Gateway countries would have to consider whether implementing stringent regulatory measures at their ports would jeopardize their popularity or upset the fragile relations currently existing between them. Prior (1997: 20) argues the case for southern gateway countries interacting and supporting one another to a greater extent. Adoption of port state jurisdiction by the Treaty parties might provide opportunities for enhanced cooperation, rather than competition, between them.

Conclusions

While the six gateway ports under discussion were developed primarily to facilitate the development of their hinterlands, all contributed marginally to the early exploration of Antarctica and exploitation of southern oceans' resources, and to varying degrees benefited from the early to mid-20th century development of Antarctic whaling. During the mid-century, as Antarctica acquired increasing political and scientific significance, all assumed Antarctic gateway roles – of relatively minor economic benefit for the three largest (Cape Town, Hobart, Christchurch/Lyttelton), more significant for the smaller and relatively undeveloped ports (Stanley, Ushuaia, Punta Arenas) of the South American sector.

The late-century advent and growth of Antarctic tourism, particularly ship-borne tourism, has provided further opportunities for all six ports. As most southern cruises are to the South American sector, Stanley, Ushuaia and Punta Arenas have become the main beneficiaries, each receiving substantial income from the enhanced flow of passengers and goods through their extended facilities. Of the three older ports, Hobart and Christchurch/Lyttelton in particular are actively presenting themselves as Antarctic gateways, and all six ports have clear reasons for promoting Antarctic tourism as strenuously as possible.

Acknowledgements

We thank Fabiana Alvaro (Antarctic Unit, IN.FUE.TUR), Horatio Ojeda (Naval Headquarters, Prefectura), Patricia Gavlan (of the shipping agency Navalia) and Marisol Vereda (National University of Patagonia San Juan Bosco) who helped our enquiries in Ushuaia; representatives of the Chilean airline DAP and the Turismo Viento Sur tourism agency, Orlando Dollenz (University of Magallanes), José Valencia (Chilean Antarctic Institute) and Jorges Guzman (Government representative) who helped us in Chile; officials in the Government Offices in Stanley – Mike Summers, Michael Hart and Michael Floyd (Customs Officer) and Mike Richardson (Foreign and Commonwealth Office); Government representatives Henry Valentine and Carole Jacobs, and Hans Van Heukelum (of the tourist agency Unique Destinations) in Cape Town; and Ben Galbraith (Antarctic Tasmania, Department for Economic Development) in Hobart.

References

Antarctic Link Canterbury (2006) *Who we are*. Antarctic Link Canterbury website; available at http://www.antarctic-link.org.nz/who.asp (accessed December 2006).

Antarctic Tasmania (2004) *Tasmania's Antarctic, Sub Antarctic and Southern Ocean Policy*. Antarctic Tasmania, Hobart, Tasmania.

ASOC (2002) Port state jurisdiction: an appropriate international law mechanism to regulate vessels engaged in Antarctic tourism. *IP 63 (Agenda Item 7), ATCM XXV*, Warsaw, 10–20 September. Antarctic and Southern Ocean Coalition, Washington, DC.

Auburn, F. (1972) *The Ross Dependency*. Martin Nijhoff, The Hague, The Netherlands.

Bauer, T. (2001) *Tourism in the Antarctic: Opportunities, Constraints and Future Prospects*. Haworth Hospitality Press, New York, New York.

Beck, P. (1988) *The Falkland Islands as an International Problem*. Routledge, London.

Bertram, E. (2005) Tourists, gateway ports and the regulation of shipborne tourism in wilderness regions: the case of Antarctica. PhD thesis, Royal Holloway, University of London, London.

Child, J. (1988) *Antarctica and Southern American Geopolitics: Frozen Lebensraum*. Praeger, New York, New York.

Crawford, A. (1982) *Tristan da Cunha and the Roaring Forties*. Charles Skilton, Edinburgh.

Cuzens, S. (2003) LOTR spin offs. *Bright: Reflecting Business Brilliance* 2 December.

Dodds, K. (1997) *Geopolitics in Antarctica*. John Wiley & Sons, Chichester, UK.

Dodds, K. (2000) Geopolitics, Patagonian toothfish and living resource regulation in the Southern Ocean. *Third World Quarterly* 2(2), 229–246.

Falkland Islands Customs and Immigration (2003) *Statistical Information*. Falkland Islands Government, Stanley, Falkland Islands.

Falkland Islands Customs and Immigration (2004) *Statistical Information*. Falkland Islands Government, Stanley, Falkland Islands.

Falkland Islands Customs and Immigration (2005) *Statistical Information*. Falkland Islands Government, Stanley, Falkland Islands.

Galimberti, D. (1996) The Antarctic unit in the tourism board of Tierra del Fuego and its role in Antarctic education. In: *Opportunities for Antarctic Environmental Education and Training, Proceedings of the SCAR/IUCN Workshop on Environmental Education and Training, Gorzia, Italy, 26–29 April 1993*. International Union for Conservation of Nature and Natural Resources, Cambridge, UK, pp. 101–103.

Government of Australia (1989) *Tourism in Antarctica: Report of the House of Representatives Standing Committee on Environment, Recreation and the Arts*. Australian Government Publishing Service, Canberra.

Hall, C.M. (2000) The tourist and economic significance of Antarctic travel in Australian and New Zealand Antarctic gateway cities. *Tourism and Hospitality Research* 2(2), 157–168.

Hart, I.B. (2001) *PESCA: A History of the Pioneer Modern Whaling Company in the Antarctic.* Aiden Ellis, Salcombe, UK.

Headland, R.K. (1989) *Chronological List of Antarctic Expeditions and Related Historical Events.* Cambridge University Press, Cambridge, UK.

Henzell, J. (2003) Canty connection. *The Press (Christchurch)* 13 December.

Herr, R.A. (1996) The regulation of Antarctic tourism. In: Vidas, D. and Stokke, O. (eds) *Governing the Antarctic: The Effectiveness and Legitimacy of the Antarctic Treaty System.* Cambridge University Press, Cambridge, UK, pp. 203–323.

IAATO (2003) Report of the International Association of Antarctic Tour Operators (IAATO) 2002–2003. *IP 78 (Agenda Item 5b and Agenda Item 10), ATCM XXVI*, Madrid, 9–20 June. International Association of Antarctic Tour Operators, Basalt, Colorado.

IAATO (2005) Overview of Antarctic tourism 2004–2005 Antarctic season. *IP 82 (Agenda Item 12), Antarctic Treaty Consultative Meeting XXVIII*, Stockholm, 6–17 June. International Association of Antarctic Tour Operators, Basalt, Colorado.

IN.FEU.TUR (2002/3) *Antarctica: Information on the transit of Antarctic tourists through Ushuaia during 2002–3.* Instituto Fueguino de Turismo, Ushuaia, Argentina.

IN.FEU.TUR (2003/4) *Antarctica: Information on the transit of Antarctic tourists through Ushuaia during 2003–4.* Instituto Fueguino de Turismo, Ushuaia, Argentina.

Johnson, P. (1997) Ushuaia port expansion. *Buenos Aires Herald* 19 June.

Kriwoken, L.K. and Williamson, J.W. (1993) Hobart, Tasmania: Antarctic and Southern Ocean connections. *Polar Record* 19(169), 93–102.

Muir, S. (2004) The role of Christchurch as an Antarctic gateway port. Thesis for Graduate Certificate of Antarctic Studies, University of Canterbury, Christchurch, New Zealand.

Peat, N. (1983) *Looking South: New Zealand Antarctic Society's First Fifty Years 1933–1983.* New Zealand Antarctic Society, Wellington.

Prior, S. (1997) *Antarctica: View for a Gateway.* Working Paper 5/97. Centre for Strategic Studies, Victoria University of Wellington, Wellington.

Selkirk, P.M., Seppelt, R.D. and Selkirk, D.R. (1990) *Sub-Antarctic Macquarie Island.* Cambridge University Press, Cambridge, UK.

Swithinbank, C. (1993) Airborne tourism in the Antarctic. *Polar Record* 29(169), 103–110.

Vereda, M. (2004) El impacto del turismo Antarctico en el desarollo de Ushuaia, Tierra del Fuego, indicadores socio-economicos. Presented at *VI jornadas nacionales de investigacion-accion en turismo 'La investigación en turismo: el desafío del nuevo milenio'*, Ushuaia, Argentina, 21–24 April.

Vicuna, F.O. (2000) Port state jurisdiction in Antarctica: a new approach to inspection, control and enforcement. In: Vidas, D. (ed.) *Implementing the Environmental Protection Regime for the Antarctic.* Kluwer Academic Publishers, Dordrecht, The Netherlands, pp. 45–69.

III Developments in Antarctic Tourism: Introduction

BERNARD STONEHOUSE

The first commercial tourist visits to Antarctica can be dated unequivocally to the late 1950s. By then the days had passed when membership of a bona fide Antarctic expedition was a privilege granted to very few. The massive post-World War II onslaughts of Byrd's Operations Highjump and Windmill, and the impressively coordinated expeditions of a dozen participant nations during the International Geophysical Year (1957/8), had brought thousands to the continent for exploration and science. But the idea that Antarctica could be used for recreational visits was not immediately acceptable to the scientists and diplomats who had now appropriated the continent. Indeed, among the scientific community 'tourist' was a term of opprobrium reserved for expedition members who failed to pull their weight, or worse – much worse – for journalists and politicians who came south for a week in summer and dined out on it for months afterwards.

That genuine tourists could now buy their way to Antarctica almost beggared belief. Some of the scientists got over it quickly; indeed a few grasped the opportunities that it provided for interdisciplinary research. Others simply hoped that it would soon go away. A few may still retain that hope, though it must now be a forlorn one. The five chapters of this third section, under the heading 'Developments in Antarctic Tourism', make it quite clear that practically every aspect of an already diverse business is growing and diversifying further as we watch.

In Chapter 9 Esther Bertram outlines the growth of ship-borne tourism, by far the most prominent of all Antarctic tourist activities and the one that seems most likely to continue its accelerating growth during the next quarter-century. Not long ago small ships carrying up to 120 passengers were the mode. Today those are disappearing, replaced by more cost-effective larger ships of up to 250, with more and more liners of 1000+ passengers appearing each year. In Chapter 10 three researchers from The Netherlands – Machiel Lamiers, Jan H. Stel and Bas Amelung – investigate the growth of adventure tourism, brought to Antarctica by both ships and aircraft. Though still a small sector of the industry as a whole, this is one that appeals to a prosperous younger generation – far more than the relatively staid, safe lectures and Zodiac tours that are standard fare of ship-borne tourists – and one that most sharply raises problems of responsibility for accidents and liability for damage and disruption.

In Chapter 11 Thomas Bauer discusses the overflights that are now a regular and popular annual feature of travel from Australia. Overflights are regularly made too from Chile, as are scenic flights with landings on King George Island, in the South Shetland Islands. The planned 'air bridge' between Hobart, Tasmania and Casey Station, Antarctica suggests possibilities of a similar development over East Antarctica. Chapter 12, by Denise Landau and John Splettstoesser, provides a viewpoint from the International Association of Antarctica Tour Operators (IAATO) on how the industry as a whole is faring, and how growth and development may continue into the future. Chapter 13, by Bernard Stonehouse and Kim Crosbie, reports retrospectively on the literature generated by Antarctic tourism over its first 50 years. Tourism involves a wide range of disciplines from social anthropology to zoology. Those who investigate it may present results in an even wider range of publications, making it difficult for new researchers (or even older ones) to come to grips with the literature. Stonehouse and Crosbie scan from tentative early reports of unlikely happenings (as the first voyages seemed at the time) to full-bodied studies of a well-established and flourishing industry.

9 Antarctic Ship-borne Tourism: an Expanding Industry

ESTHER BERTRAM

Royal Holloway, University of London, Esher, Surrey, UK

Introduction

This chapter explores the development of Antarctic tourism during its first 50 years, identifies its current forms and recent diversification of activities, summarizes the impacts of such activities, and assesses the effectiveness of regulations available for their management. In conclusion it draws attention to shortcomings in environmental assessment criteria. Though concerned mainly with ship-borne tourism, for comparative purposes it deals also with land-based and air-borne activities.

Tourism and Its Uses of Antarctica

Bertram (2005: 22) defines Antarctic tourists as:

> Recreational visitors to the Treaty area who are not affiliated in any official capacity with an established governmental or NGO program, or involved in independent research or journalistic or artistic documentation of the continent.

Currently some 30,000 tourists visit Antarctica yearly, of whom over 95% travel by ship. Most visit the Maritime Antarctic, the sector south of South America including the Antarctic Peninsula and southern islands of the Scotia Arc. Milder throughout the year than continental Antarctica and with longer summers, this sub-region is scenically and biologically more attractive than continental Antarctica, and can be reached in 2 days or less by sea from South America, the closest continental landmass (Fig. 9.1). Much of it is relatively free of sea ice between early November and March. Accessibility has prompted several national scientific programmes to establish research stations in the area.

Continental Antarctica is colder, more difficult of access and, with the notable exception of the Victoria Land coast of the Ross Sea, less interesting to recreational visitors. Inland Antarctica is accessible mainly by air, but air-borne tourism remains a costly and limited alternative to ship-borne tourism.

The International Association of Antarctica Tour Operators

The International Association of Antarctica Tour Operators (IAATO), the organization that represents and coordinates commercial tourism in Antarctica, is discussed in detail by Landau and Splettstoesser (Chapter 12, this volume). Established in 1991 as a consortium of ship operators, it has grown and diversified to accommodate new developments within the industry. Membership is voluntary: not all Antarctic tour operators have joined, but most see advantages in conforming to its requirements. Up-to-date listings of members are available on the IAATO website (www.iaato.org). Membership

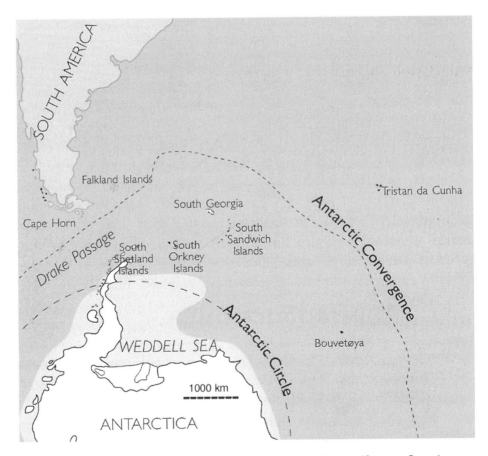

Fig. 9.1. Antarctica, South America and the southern Atlantic Ocean. (Source: Stonehouse, 2006.)

involves responsibilities; for example, to use appropriate forms of transport, hire qualified staff and limit landings to 100 passengers ashore at a time with a 1:20 staff-to-passenger ratio. Provisional and probationary operators seeking full membership carry approved observers on their first visits to the region.

IAATO's original bylaws, adopted in 1991, required member companies to carry not more than 400 passengers per voyage, a limit determined by the capacity on Antarctic cruises of the largest ship operating at the time. The advent of larger ships made it necessary for the limit to be stretched to 500 passengers on ships that make landings; no capacity limit has been imposed on the still-larger liners now operating, that do not seek to land passengers.

IAATO members in Antarctica provide a valuable 'first response' network of ships, available to all ships and stations in emergencies. Given the very limited response infrastructure otherwise present in Antarctica, this is a most valuable and important service. Members have voluntarily, in some instances heroically, assisted in medical evacuations, and provided support to ships in difficulties and emergencies. They have also provided transport and logistic support for scientific research – a role greatly appreciated by all who have benefited from it.

Ship-borne Tourism

Ship-borne tourism began in 1957–1959 with four visits by Argentinean and Chilean naval transports, which accommodated tourists whose fares helped to pay costs of servicing the national expeditions (Reich, 1980: 205). Snyder (personal communication, 2006) ascribes its start at this time to a unique combination of favourable world publicity associated with the International Geophysical Year 1957/8 (IGY) and the availability of commercial jet airliner and modern cruise-ship transport. World attention was directed to Antarctica by well-publicized explorations linked with the names of Richard Byrd, Vivian Fuchs and Edmund Hillary. Critically important mapping of the continent by aerial surveys, and the subsequent distribution of maps and photographs, brought favourable attention to the continent: the US National Geographical Society alone distributed 2,270,000 copies of new maps to its international membership. The remarkable cooperation among the international scientific community in establishing 46 IGY research stations made clear the availability of the continent to modern ships and aircraft. It required only entrepreneurial operators to initiate Antarctic tourism.

The initial voyages were quickly superseded by dedicated cruises in small 'expedition' ships carrying 50–120 passengers, which for many years have largely dominated the trade (Fig. 9.2). Small numbers of passengers continue to be carried on naval transports. Not being registered with IAATO, these voyages are not included in annual statistics and exact figures are unknown. The first larger cruise ship to enter the field, *Ocean Princess* in 1990–1993, had a capacity of up to 480 passengers, but carried only 250–400 on its annual Antarctic voyages. Several larger ships carrying 400–500 passengers have since been involved. The most recent development has been the advent from 2000 of liners certified to carry 800–3000 passengers, usually as part of longer South American or worldwide cruises (Fig. 9.3).

Expedition ship and small sailing vessel operations

The most important characteristic of small-ship operations is the facility for landing passengers at selected sites, using fleets of inflatable boats with outboard engines. Arguably the most invasive of all tourist activities, landings are particularly subject to IAATO regulations, developed over many years (Chapter 12, this volume). Small cruise ships accommodating 60–120 passengers began, and have for long upheld, a traditional 'Lindblad pattern' of management (Stonehouse and Crosbie, 1995: 221) which preceded the advent of IAATO but became the basis of IAATO guidelines and management practices. Passengers receive on-board lectures about the environment, wildlife, history and IAATO guidelines on comportment ashore. One, two or three landings are made daily. Visitors are taken ashore in inflatable craft with outboard engines, and stay ashore for 2 to 3 h with well-informed naturalists on hand to oversee and provide information.

Under IAATO rules not more than 100 passengers are allowed ashore at a time – a restriction imposed primarily by the need to return shore parties aboard quickly should the weather change. Ships carrying 300 to 500 passengers necessarily take much longer than small ships to accomplish a landing. Cruise liners with over 1000 passengers lack inflatable boats, and would find the logistics of landing difficult or impossible to deal with.

Fig. 9.2. *Endeavour*, a cruise ship that carries up to 120 passengers, landing them from Zodiacs (inflatable boats). (Photo: E. Bertram.)

Fig. 9.3. *MV Crystal Symphony*, a cruise ship that carries up to 1000 passengers, but makes no landings on Antarctic cruises. (Photo: E. Bertram.)

Among the small sailing vessels and yachts that visit Antarctica, only those that operate commercially (i.e. for charter or advertised cruises) are members of IAATO. Others (it is not clear how many) operate privately and, as far as is possible to tell, without reference to regulations or regulatory bodies. Those that are IAATO members carry up to a dozen passengers and tend to follow the Lindblad style of management, including lectures and landings accompanied by naturalists. Day-to-day management aboard is less formal and more adventurous than on the small cruise ships, appealing particularly to small groups with special interests, such as filming, climbing or bird watching.

Expedition leaders and accompanying teams of lecturers and guides traditionally receive support from the passengers, who willingly accept the spirit of the IAATO guidelines. Occasional individuals or groups of passengers respond negatively, feeling strongly that, as they are paying substantially for the experience, they resent attempts to restrict their activities on landing. From my own observations, experienced staff can usually deal with such situations, but ultimately lack the authority or powers of enforcement of, for example, national park rangers.

In the 2004/5 season, 36 IAATO and non-IAATO expedition ships and small sailing vessels together made a total of 207 voyages to the Antarctic region, carrying 22,297 passengers. Non-IAATO visits in this category included 17 sailing vessels/yachts, for which no further data are known (IAATO, 2005: 3–5).

Larger ship operations with landings

IAATO admits operators of ships with a capacity of over 200 passengers to make landings in Antarctica only if carrying fewer than 500 passengers. They may land no more than 100 at a time, and in all other respects follow the Lindblad pattern and IAATO guidelines as closely as possible (B. Riffenburgh, Expedition Leader *Marco Polo*, personal communication, 2003). Taking larger numbers of passengers ashore imposes restrictions. Choice of sites is limited to a small number where landing is relatively easy. Passengers' time ashore is limited, usually to 1 h, within restricted boundaries (sometimes marked out by traffic cones). Each landing takes much longer, so the overall number of landings is limited to four or five per voyage. Nevertheless many passengers appreciate the relative comfort and cheapness of larger ships, their stability in rough weather, and the alternative attractions on board including casinos and evening shows. In the 2004/5 season three IAATO-member ships in this category made a total of 13 voyages to the Antarctic region, carrying 3768 passengers. Two non-IAATO ships in this size range made eight voyages during the season, carrying a total of 4088 passengers between them (IAATO, 2005: 2).

Passenger liner operations

The first visit of a large passenger liner, *MS Rotterdam*, to Antarctica occurred in January 2000, marking a significant new development in the Antarctic tourism industry. Largest of the Holland America fleet, this ship's size and seeming vulnerability in ice-strewn waters led to many misgivings. In presenting their case for a permit to the US licensing agency, the company quoted its experience of operating in ice under the stringent environmental requirements of the US National Park Service in Glacier Bay, Alaska. It drew attention to the lack of firm operational requirements in Antarctic waters, undertaking instead to meet the rigorous requirements imposed in a sensitive and well-protected area of Alaska. It disarmed early critics of the voyage by not seeking to make landings. Similar cases have since been put by other operators who have subsequently followed Holland America's lead.

Stonehouse and Brigham (2000: 347–349), who observed procedures during the first cruise, pointed out possible dangers from tight scheduling (which might encourage faster-than-safe travel in ice-strewn waters) and the inadequacy of standard lifeboats and rafts to protect passengers and crew under Antarctic conditions. Should such a ship founder in Antarctic waters, up to 1800 passengers and crew would need to be rescued as quickly as possible. Their suggestion that:

> there is a strong case for scheduling a second large ship either in the same area, or at least within two days' travel (for example around southern South America), which could make an immediate response and provide the emergency facilities required

represents a desirable situation that is fast approaching as more large liners enter the trade. In the 2004/5 season four large non-landing liners made four such cruise-only visits to the Antarctic. Operators for three were IAATO members, and the ships carried a total of 5027 passengers (IAATO, 2005: 4).

Large-liner cruising is currently the fastest-growing sector of Antarctic ship-borne tourism.

Statistical records of tourist numbers, particularly during the pre-IAATO era, significantly underestimated the total numbers of non-scientific visitors landing in Antarctica. Certain tourist industry segments and industry-related personnel were not consistently reported. Enzenbacher (1992a: 17) states:

> The total number of tourists who have visited Antarctica is difficult to determine with certainty, due to lack of uniformity in reporting procedures. Scheduled commercial cruises or flights are normally reported to home governments by respective tour operators, and data are later exchanged under the information provision of Para VII (5) of the Antarctic Treaty. Precise numbers of visits made by small or non-commercial expeditions to Antarctica are more difficult to obtain: many visits may never be reported.

IAATO's imposition of improved methods for collecting and collating statistics has to some degree overcome this problem, but there are still difficulties over reporting numbers of naturalists, boat handlers, crew members and cruise-ship administrative staff. These are not normally counted as 'tourists', but frequently go ashore with the tourists and present as few or as many challenges to the environment as those whom they serve.

Among other groups who are difficult to account for is the fast-growing number of independent yacht operators and their passengers, many (but not all) operating from South America, who have not seen advantage in joining IAATO and pursue their own courses in Antarctica. Numbers of independent adventure travellers (discussed more fully in Chapter 10, this volume) are also increasing, whose recreational pursuits tend to place emphasis on extensive site utilization, elaborate logistical support and a potential need for emergency evacuation services.

Flight/cruise operations

The voyage from South America to Antarctica across Drake Passage can be rough and uncomfortable, and dangerous for elderly or part-disabled passengers. From time to time efforts have been made to eliminate it by flying passengers from Tierra del Fuego to the Chilean airstrip Teniente Marsh, on King George Island, South Shetland Islands. The first such operation took place in January 1982, involving a Fuerza Aérea de Chile (FACH) C-130, for immediate transfer to a cruise ship lying at anchor offshore. Flight-cruises have operated spasmodically since then, hampered and made unreliable by frequent bad weather and poor landing facilities at the Antarctic end. Their reliability has recently increased, notably by use of Dash-7 aircraft, and regular scheduled flights are now made. In 2004/5 the Chilean airline company of Patagonia (DAP), a provisional member of IAATO, made 29 flights from Punta Arenas to King George Island, carrying a total of 657 passengers. These visits included, as well as an Antarctic cruise, overnight stays at Teniente Marsh Station and guided tours of other local stations and wildlife (IAATO, 2005: 9).

Ship-borne tourism: summary

Figure 9.4 illustrates the development of Antarctic ship-borne tourism between 1956 and 2005. Four visits by Argentinean and Chilean naval vessels in 1957–1959 preceded the start of regular commercial cruises from 1966. These were dominated by small ships (for example *MS Explorer*, capacity 96 passengers), with occasional visits from the larger Argentine ships *Libertad* (400 capacity) and *Rio Tunyan* (394 capacity) between 1968 and 1976. Six voyages by the Argentine-operated *Regina Prima* (474 passenger average) and a single cruise by the Spanish *Cabo San Roque* (841 capacity) caused the 1974/5 peak (Reich, 1980: 205–206).

Numbers of visiting ships increased from three or fewer between 1957 and 1987 to ten in 1991/2 and 12 in 1992/3 (Enzenbacher, 1992b: 260, 1994: 108). Through the 1990s there followed an irregular annual increase, due in part to an influx from 1991 of small, ice-strengthened Soviet research vessels, which became available for charter to tour companies. *Marco Polo* (800 passenger capacity, but carrying only about 500 in the Antarctic) increased total numbers of tourists landing from 1993, and has been visiting continuously ever since.

A small rise in tourist numbers over the millennium year was due to the single-season advent of two large vessels, *Ocean Explorer 1* (850 capacity) and *Aegean 1* (630 capacity), each making two voyages with landings. 1999/2000 also marked the first visit of the much larger cruise liner, *MV Rotterdam* (1200 capacity), making no landings. Such visits have been a

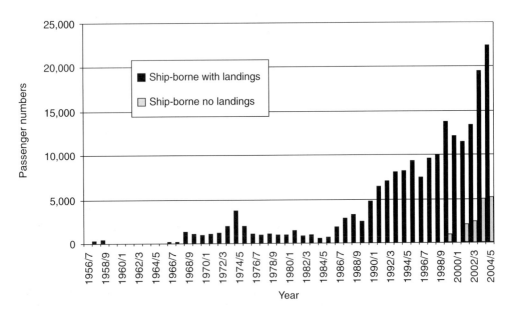

Fig. 9.4. Approximate numbers of ship-borne tourists visiting the Antarctic, 1956–2005, showing ship-borne with landings and cruise-only visitors. (Sources: Reich, 1980; Enzenbacher, 1992a, 1994; IAATO, 1999, 2000, 2001, 2002, 2003, 2004, 2005.)

continuous feature since the 2001/2 season. After 2000, numbers of ship-borne visitors making landings decreased, but then rose again with the advent from 2004/5 of four more large vessels (*Saga Rose*, *Vistamar*, *Discovery* and *Nordnorge*). More large non-landing liners are now involved, including several with capacity of over 1000, and capable of carrying over 3000.

A further trend, significant though not apparent from Fig. 9.4, is the ageing and withdrawal from Antarctic service of some of the earlier small ships and their gradual replacement by ships of larger passenger capacity. It will be interesting to see if smaller vessels, which formerly dominated the scene, will remain among the larger ones as a niche market offering specialized services.

Air-borne Tourism: a Comparison

Although sea-borne and air-borne tourism to the Antarctic began within a year of each other in the late 1950s, air-borne tourism has since been far outstripped, due mainly to high costs and limitations imposed by unreliable weather and lack of land-based accommodation for visitors. Overflights without landings began with the pioneering flight of a DC6B of the Chilean national airline in December 1956 (Reich, 1980: 209). Flights with landings have given rise to land-based adventure operations, involving climbing, skiing and visiting out-of-the-way corners of Antarctica. Developments in Antarctic air-borne tourism have been detailed by several authors (see for example Reich, 1980; Swithinbank, 1989, 1990, 1991, 1992, 1993a,b, 1994, 1995, 1996, 1997a,b, 1998, 1999, 2000; Tracey, 2001).

Following two early flights in 1956 and 1957, commercial landings in Antarctica were rare until 1984, when the Chilean government began marketing short-stay trips to their complex of bases on King George Island, South Shetland Islands. Soon afterwards a private company, Adventure Network International (ANI), began operating between South America and the Antarctic Peninsula. Antarctic Logistics & Expeditions (ALE) purchased ANI in the 2003/4 season. ANI/ALE caters for adventure tourism mainly in the continental interior. Clients include mountaineers, trekkers and visitors who want to experience Antarctica remote from the cruise-ships' beaten track. It also provides emergency rescue services for other tourist activities, and is available to support other private or national expeditions. In the 2004/5 season, ANI/ALE transported and supported 221 tourists on private expeditions or visiting inland areas. Aside from their tourism activities ANI/ALE has also undertaken support for governmental activities (IAATO, 2005: 8).

Helicopters are increasingly used for tourist enjoyment in Antarctica. FACH helicopters normally employed in taking scientists from Teniente Marsh to other points on the South Shetland Islands may be hired for transporting visiting tourists. Helicopters carried by some of the larger cruise ships (notably Russian icebreakers) to reconnoitre ice conditions are frequently used to fly passengers ashore, for sightseeing excursions over the sea ice, or to visit emperor penguin colonies that are otherwise inaccessible.

Antarctic overflights to Victoria Land began in 1977 from New Zealand and Australia, but stopped in 1980 after a tragic accident in which 257 passengers and crew were killed. They were resumed in 1994/5 from Melbourne and Sydney (Headland and Keage, 1995: 347). Similar but shorter flights are made from Punta Arenas over the Antarctic Peninsula and the South Shetland Islands, often coordinated with the arrival in port of an Antarctic cruise ship, offering a preview of the area into which the ship is heading. Australian operators Croydon Travel and Chilean operators LanChile are associate members of IAATO, which estimates that, in 2004/5, 2030 passengers visited Antarctica on overflights (IAATO, 2005: 9–10).

Table 9.1 illustrates the much smaller numbers of passengers involved in air-borne than in ship-borne tourism, and the relatively tiny numbers landing.

Landing Sites

How are landing sites selected for use, and what determines a popular site? Stonehouse (1992: 217) identifies 15 factors that make sites attractive to operators, ranging from offshore sea room for ships to the presence of penguin colonies and relics of earlier human occupation. Sites vary considerably in their attractiveness, but tend to retain their popularity ratings from year to year. Stonehouse and Crosbie (1995: 206) comment further:

Table 9.1. Passengers landed and carried in ships and aircraft, 1992–2005. (Sources: IAATO, 2004, 2005.)

Year	Passengers landed from ships	Passengers carried in non-landing ships	Passengers landed from aircraft	Total passengers landed	Passengers carried in overflights	Total passengers
1992/3	6,983	0	84	7,067	2,134	9,201
1993/4	7,957	0	69	8,026	2,958	10,984
1994/5	8,098	0	104	8,202	3,301	11,503
1995/6	9,212	0	169	9,381	3,146	12,527
1996/7	7,323	0	110	7,433	3,127	10,560
1997/8	9,473	0	131	9,604	3,412	13,016
1998/9	9,857	0	79	9,936	2,041	11,977
1999/2000	13,687	936	139	13,826	2,412	17,174
2000/1	12,109	0	150	12,259	1,552	13811
2001/2	11,429	2,029	159	11,582	2,412	17,457
2002/3	13,263	2,424	180	13,443	1,552	17,547
2003/4	19,369	4,949	517	19,886	2,827	27,662
2004/5	22,297	5,027	878	23,175	2,030	30,232

For tourists, sites without wildlife or human artefacts have little appeal, while those in spectacular settings of glaciers and mountains are particularly favoured. However, it is cruise directors who ultimately choose where landings are made. Each has a list of particular sites acceptable on grounds of safety, known to have proved popular before, and often favoured especially for such factors as reliability of weather or ease of landing. The most popular sites are those that feature on most of the cruise directors' lists.

How often are individual sites visited? In a contribution to the VIII International SCAR Biology Symposium, Stonehouse and Bertram (2001) analysed IAATO statistics for the preceding 10 years that provided a perspective on site use. Approximately 270 sites had by then been identified as landings in the Antarctic Treaty area, by far the majority in the Maritime Antarctic. Of these, many had been visited only a few times since their discovery, and very few new sites were appearing in annual listings. In the single season 2000/1 (which was typical of its period) 18 cruise ships made 116 voyages during a season of approximately 20 weeks, carrying a total of some 12,000 visitors. Though cruise directors were free to land at any of the known sites, or indeed to find new sites if they so desired, landings in that season were made at only 122 (45%) of the known sites. Each visit lasted 2 to 3 h and involved on average 90 passengers. Table 9.2 shows the distribution of visits among these sites.

Among the least-visited sites, 71 (58%) received only one or two visits in the season, and 97 sites (79%) were visited fewer than ten times – on average less than once every second week. Among the more popular sites, only nine

Table 9.2. Numbers of visits paid to 122 sites where passengers were landed from cruise ships during the season 2000/1. (Source: IAATO, 2001.)

No. of sites visited	No. of visits per season	% of sites in each group	Frequency of visits in a 20-week season
1	>61	0.8	Over three per week
4	51–60	3.3	
4	41–50	3.3	Over two per week
3	31–40	2.5	
4	21–30	3.3	Over one per week
9	11–20	7.4	
11	6–10	9.0	Over one per four weeks
5	5	4.1	One per four weeks
6	4	4.9	
4	3	3.3	
19	2	15.6	One per 10 weeks
52	1	42.6	One per 20 weeks
122		100	

(7.4%) received more than 40 visits (two visits per week) and only 16 were visited on average more than once per week. The most popular of all (Whalers Bay, Deception Island, South Shetland Islands) rated 79 visits, almost four per week (Stonehouse and Bertram, 2001; cited in Bertram, 2005: 310–316).

All but one of the most popular sites lay within the Maritime Antarctic; the exception was Commonwealth Bay, the site of Mawson's Australasian Antarctic Expedition (1911–1914) hut, which received ten visits during the season.

It is relevant to note that numbers of ship-borne passengers landing have almost doubled since this analysis, but the tendency towards concentration of landings at a relatively small number of sites has continued. IAATO figures reported for the 2005/6 season (IAATO, 2006) indicate that 30 ships visited (an increase of three over the previous year) and that, as in years past, 30 sites received 85% of the visits, with only six sites receiving 37% of all visits (IAATO, 2006: 1). The most recent Antarctic Treaty Consultative Meeting is now preparing to adopt specific-site guidelines for the most heavily visited sites (United Kingdom, 2006), though not yet to introduce monitoring and assessment measures that might provide a fuller understanding of impacts and management issues involved.

Expansion and Diversification of Antarctic Tourism

Figures 9.4 and 9.5 illustrate the expansion of Antarctic tourism during its first 50 years; Table 9.3, showing in more detail the accelerated expansion since 1992, raises the following points:

1. Between 1992 and 1999 the number of passengers landing from ships expanded by 141%; thereafter it expanded more rapidly with a substantial increase in the number of ships and voyages. It currently shows an increase of 319% overall since 1992.
2. The number of passengers landing from aircraft fluctuated widely during the first half of the period, with no overall tendency to increase. Thereafter it showed a steadier increase, indicating the recent success of Category 6 air/ship operations, and overall has increased more than tenfold since 1992.
3. During the five seasons in which increasing numbers of non-landing liners have operated, passenger numbers carried by these ships have increased over fivefold.
4. Numbers of passengers carried in overflights fluctuate widely from year to year, but show no overall tendency to increase.
5. Overall, the total number of passengers landing in Antarctica, and the total number visiting the continent, has increased respectively by 328% and 329%.
6. Of all passengers landed in Antarctica in 2004/5, 96% arrived by ship – a slight reduction on the overall mean, again due to the recent growing success of air/ship operations.

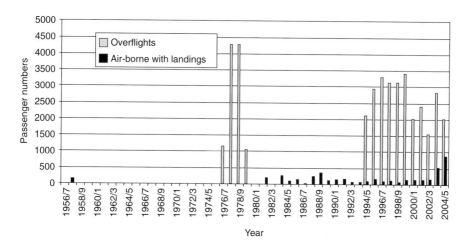

Fig. 9.5. Air-borne Antarctic tourism, 1956–2005, showing passengers carried on overflights and passengers landing. (Sources: Reich, 1980: summary of 1977–1980 flights, from which the numbers in figure have been estimated; Swithinbank, 1989, 1991, 1992, 1993a,b, 1994, 1995, 1996, 1997a, 1998, 1999; IAATO, 2000, 2001, 2002, 2003, 2004, 2005; Bauer, 2001.)

Table 9.3 presents in more detail the growth of ship-borne tourism during the 1992–2005 period. During the first six seasons, numbers of operators, ships, voyages and landings remained fairly constant, at levels consistent with those of the preceding decade. Following slight increases in 1998/9 and

Table 9.3. Ship-borne tourism: numbers of operators, voyages and landings, 1992–2005. (Source: IAATO, 2004: 3, 2005.)

Year	Total ship-borne passengers	Operators	Ships	Voyages
1992/3	6,983	10	12	59
1993/4	7,957	9	11	65
1994/5	8,098	9	14	93
1995/6	9,212	10	15	113
1996/7	7,323	11	13	104
1997/8	9,473	12	13	92
1998/9	9,857	15	15	116
1999/2000	14,623	17	21	154
2000/1	12,109	15	32	131
2001/2	13,458	19	37	117
2002/3	15,687	26	47	136
2003/4	24,318	31	51	180
2004/5	27,324	35	52	207

1999/2000, and a slight decline following the well-advertised cruises of the millennium year, a more dramatic increase began in 2002/3, continuing into the most recent season of 2004/5. During these past three years, numbers of operators, ships and voyages have all achieved records.

Rising popularity of larger ships

Table 9.4 indicates an irregular but positive increase in numbers of passengers travelling to Antarctica in larger ships over the last 7 years. Argentine and Chilean vessels operating in the 1960s and 1970s were in the 400 capacity range. After that time, however, up to the end of the 1990s, Antarctic tourism was dominated by small ships of up to 120 passenger capacity. *Marco Polo*, operating from 1991, was the first to challenge the original 400-passenger limit. Several other larger ships have since appeared and proved popular, greatly increasing the proportion of passengers making use of larger ships.

Nationalities of Antarctic tourists

Between 1994 and 2005 tourists to Antarctica originated mainly from the USA, Germany, the UK, Australia and Japan (Fig. 9.6). Of these, US tourists visited in the greatest numbers every year, ranging from 34% of all visitors in 1994/5 and peaking at 52% in the millennium year, before returning to about 40% over the last 4 years. Japanese numbers have been low but steady throughout the period; the spectacular expansion of the last 2 years has been shared almost equally by the other four nations.

Table 9.4. Numbers of Antarctic tourists travelling on ships of 250 capacity or more, 1998–2005. (Sources: IAATO, 1999, 2000, 2001, 2002, 2003, 2004, 2005.)

Year	Total passengers visiting by ship	Passengers on ships 250 capacity or more	% of total
1998/9	9,857	2,177	22
1999/2000	14,623	6,152	42
2000/1	12,109	2,642	21
2001/2	13,458	5,187	38
2002/3	15,687	6,636	44
2003/4	24,398	12,666	52
2004/5	27,324	13,164	48

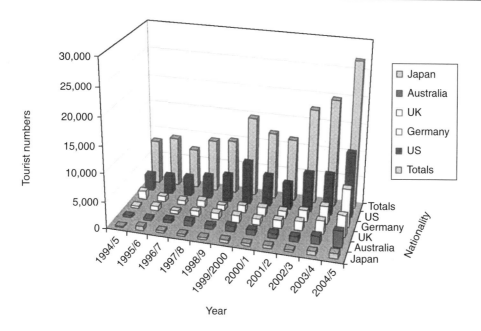

Fig. 9.6. The five countries from which the greatest number of Antarctic tourists originated, 1994–2005. (Source: IAATO, 2005.)

In 2004/5, of almost 23,000 visitors of known nationality, 36.5% were from the USA, 15.9% from the UK, 13.7% from Germany and 10.0% from Australia. In that year almost equal numbers originated in Canada (3.7%) and Japan (3.2%); the next most frequent came from Switzerland (1.9%) and the Netherlands (1.8%). Conspicuously absent from these statistics, in view of their proximity and strong political interest, are citizens of Argentina, Chile or any other South American nation: the few who visit may find it easier and cheaper to travel as passengers in ships of their own navies.

With Antarctic tourism continuously dominated by American tour operators (of current members of IAATO, 46.9% are US-based), it might be assumed that the decline in both air-borne and sea-borne travel that followed the attack on New York of 11 September 2001, indicating a reluctance by US citizens to travel outside the USA, would have adversely affected numbers of Americans travelling to Antarctica. The continuing increase in Antarctic tourism (which remains dominated by Americans) suggests that Americans regard Antarctica as a relatively safe destination for travel. This is in line with worldwide figures for cruise-ship tourism, which appears to have remained only marginally affected by world events. In fact, in the last decade cruise tourism has been the tourist sub-sector with the highest growth rate (WTO, 2003). Overall more than 250 cruise liners carried 10 million people to different areas of the world in 2003.

Visits to landing sites

The main points of contact between ship-borne tourists and Antarctica are the landing sites, of which over 270 have been identified as having been used during the time since IAATO records began (see 'Landing sites' above). Table 9.5 shows the numbers of sites in Antarctica at which ship-borne tourists landed during the period 1994–2005 in both maritime and continental regions. Data include only passenger landings from ships: they do not include Zodiac cruises without landing or helicopter landings (which are rare). Nor does it include such recently developed innovations as site visits involving prolonged stays for camping or scuba diving – new activities that are now collated separately on the IAATO website (http://www.iaato.org/tourism_stats.html) and are particularly significant for site management.

Table 9.5 illustrates the many more sites used in maritime than in continental Antarctica, a factor reflected also in the numbers of ship visits and passengers landed in the two regions. In both regions, numbers of sites visited, numbers of ship visits and numbers of passengers landing all show irregular increases during the period under review. The number of ship visits and the total numbers of passengers landing show a very strong positive mutual correlation (Spearman's rank correlation coefficient 0.92). The table shows also that, within the past decade, the number of sites visited by tourists has increased by 50%, the number of visits to sites (essentially Zodiac trips) has slightly more than doubled, and passenger landings have almost doubled. Current reliable figures represent enormous advances over the less precise, but none the less credible, figures for the 1970s and 1980s.

Table 9.5. Number of peninsula and continental sites visited, ship visits and total numbers landed at all sites[a], 1994–2005. (Sources: Reich, 1980; Enzenbacher, 1992a, 1994; IAATO, 1999, 2000, 2001, 2002, 2003, 2004, 2005.)

Year	Passengers	Sites visited			Ship visits			Passengers landed		
		Mar	Cont	Total	Mar	Cont	Total	Mar	Cont	Total
1994/5	8,098	82	8	90	615	12	627	53,453	1,411	54,864
1995/6	9,212	75	8	83	786	10	796	60,530	1,349	61,879
1996/7	7,323	89	21	110	779	39	818	54,975	3,866	58,841
1997/8	9,473	80	16	96	723	35	758	66,815	3,978	70,793
1998/9	9,857	94	20	114	866	26	892	75,216	4,136	79,352
1999/2000	13,687	101	24	125	1,129	58	1,187	92,555	5,195	97,750
2000/1	12,109	126	27	153	1,064	83	1,147	92,814	5,015	97,829
2001/2	11,429	94	28	122	874	57	931	78,180	3,413	81,593
2002/3	13,263	122	26	148	1,274	29	1,303	105,282	2,512	107,794
2003/4	19,369	120	26	146				134,572	3,370	137,942
2004/5	22,297									

[a]Numbers do not include Zodiac visits, helicopter landings, camping or scuba diving activities. Mar, maritime sites; Cont, continental sites.

The presence ashore of tens of thousands of tourist visitors each summer has led to misgivings about impact from sheer weight of numbers. Headland (2002: 1) provides a perspective by comparing 'person-hours or days' spent by scientists and tourists in the Antarctic during 2001/2. For ship-borne tourists he used average times at landing sites, adding to the total nights spent in the region by those involved in air/land operations. This he compared with the average time spent by scientists and supply ship/plane personnel in the region. He concluded that, in that particular year, though tourists greatly outnumbered scientists, they accounted for only 1.07% of the total person-days spent ashore. This calculation does not acknowledge that ship-borne tourists spend all their time ashore in the most sensitive wildlife breeding areas, whereas the station personnel notch up their hours mainly within their stations. However, Headland (personal communication, 2003) points out that many stations are built in ice-free areas which are also the breeding sites of much of the wildlife.

Not illustrated in Table 9.5 is a factor of diversification in the use of landing sites. Traditionally passengers land for the experience of landing in an Antarctic setting, to stroll within a few hundred metres of the point of landing, visit wildlife and take photographs under guidance, then to return to the ship. During the past few seasons particular operators have varied ship-borne landings by providing for camping ashore, climbing, marathon running, scuba diving, canoeing and other diversions that far exceed the self-imposed limits set by the early operators. Though these varied activities have been validated by Initial Environmental Evaluations, there is no evidence of their having been subject to scrutiny, other than by the operators themselves, to ensure that the requirements and intentions of the Environmental Protocol are being met.

Tourism Management

With Antarctic tourism, particularly ship-borne tourism, clearly growing at an accelerating rate, it becomes increasingly important that sufficient regulation is in place to manage the activities of the industry. In the absence of sovereign governments, all human activities in Antarctica are regulated by a single management authority, the Antarctic Treaty System, of which the relevant instrument is the 1996 Environmental Protocol. Though claims to Antarctica are placed on hold under Article 4 of the Treaty, sovereignty still plays a crucial role in tourist matters. As the Protocol is given effect through domestic legislation, the sovereignty of the national authority to which an operator applies for permission to conduct operations is crucial (Kriwoken and Rootes, 2000: 142). Different states that are signatories to the Treaty have different legal systems, providing variation in how the Protocol is interpreted. Boczek (1988) comments on differences in interpretation due to the vague language of the Antarctic Treaty, allowing a lack of conceptual rigor in the regulations that apply to tourists under the different regimes. Tracey (2001: 195–206) provides a comprehensive country-by-country overview of the methods used to implement the Protocol.

Currently, Initial Environmental Evaluations (IEEs) are required for ship-borne operations with and without landings, overflights and all other tourist-related operations in Antarctica. At the first implementation of the Protocol in 1996, 'programmatic' IEEs were devised by IAATO covering standard landings, making it easier both for ship-borne operators to apply for permits and for permitting authorities to grant them. They avoided, however, the important issue of whether all sites are equally vulnerable to the standard activities specified. As Hemmings and Roura (2003: 22) point out, this provision allowed operators to maintain a degree of flexibility in choosing their landing sites, but limited the potential for monitoring activities.

Once a permit has been granted, in the absence of practical external regulation operators regulate their own ongoing activities, bound only by their own interpretations of the IAATO guidelines. Article 14 of the Protocol requires parties to appoint and accommodate observers, which has mainly been implemented for station inspections, very little for the observance of tourist activities. Provisions for monitoring under the Protocol have not so far been implemented, and there is no general provision for management plans which are generally considered essential premises for monitoring procedures.

There appears to be little agreement between states on procedures and standards in assessing IEEs, perhaps because few countries have arranged for coordination and 'across-the-board' implementation. Article 6 of the Protocol requires parties to coordinate activities and help each other in the preparation of impact assessments. However, at present there are no quantitative standards among the countries that assess tourism activities to assist tour operators in determining precise measures of impact (Kriwoken and Rootes, 2000: 148). These authors comment that the current system relies on a case-by-case reactive process of environmental impact assessment supported by the adoption of IEE, with no comprehensive system of across-the-board standards for implementation or evaluation. Without an external body overseeing these assessments, there is the potential for tour operators to create IEEs to fit their own business interests.

Presuming that the challenges associated with the IEEs can be surmounted, there remains the issue of cumulative impact evaluation. The single tour operator and their passengers may behave admirably within the scope of their IEE, but the repetitive use of a site by several operators can result in unintended consequences that no one desires. Individual tour operators cannot be expected to perform this type of analysis, and institutional relationships are not sufficiently vested to accomplish this function either. As a starting point, consensus-based management goals and objective will have to be defined – a daunting first step indeed.

The impact of ship-borne tourists making landings is one concern raised by critics of Antarctic tourism, in view especially of increasing numbers of ships and ship-borne operations. There is also concern about large-scale pollution and search-and-rescue implications caused by any visiting ships, but especially by the large liners that are visiting in increasing numbers. Little research has been undertaken to date on the overall impacts of well-managed ships visiting Antarctic waters without untoward incidents. Little evidence has appeared to suggest that emissions of smoke, oil or other contaminants from such ships have measurable effects: the proliferation of cruise ships cannot in itself be regarded as an environmental threat.

IAATO ships follow IAATO guidelines by minimizing their everyday impacts. Some crew members are employed purely for sorting waste materials, including paper, cardboard and glass, for delivery to their gateway ports. Human waste is treated according to strict guidelines. Drinking water is produced on the ships by treating seawater. Food wastes are broken down and treated before being voided at sea. Nothing is released from ships while in the Antarctic Treaty area. During my own observations and those of colleagues, IAATO guidelines have been followed precisely.

Summary and Conclusions

Antarctic ship-borne tourism at the end of its first half-century is showing every sign of rapid growth and development. Formerly restricted to small passenger ships operating the successful, environmentally sound but limited Lindblad pattern of operations, it has now diversified to include larger ships landing passengers, still-larger liners that make no landings, flight/cruise operations, and landings with a diversity of onshore and near-shore activities – all of which appear to be expanding. The programmatic IEEs covering landing operations provide for non-specific permitting, both for ease of operation and to allow operators the right to select landings for their own convenience. However, they make no reference to individual vulnerabilities of landing sites and are not implemented equally by the different countries involved in permitting. Under IAATO guidance, Antarctic tourism remains largely a self-governing industry with every prospect of further expansion.

References

Bauer, T.G. (2001) *Tourism in the Antarctic: Opportunities, Constraints, and Future Prospects.* The Haworth Hospitality Press, New York, New York.

Bertram, E. (2005) Tourists, gateway ports and the regulation of shipborne tourism in wilderness regions: the case of Antarctica. PhD thesis, Royal Holloway, University of London, London.

Boczek, B.A. (1988) The legal status of visitors, including tourists, and non-governmental expeditions in Antarctica. In: Wolfrum, R. (ed.) *Antarctic Challenge III.* Duncker and Humblot, Berlin, pp. 455–490.

Enzenbacher, D.J. (1992a) Tourists in Antarctica: numbers and trends. *Polar Record* 28(164), 17–22.

Enzenbacher, D.J. (1992b) Antarctic tourism and environmental concerns. *Marine Pollution Bulletin* 25(9–12), 258–265.

Enzenbacher, D.J. (1994) Antarctic tourism: an overview of 1992/93 season activity, recent developments and emerging issues. *Polar Record* 30(173), 105–116.

Headland, R.K. (2002) Effects of tourists visiting Antarctic regions proportional to all human effects. Paper presented at *IAATO Meeting*, Cambridge, UK, 1–4 July.

Headland, R.K. and Keage, P.L. (1995) Antarctic tourist day-flights. *Polar Record* 31(178), 347.

Hemmings, A. and Roura, R. (2003) A square peg in a round hole: fitting impact assessment under the Antarctic Environmental Protocol to Antarctic tourists. *Impact Assessment and Project Appraisal* 21(1), 13–24.

IAATO (1999) Overview of Antarctic tourism activities 1998–9. *Antarctic Treaty Consultative Meeting XXIII*, Lima, 24 May–4 June. International Association of Antarctica Tour Operators, Basalt, Colorado.

IAATO (2000) Report of the International Association of Antarctic Tour Operators (IAATO) 1999–2000. *SATCM/IP (Agenda Item 8), Special Antarctic Treaty Consultative Meeting*, The Hague, The Netherlands, 11–15 September. International Association of Antarctica Tour Operators, Basalt, Colorado.

IAATO (2001) Report of the International Association of Antarctic Tour Operators (IAATO) 2000–2001. *ATCM XXIV/IP (Agenda Item 5b), Antarctic Treaty Consultative Meeting XXIV*, St Petersburg, Russian Federation, 9–20 July. International Association of Antarctica Tour Operators, Basalt, Colorado.

IAATO (2002) Report of the International Association of Antarctic Tour Operators (IAATO) 2001–2002. *IP 74 (Agenda Item 5b), Antarctic Treaty Consultative Meeting XXV*, Warsaw, 9–20 September. International Association of Antarctica Tour Operators, Basalt, Colorado.

IAATO (2003) Report of the International Association of Antarctic Tour Operators (IAATO) 2002–2003. *IP 78 (Agenda Item 5b and Agenda Item 10), Antarctic Treaty Consultative Meeting XXVI*, Madrid, 9–20 June. International Association of Antarctica Tour Operators, Basalt, Colorado.

IAATO (2004) Report of the International Association of Antarctic Tour Operators (IAATO) 2003–2004. *IP 68 (Agenda Item 4b and Agenda Item 11), Antarctic Treaty Consultative Meeting XXVII*, Cape Town, South Africa, 24 May–4 June. International Association of Antarctica Tour Operators, Basalt, Colorado.

IAATO (2005) IAATO overview of Antarctic tourism 2004–2005 Antarctic season. *IP 95 (Agenda Item 12), Antarctic Treaty Consultative Meeting XXVIII*, Stockholm, 6–17 June. International Association of Antarctica Tour Operators, Basalt, Colorado.

IAATO (2006) Brief update on the Antarctic Peninsula landing site visits and site guidelines. *IP 66 (Agenda Item 12), Antarctic Treaty Consultative Meeting XXIX*, Edinburgh, 12–23 June. International Association of Antarctica Tour Operators, Basalt, Colorado.

Kriwoken, L. and Rootes, D. (2000) Tourism on ice: environmental impact assessment of Antarctic tourism. *Impact Assessment and Project Appraisal* 18(2), 138–150.

Reich, R.J. (1980) The development of Antarctic tourism. *Polar Record* 20(126), 203–214.

Stonehouse, B. (1992) Monitoring shipborne visitors in Antarctica: a preliminary field study. *Polar Record* 28(166), 213–218.

Stonehouse, B. (2006) *Antarctica from South America*. Originator Publishing, Great Yarmouth, UK.

Stonehouse, B. and Bertram, E. (2001) Monitoring and assessment of impacts at Antarctic tourist landing sites. Paper presented at *Antarctic Biology in a Global Context, VIII SCAR International Biology Symposium*, Amsterdam, 27 August–1 September. Abstract S6003.

Stonehouse, B. and Brigham, L. (2000) The cruise of *MS Rotterdam* in Antarctic waters. *Polar Record* 36(199), 347–349.

Stonehouse, B. and Crosbie, K. (1995) Tourist impacts and management in the Antarctic Peninsula area. In: Hall, C.M. and Johnston, M. (eds) *Polar Tourism: Tourism in the Arctic and Antarctic Regions*. John Wiley & Sons, Chichester, UK, pp. 217–233.

Swithinbank, C. (1989) Non-government aircraft in the Antarctic 1988–1989. *Polar Record* 25(154), 254.

Swithinbank, C. (1990) Non-government aircraft in the Antarctic 1989–1990. *Polar Record* 26(159), 316.

Swithinbank, C. (1991) Non-government aircraft in the Antarctic 1990–1991. *Polar Record* 28(164), 66.

Swithinbank, C. (1992) Non-government aircraft in the Antarctic 1991–1992. *Polar Record* 28(166), 232.

Swithinbank, C. (1993a) Airborne tourism in the Antarctic. *Polar Record* 29(169), 103–110.

Swithinbank, C. (1993b) Non-government aircraft in the Antarctic 1992–1993. *Polar Record* 29(170), 244–245.

Swithinbank, C. (1994) Non-government aircraft in the Antarctic 1993–1994. *Polar Record* 30(174), 221.

Swithinbank, C. (1995) Non-government aircraft in the Antarctic 1994–1995. *Polar Record* 31(178), 346.

Swithinbank, C. (1996) Non-government aircraft in the Antarctic 1995–1996. *Polar Record* 32(183), 355–356.

Swithinbank, C. (1997a) Non-government aircraft in the Antarctic 1996–1997. *Polar Record* 33(187), 340.

Swithinbank, C. (1997b) New intercontinental air route: Cape Town to Antarctica. *Polar Record* 33(186), 243–244.

Swithinbank, C. (1998) Non-government aircraft in the Antarctic 1997–1998. *Polar Record* 34(190), 249.

Swithinbank, C. (1999) Non-government aircraft in the Antarctic 1998–1999. *Polar Record* 36(196), 51–52.

Swithinbank, C. (2000) Non-government aircraft in the Antarctic 1999–2000. *Polar Record* 36(198), 249.

Tracey, P. (2001) Managing Antarctic tourism. PhD thesis, Institute of Antarctic and Southern Ocean Studies, Hobart, Tasmania.

United Kingdom (2006) Site guidelines for Goudier Island, Port Lockroy. *WP (Agenda Item CEP 7), Antarctic Treaty Consultative Meeting XXIX*, Edinburgh, 12–23 June.

WTO (2003) *Worldwide Cruise Tourism*. World Tourism Organization, Madrid.

10 Antarctic Adventure Tourism and Private Expeditions

MACHIEL LAMERS, JAN H. STEL AND BAS AMELUNG

International Centre for Integrated Assessment and Sustainable Development, University of Maastricht, Postbox 616, 6200 MD Maastricht, The Netherlands

Introduction

Tourist arrivals in Antarctica have increased sharply over the last decade. Modern transport technologies have improved the accessibility of this remotest of continents, creating new opportunities for both commercial and private expeditions. As a result, the portfolio of tourism activities has gradually become more diverse; a broad range of market segments is now catered for, including luxury tours and adventure tourism. The classic Antarctic expedition cruises involving small to medium-sized ships and landings in inflatable boats are now complemented by, for example, eclipse-viewing trips reported by the *Antarctic Non-government Activity Newsletter* (ANAN, 2003: 96/05, 95/01), fly–sail cruises, overflights and cruises on very large cruise liners, as well as by land-based adventurous activities such as mountain climbing, cross-country skiing and marathon running (IAATO, 2004b). The trend towards diversification has given rise to concerns about the desirability and appropriateness of certain tourist activities in an Antarctic setting. In policy circles, 'adventure tourism' has become a catchall term for all insufficiently prepared trips to the Antarctic. There is thus a lack of clear terminology and definition, which hampers the identification, analysis and possible solution of any tourism-related problems in Antarctica (Murray and Jabour, 2004). However, terminology is just one side of the issue. Key concepts in understanding the commotion around adventure tourism are risk and impact, and these are only loosely linked to tourism typologies. This chapter explores the risks of adventure tourism in Antarctica and their determinants, in conclusion exploring the implications of our findings for its current and future development and governance.

Adventure Tourism in Antarctica

The process of negotiating and ratifying the Environmental Protocol in the 1990s temporarily reduced attention on tourism, but the issue re-emerged on the agenda of the first Antarctic Treaty Consultative Meeting (ATCM) of the new millennium (ASOC, 2001). Since then the discourse has focused on specific and technical issues, such as ship sizes, site-specific guidelines, cumulative impacts and 'adventure tourism'. This latter issue has recently been addressed by the Council of Managers of National Antarctic Programs (COMNAP) and the International Association of Antarctica Tour Operators (IAATO). Various policy documents report on incidents involving adventure tourists that required intervention from National Antarctic Programs (NAPs) in terms of search and rescue (SAR), medical support, accommodation and transportation (COMNAP, 2002). Such assistance is typically very expensive and risky, and disruptive for the stations' usual activities. Its effects on science programmes that may have taken years of planning cannot be easily undone by financial compensations of the direct costs incurred (Chiang, 2000: 27; ANAN, 2003: 95/03).

The financial and operational consequences of several serious incidents have given adventure tourism a negative connotation in policy circles and raised questions about its desirability and options for regulation. Some NAPs have even established stringent policies regarding station visits and services such as accommodation, food and fuel. At some frequently visited stations, adventure tourists are informed upon arrival that they are basically entitled to nothing, a point confirmed in many written accounts by adventure tourists. However, in practice such hostility hardly ever leads to serious conflicts; the station personnel often seem happy to have guests and treat them hospitably. Nevertheless there have been serious conflicts between NAPs and adventure tourists, up to the highest political and diplomatic levels (New Zealand and United States, 2004); see Table 10.1 and its discussion below.

The Environmental Protocol currently regulates activities in Antarctica from an environmental perspective by demanding initial environmental assessments and notifications. However, since its adoption, some Treaty parties and authors have expressed concerns about its limitations in regulating tourism (Richardson, 2000: 77). COMNAP has stated that 'high risk adventure tourism' often cannot be regulated within this environmental legal framework, since risks to the environment are generally considered low in this type of tourism. As a consequence, parties do not have legal powers to require adventurers to undertake contingency planning or carry insurance to cover SAR costs in case of emergency (COMNAP, 2002, 2003). Meanwhile IAATO has made it very clear that its member organizations do not cause any problems (IAATO, 2003c): all are subject to a list of strict personal, financial and operational requirements. Furthermore, IAATO members have been called in where necessary to rescue independent adventure tourists in difficulties.

In the academic literature the significance and scope of the notion of adventure tourism are heavily debated. The many definitions differ in focus, depending on the research context (Swarbrooke *et al.*, 2003). Many definitions stress tourist motivation as a distinguishing factor: adventure tourism is distinguished from other types of tourism by differences in the participants' intent. For our purposes, the relevance of this type of definition is limited, because they are unrelated to the risks and impacts that are imposed on others.

A very general observation about adventure tourism is that it entails an interaction between a participant and an environment in which the outcome is uncertain (Hall, 1992; Priest, 2001: 112; Swarbrook *et al.*, 2003: 9). The inherent uncertainty often translates into risks for participants. In the Antarctic context, these risks are amplified by the continent's inhospitable climate and general lack of facilities.

Paradoxically, these very uncertainties and real or perceived risks are among adventure tourism's main attractors – very different from scientific activities and mainstream tourism in Antarctica, where intentions are to exclude or avoid as many risks as possible. The role of risks may well constitute one of the main distinctive features of adventure tourism.

Adventure tourism in Antarctica can be characterized as a broad spectrum of self-initiated or commercially provided journeys, or single activities, with a challenging or innovative nature, to or within the Antarctic continent. In the following sections we discuss activities that match these general characteristics. An important distinction is made between commercially provided adventure tourism activities and those that are independently pursued. Specifically, IAATO-member companies organize various types of adventure tourism and support private expeditions in terms of transport and backup. A distinction is also commonly made between ship-based and land-based tourism. We discuss below: (i) ship-based tourism, consisting of activities in the coastal zones that are operated from tourist ships; (ii) land-based tourism, involving activities away from the coasts, operated from tented camps serviced by aircraft; and (iii) forms of adventure tourism that do not belong to either category, for example independent adventurers arriving by private boats or aircraft.

Ship-based adventure tourism

Ship-borne tourism has a history of decades in the Antarctic. The Antarctic coastal zones are now sailed for tourism purposes by a variety of ship types and sizes, from small yachts and expedition ships to sailing ships and cruise liners. Most ships visit the Antarctic Peninsula for traditional activities, such as cruising in inflatable boats and wildlife viewing. New generations of travellers show a keen interest in more active elements. As a result, the range of activities has broadened over the past decade to include sea kayaking, scuba diving, snorkelling, ice camping and climbing (IAATO, 2002). In general most of these additional adventure activities are organized from the smaller ships (i.e. yachts and expedition ships).

According to IAATO (2003c), the passengers participating in these ship-based adventure packages account for less than 0.5% of all tourists travelling to Antarctica. Based on this statement, the number of adventure tourists can be estimated at 120 for the 2003/4 season and at 150 for the 2004/5 season.

IAATO members that organize these new activities have developed guidelines and operating procedures for passengers, staff and crew (ANAN Archive, 2002: 77/08) that address the activities' specific risks. Prior to any trip, tour operators screen the passengers for physical and mental competence, and for experience in the particular sports involved. In addition, participants are required to declare in writing that they accept the risks involved in the activity. Appropriate and qualified staff facilitates the activities (IAATO, 2003c). IAATO ensures that member companies are well insured and that they are capable of dealing with incidents without much reliance on facilities and support from uninvolved national programmes. The network of IAATO-member companies active in the Antarctic region and the pre-established contacts with NAPs anticipates this need. Despite safety regulations, fatal incidents have occurred among scuba divers (IAATO, 2003a, 2005a).

Land-based adventure tourism

Land-based tourism activities rely on air links for transportation into and out of the Antarctic interior. Currently, two commercial airlines are operational. The Adventure Network International/Antarctic Logistics and Expeditions (ANI/ALE, formerly Adventure Network International), registered in the USA, has been operating in Antarctica for 17 consecutive seasons. From Punta Arenas and Cape Town ANI/ALE services several tented camps and landing strips in Antarctica. The Chilean Aerovias DAP is an airline with a long history of flying between Punta Arenas and King George Island in the Antarctic Peninsula. DAP operations are limited to station visits and fly–sail operations, and are thus of little interest to us. As a result, our overview is limited to the ANI/ALE activities.

ANI/ALE was established by the late Gilles Kershaw, a well-known Antarctic pilot and one of the pioneers of landing a wheeled aircraft on natural blue-ice fields. Blue-ice runways were vital in the establishment of ANI/ALE because it opened up the Antarctic interior and created opportunities for adventurers and even NAPs (Swithinbank, 1998a). Another important factor has been the small but steady market of mountaineers and polar adventurers desiring to go to the Antarctic interior. Initially these adventurers were mostly the world's top mountaineers climbing Antarctica's tallest mountain, the Vinson Massif, as part of the seven highest summits on each continent. Later these adventurers also included South Pole cross-country skiers using kites and sledges, mountaineers climbing the numerous unclimbed peaks, and single activities such as skydiving.

Private expedition teams are constantly in search of innovative and challenging activities. As a result, the composition of private groups varies widely, as do the activities and locations chosen. ANI/ALE provides services such as transport, contingency planning, SAR services and medical evacuations. Typically, mountain climbers and South Pole skiers only use the transport and rescue services. In addition, ANI/ALE organizes a number of land-based adventure activities and less-demanding packages. The adventure packages include skiing trips to the South Pole, mountaineering trips and marathons. The less-demanding packages include the South Pole fly-in, the emperor penguin trip and small-scale activities around the tented camp, such as driving snowmobiles and short skiing trips. Finally, besides supporting private expeditions and operating their own tourist itineraries, ANI/ALE provides logistics and backup support for several tourist vessels and NAPs in Antarctica.

Adventure tourism in the Antarctic interior is a very small niche market. Contrary to the general belief, average growth rates are not high either, as Fig. 10.1 shows. Nevertheless, the trend has been upwards over the last few years, interrupted only by a dip in 2003/4 when operations were limited as ANI was taken over by ALE. Figure 10.1 must be interpreted cautiously as a significant share of the passengers does not seem to qualify as adventure tourists. Arguably, the only real adventure tourists are the ones who take part in private expeditions, mountaineering activities and ANI/ALE's own adventure tourism activities. These visitors and adventurers are independent and self-reliant for part of the expedition, or move away from the base camps into difficult terrain, with or without experienced guides. People in the category of 'other passengers' did not perform any of the adventurous activities described above, but were visitors participating in the less-demanding packages, ANI/ALE personnel and other unspecified passengers.

Being a founding member of IAATO, ANI/ALE applies stringent requirements to its own trips and the private expeditions it supports. These requirements relate to client acceptance and risk management and include appropriate insurance in case of an emergency; environmental impact assessment for the proposed journey; permission from the relevant national government departments; the quantity of food that is taken and the caloric value of each meal; information on the route taken and the equipment used (camping, communication, medical); experience of team members in polar regions and with the proposed activity; and backup planning in case of technical problems. As an extra safety measure, ANI/ALE automatically dispatches an aircraft to pick up expedition teams if no communication can be established for 48 h (IAATO, 2003c).

In 1997 the only fatal incident within the operational history of ANI/ALE occurred. On the first Antarctic skydiving attempt above the South Pole, three of the participants died when their parachutes failed to open (IAATO, 1998). It was argued that despite significant polar experience in the group, they lost their sense of perception looking down on the white expanse of the ice cap. Further, no emergency deployments were in place (Chiang, 2000: 26). This expedition resulted in an incident that triggered a discussion about the

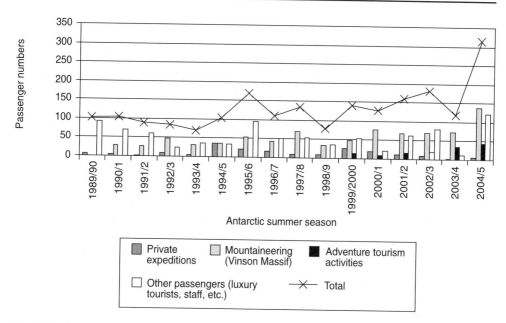

Fig. 10.1. Numbers of passengers carried by Adventure Network International/Antarctic Logistics and Expeditions (ANI/ALE) per austral season, 1989/90 to 2004/5. (Sources: Swithinbank, 1989, 1990, 1992a,b, 1993, 1994, 1995, 1996, 1997, 1998b, 1999, 2000; IAATO, 1997, 1998, 1999, 2000, 2001, 2002, 2003a,b,c, 2004a,b, 2005a,b.)

appropriateness of skydiving in Antarctica (Outside Online Dispatches, 1998). In spite of this, another skydiving expedition (the Millennium Expedition 1999/2001) was organized by a Russian polar outfitter in 2001; no major problems occurred. Nevertheless, ANI/ALE has declared that it will not organize any further skydiving activities in Antarctica (D. Rootes, personal communication, 2005).

Other forms of adventure tourism

There is more to adventure tourism than the activities conducted and supported by IAATO-member companies. In the history of Antarctic adventure tourism there have been cases of NAPs supporting private expeditions in terms of transport, food and other services. In fact, the first private expeditions and adventure tourism programmes were a direct spin-off from large-scale governmental expeditions, involving the same people and using the routes and experiences gained (Swithinbank, 1998a). There are numerous examples of private mountaineering and skiing expeditions that were supported in terms of logistics and backup by NAPs. These services were either voluntarily granted to adventure travellers or provided in emergency cases.

Also, more recently private adventurers sometimes try to make arrangements with NAPs to arrange particular services to cut the costs of their expeditions. Some NAPs have protested against these 'hopping-and-shopping' practices and established stringent permit requirements, regulations for station visits, or called for improved communication between the NAPs. However, the negative attitude towards Antarctic tourists is not shared by all NAPs (IAATO, 2003c).

Recently, Antarctic Logistics Centre International (ALCI) started operating a new air link between Cape Town and Dronning Maud Land (DML). While initially intended as a transport link for NAPs operating in the DML area, several reports show that paying passengers have been taken on board. Moreover, several proposals exist for mountaineering expeditions (ANAN Archive, 2001: 83/01, 61/01; the poles.com, 2006). ALCI has not provided any information on the exact passenger numbers that have been carried to and from the Antarctic and their operational procedures (IAATO, 2003a,b,c). In the 2004/5 season the Uruguayan national programme has reported to have taken paying passengers on their operational flights to Artigas Station on King George Island (IAATO, 2005a; Uruguay, 2005). Paying tourists can provide a welcome source of income for NAPs with limited budgets.

In addition, personnel of national programmes are known to participate in mountaineering and skiing activities (Murray and Jabour, 2004: 311), and may sometimes even attempt to traverse the ice cap (ANAN Archive, 2001: 41/02). No official statistics are available for these activities, so that the scope and nature of these arrangements are difficult to assess.

A final group of travellers are the so-called 'one-off' adventurers. These are independent groups or individuals that rely on their own transport into the Antarctic Treaty area. Their means of transport include privately owned or rented yachts and aircrafts. For these one-off expeditions the challenge typically lies in 'getting there', rather than in performing particular activities in Antarctica. Because of the independent nature of these activities, statistics are hardly available. It is very difficult to estimate how many travellers are involved, what activities are carried out, where travellers are located, what kind of equipment is used and how well trips are planned. In fact, such information is gathered only when an accident occurs.

Recent Incidents That Have Caused Concern

Chiang (2000) reports that, from 1979 to 2000, 26 incidents were recorded in adventure tourism expeditions that have caused policy-makers to express concern for the development and effects of adventure tourism; in Chiang's view, the number of such incidents is growing. Murray and Jabour (2004) in contrast reject this focus on adventure tourism by pointing to the notion's fuzzy definition. We approach the issue by discussing the main factors involved in incidents recorded during five recent seasons 2000–2005 (Table 10.1).

Table 10.1. Incidents during Antarctic adventure tourism expeditions in 2000–2005 that have caused concern.

Season	Incident
2000/1	In 2001 Rolf Bae and Eirik Sonneland, two Norwegian adventurers, arrived at Scott Station (New Zealand) after a 2900-km, 107-day unsupported skiing trek from Troll Station (Norway) in DML to Ross Island. Bae and Sonneland started off after wintering at Troll Station with no formal arrangements on transport or SAR. They were not able to communicate their circumstances and position because of failing communication equipment. Furthermore, they surprised everyone by taking off for Ross Island after arriving at the South Pole, since this was not in their initial plans. They arrived at Ross Island with little food and were accommodated and fed at Scott Station. They were able to leave Antarctica on a tour vessel (ANAN Archive, 2001: 41/02). In the same season two Australian adventurers, Peter Bland and Jay Watson, ran into difficulties in an attempt to cross the Antarctic Peninsula after Bland was injured in an avalanche. Despite the negative advice while obtaining permits by the Australian authorities, Bland and Watson chartered a private yacht and set off for the Peninsula. The activities involved in this Antarctic Peninsula trek included kayaking, mountain climbing and skiing. It took the combined efforts of the yacht's crew, a nearby tour ship and the Chilean national programme to save Bland in a very difficult and dangerous rescue attempt. Bland and Watson had no official permit for their expedition and no formal SAR plan (ANAN Archive, 2001: 41/01).
2001/2	In 2002 a group of Russian government officials and tourists travelling with Cerpolex, a French company, were stranded for 2 days at the South Pole because their aircraft failed to start. They were accommodated and catered for at Amundsen–Scott Station (USA) and eventually flown out at their own expense (ANAN Archive, 2002: 79/05). In 2005 their aircraft, an Antonov-3, was successfully recovered by combined efforts of the USA and Russia (Amundsen–Scott South Pole Station, 2005).
2002/3	In 2002 the French pilot Henri Chorozs made an unexpected emergency landing on Marion Island (off the coast of DML) in an attempt to become the first to fly around the world via both poles in a single-engined aircraft. After a hard landing, Chorozs was quickly pulled from his aircraft by a South African rescue team. He stayed at Marion Island (South Africa) for 10 days, then was transported off the island by a French naval vessel (ANAN Archive, 2002: 88/01). In early 2003 a UK-registered helicopter crashed into the ocean near the South Shetlands in the Antarctic Peninsula. The Chilean navy rescued two British pilots from a life raft. Apparently the British authority was not aware of this expedition prior to the incident (ANAN, 2003: 91/01, 93/03). A few days later a scuba diver died while making a check dive with the Netherlands-based tour company Oceanwide Expeditions. The victim was part of a group of nine Latvian scuba divers on board the tourist vessel *Gregory Mikheev*. Despite resuscitation attempts by the ship's doctor and the help offered by a nearby Brazilian research ship, he died (ANAN, 2003: 91/02).

Continued

Table 10.1. — *Continued*

Season	Incident
2003/4	Below are the season's three most notable incidents involving private aviators: British pilot Polly Vacher had to abort her attempt to cross the Antarctic by aircraft because of bad weather conditions. She had made several arrangements with the national programmes of New Zealand and the UK and a tourist vessel for services (ANAN Archive, 2002: 64/11). The expedition was cancelled (IAATO, 2004a: 21). Australian aviator Jon Johanson landed at Ross Island after becoming the first person to fly across the South Pole in a homemade aircraft. He had no fuel left to return to New Zealand and eventually got a refill from the Vacher expedition's fuel dump that was stored at McMurdo Station (USA) (IAATO, 2004a; New Zealand and United States, 2004). British helicopter pilots Jennifer Murray and Colin Bodill planned to circle both poles when their helicopter crashed near Patriot Hills. They were rescued by ANI/ALE according to their contingency plans (IAATO, 2004a).
2004/5	In early 2005 UK sailor Stephen Thomas died after falling into a crevasse near Port Lockroy in the Antarctic Peninsula. Thomas and his crew reached Antarctica by private yacht after having previously sailed to the Arctic region. Despite his considerable experience in mountaineering, Thomas was unaware of the specific Antarctic conditions. He was retrieved by his yacht crew, examined by a medical doctor on a nearby cruise ship and pronounced dead (BBC News, 2005; IAATO, 2005b).

DML, Dronning Maud Land; SAR, search and rescue; ANI/ALE, Adventure Network International/Antarctic Logistics and Expeditions.

Coinciding with general visitation patterns, most of the incidents occurred in the Antarctic Peninsula and in the area around the geographical South Pole. The Peninsula is relatively easily accessible, while the South Pole is attractive because of its mythical qualities. Most of the incidents with private aviators happened near the airstrips of research stations. Incidents also occurred in other locations, such as the ANI/ALE's Patriot Hills camp and Marion Island, off the coast of DML. Remoteness is a key risk factor in Antarctica. The Russian party of 17 in 2002, and other smaller groups, would have been in serious trouble had they not ended up near a station or ship. Good information on the environmental conditions and the available infrastructure are also vital, as the incidents of Henri Chorozs and Stephen Thomas show.

Incidents occurred in expeditions of varying group composition, and of varying physical and mental abilities and levels of experience. Aviators usually travelled alone, whereas the land-based activities that failed both consisted of parties of two or more. Adventure tourists travelling with established companies were typically better prepared for the conditions in the Antarctic interior than the members of one-off expeditions. In 2005 Stephen Thomas was well prepared for crossing the Southern Ocean in his yacht, but not for trekking on the islands of the Peninsula. However, even experienced trekkers, aviators and sailors can underestimate the dangers of the Antarctic environment, or be simply overwhelmed by them.

The incidents occurred with a broad range of activities, including cross-country skiing, kayaking, scuba diving, mountain climbing and aviation. For some expeditions, trekking across a part of the continent is the main challenge, whereas for others this challenge lies in simply 'getting there'. In 2002, the Russian group managed to achieve this latter objective, but had paid insufficient attention to the issues of 'staying there' and 'getting away if something goes wrong'. Between 2002 and 2004 adventurers took on the new challenge of flying around the world over the poles in their own aircraft. The competitive spirit made adventurers rush their preparations and consequently many failed.

Activities by IAATO-member companies are well-organized from the perspective of safety regulations and minimizing risk and uncertainty for third parties. Nevertheless, incidents have occurred lately, such as the Polar First expedition in 2004 and the Oceanwide diving expedition in 2003. Because of elaborate contingency planning and pre-arranged backup, no request for help from external parties was needed. Most of the other activities were not announced to other parties in the Antarctic before it was too late. No incident occurred in the case of Bae and Sonneland in 2001, but third parties were worried because of the unexpected behaviour of the adventurers and the lack of pre-arranged backup. In other cases, assistance was requested from scientific stations and tourist ships nearby. These interventions are known to be costly and to incur additional risks upon the rescue personnel, but so far the impacts have not been estimated quantitatively.

Despite the lack of backup plans and SAR contracts, most of the parties managed to obtain a permit from their respective national authorities. Exceptions are Bland and Watson in 2001, who required a permit but set off without one, and the two British helicopter pilots in 2003, who did receive a permit but from the Chilean authorities through established contacts. Obviously, it is very difficult to manage and control travellers arriving in the Antarctic by their own means of transport. However, by making the issuing of permits conditional upon the development of contingency plans, the impacts of accidents can be limited. The SAR and insurance contracts guarantee that the costs of any emergency operation can be recovered.

A number of risk determinants emerge from the analysis of these incidents. First of all, the level of risk depends on the activity that is being undertaken, i.e. the physical and mental challenge that the activity demands or the innovative nature of the activity. Extraordinary single activities, such as scuba diving, have risk profiles that differ from journeys or expeditions venturing deep into the continent. Remoteness composes a second risk factor. In contrast to mainstream tourism, adventure activities are typically set in remote and exotic places, in this case the Antarctic polar environment (Swarbrooke *et al.*, 2003). Remote polar environments can pose great risks, such as crevasses and unreliable weather conditions, especially when the tourist vessel, tented camp or other human environment is far away. A third determinant of the level of risk is the nature of the participants. These adventurers can be solo adventurers or groups, endowed with different levels of experience regarding the activity or the environment, and with different physical and mental abilities. Typically, adventure tourists have to perform at the very limits of their capabilities in order to succeed. The fourth element is organization. Adventure tourism can take the form of a self-initiated, independently organized expedition or a commercially offered tourist itinerary. In the first category the expedition members are responsible for possible financial, environmental and safety risks. In the second category commercial tour companies provide an adventure experience for paying clients, manage the risks that are involved and carry the responsibility for the outcome. In practice, many expeditions are combinations of these two extremes, with commercial tour operators taking care of just a few aspects of the otherwise private expeditions.

Figure 10.2 is a graphical representation of the four factors that together determine the level of risk associated with a particular activity, i.e. activity (what), environment (where), participants (who) and planning (how). Ewert and Hollenhorst (1989) and Weber (2001) propose that the final level of risk depends on the interplay between individual attributes (e.g. experience and skill) on the one hand, and activity and environment attributes on the other. Bentley *et al.* (2001: 334) contend that a single risk factor rarely leads to an accident or crisis situation; usually combinations of multiple factors create the conditions in which a mishap can occur. If the four factors are not well balanced, relatively small injuries (like a cut) can lead to a chain reaction of other mishaps or incidents. Our analysis also points at the importance of proper management and organization, and at the desirability of being independent from third parties for assistance. Examples of private expeditions and adventure tourism activities show that mishaps do not necessarily have to lead to incidents involving third parties, provided that sufficient contingency plans are in place.

Policy Development

Following discussions on tourism at the ATCMs of Warsaw (2002) and Madrid (2003), an Antarctic Treaty Meeting of Experts on Tourism and Non-Governmental Activities (ATME) was organized in Norway in 2004. Concerning adventure tourism, the meeting yielded two types of proposals. The first was the adoption of additional guidelines and requirements for organizers of activities within the Antarctic Treaty consultative parties' (ATCPs) national permit systems – guidelines that would increase self-sufficiency and mitigate human risks (United Kingdom, 2004a; United States, 2004). The second was to improve the exchange of information and the coordination of activities between different ATCPs and stakeholder groups. The idea behind this proposal is that a centralized and open database of tourist activities and non-governmental expeditions would take away much of the uncertainty that surrounds adventure tourism (COMNAP, 2004; New Zealand and United States, 2004; United Kingdom, 2004b). Further,

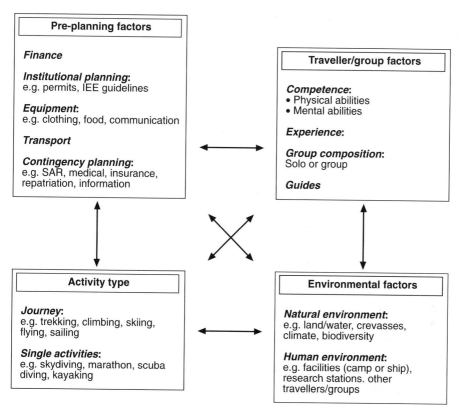

Fig. 10.2. Risk determinants in Antarctic adventure tourism. (IEE, Initial Environmental Evaluation; SAR, search and rescue.)

'hopping-and-shopping' practices for permits and support of adventurers could be brought to light and targeted.

In response to the ATME's proposals, at the 2004 ATCM in Cape Town, a resolution and a measure were adopted containing new guidelines for tourist activities. First, Resolution 4 was adopted on 'Guidelines on Contingency Planning, Insurance and Other Matters for Tourist and Other Non-governmental Activities in the Antarctic Treaty Area'. This resolution was strengthened by a measure agreed upon with immediate voluntary effect under the same title and contents (ATCM, 2004: 27–28). Major elements in these guidelines are contingency plans, including SAR, medical provisions, insurance and liability issues. In addition two resolutions were adopted to cope with the uncertainty aspect and the lack of information about activities taking place in the Antarctic Treaty Area: Resolution 3 on 'Tourism and Non-governmental Activities: Enhanced Co-operation Amongst Parties'; and Resolution 5 on the 'Establishment of an Inter-sessional Contact Group to Improve Exchange of Information Between Parties'.

Will these measures that were agreed upon at the Cape Town ATCM solve the problem of risk management in Antarctic adventure tourism? So far, policy decisions only include recommendations for ATCPs and voluntary measures. Moreover, it may still be easy for adventurers to find gaps in the regulatory systems, especially by establishing contacts and arranging support in countries with different permitting systems. Also, private adventurers from non-Treaty countries do not have to keep to these rules.

Remarkably, policy proposals and agreed measures focus solely on the pre-planning phase of the risk determinant model, i.e. making sure that adventurers depart well-prepared, self-sufficient, well-insured and that ATCPs, NAPs and other stakeholders are well-informed. The analysis has shown that the pre-planning phase, especially contingency planning and backup, is of crucial importance for the success or failure of a private expedition or adventure tourism itinerary. Although important, the analysis has also shown that not all risks can be anticipated, especially since vital information on the harsh and unreliable environmental conditions are generally missing. Moreover, earlier we learned that adventurers might be deliberately seeking these risks in their expeditions.

The risk determinant model proposes a number of other strategies to mitigate human risks in Antarctica. An important role in checking permits and contingency plans, providing information about local environmental risks and informing third parties could be allocated to the Antarctic gateway cities, especially with regard to 'one-off' yachting trips or expeditions involving aircraft. Further, one could think of closing parts of the Antarctic continent for particular types of adventure that are not considered suitable because of ecological vulnerability, high human risk, scientific research, or the impossibility of possible SAR or medical evacuations in case something goes wrong. One could think of the establishment of a 'dead zone' similar to that at Mount Everest. On the other hand, evidence from the Himalayas shows that decades of high-risk activities can have a significant cumulative impact in remote and cold climates. Another logical improvement in mitigating human risk is the establishment of permanent rescue facilities in the areas where most of the activities take place, in this case the Peninsula. On the other hand, much resistance exists for the establishment of permanent land-based tourism infrastructures in Antarctica. There is no consensus so far on this issue.

Risks involved in adventure tourism and private expeditions depend to a very large extent on the characteristics and prior experiences of the individual. It proves to be very difficult to assess whether an adventurer is well-prepared, especially for policy executers who have never been to the Antarctic or participated in similar activities. A possible way to overcome this problem is to review permit applications by a variety of experienced experts. Another option is to have the right to refuse a permit if an adventurer has caused problems in the past. One could also think of an arrangement whereby one-off expeditions are obliged to collaborate with a certified, well-established company. This, on the other hand, would stimulate the creation of monopolies in the Antarctic tourism industry.

Finally, the skydiving incident described earlier has triggered a discussion about the appropriateness of particular activities in the Antarctic. Some activities or developments could be considered in conflict with intrinsic Antarctic values, such as the designation of Antarctica as a wilderness. One can think of the use of motorized vehicles, or adventure activities that are not considered 'Antarctic' such as bungee jumping, paintball or even 'extreme ironing' (a fairly new phenomenon whereby participants try to iron under extreme conditions and provide photographic evidence; for more information see www.extremeironing.com). A possible policy option could be to ban those activities that are jeopardizing the Antarctic intrinsic values, despite the level of pre-planning, commercial involvement or experience of participants.

We must understand that adventure tourism does not develop in isolation. Rather, it is related to and co-evolves with other issues, such as the growth of commercial tourism, scientific programmes (such as the International Polar Year in 2007/8) and the creation of land-based facilities (see for example New Zealand, 2004, which argues for a prohibition of the development of land-based tourism facilities – it is thought that such facilities have irreversible environmental effects and create possibilities for tourism development, including adventure tourism). Also involved are newly established air links (for example the recently established Cape Town–DML link, and an air link currently under consideration by the Australian Antarctic Division); the establishment of an environmental liability regime; raising of awareness among adventurers and yachting people who may not fully appreciate the climatic and other dangers (ANAN Archive, 2002: 82/07); and the role of port city harbours and airports (ANAN Archive, 2002: 82/04). These interdependencies should be acknowledged, and merit more detailed investigation.

Conclusions

Currently, adventure tourism and private expeditions in Antarctica are a very small niche market. Just several hundreds of adventurers take part in them annually, and the number of problematic expeditions is limited to just a few per season. One may wonder whether adventure tourism and private expeditions are a problem at all, with such small numbers of participants and incidents. Yet, with an eye on the improving accessibility of Antarctica and the trend towards increased diversification, adventure tourism may be at the eve of a period of rapid growth. The level of attention that adventure tourism gets from policy-makers may seem unjustified at first sight, but warranted from a longer-term perspective, particularly in view of the slowness of Antarctic decision-making processes on tourism.

Murray and Jabour (2004) claim that independent expeditions have quite wrongly been depicted as the opposite of mainstream, organized tourists and other (primarily scientific) human operations in Antarctica. We must acknowledge that these well-established Antarctic institutions also once started as adventurous expeditions, including risks and incidents. The adventure spirit has been an intrinsic part of Antarctica from the moment man set foot on the continent. Adventure tourism and private expeditions highlight the uncertainty of possible outcomes in Antarctic tourism activities and the unpredictable course of its development. Further, independent expeditions benefit the Antarctic community by introducing creative innovations in polar technology, transport and geographical knowledge. As a result, many authors agree that adventure tourism should have a place in Antarctica, provided that it is organized properly (IAATO, 2003c).

The management of tourism activities in Antarctica is largely left to the tourism sector. The IAATO network issues guidelines and procedures to tackle any problems in a technical way. As a result, fundamental decisions about Antarctic tourism, such as structural limitations or its moral implications, are left untouched. More and more authors claim that the pace and direction of the current tourism trends call for the development of a more proactive and comprehensive policy (Bastmeijer and Roura, 2004). In the previous section we have seen that for adventure tourism plenty of management and policy options are available. However, diverging interests regarding tourism in the short term may hamper consensus building in the longer term. In order to create such a proactive comprehensive tourism policy, a common future vision between the different stakeholder groups and the ATCPs is required.

References

Amundsen–Scott South Pole Station (2005) *Russian aircraft adventures 2004–05.* Amundsen–Scott South Pole Station website; available at http://www.southpolestation.com/trivia/00s/russians/ilyushin.html (accessed January 2007).

ANAN (2003) *Antarctic Non-government Activity News, 2003 Issues.* Australian Antarctic Division, Kingston, Tasmania; available at http://www.aad.gov.au/default.asp?casid=4974 (accessed January 2007).

ANAN Archive (2001) *Year 2001 Antarctic Non-government Activity News Archive.* Australian Antarctic Division, Kingston, Tasmania; available at http://www-old.aad.gov.au/goingsouth/tourism/News/2001/default.asp (accessed January 2007).

ANAN Archive (2002) *Year 2002 Antarctic Non-government Activity News Archive.* Australian Antarctic Division, Kingston, Tasmania; available at http://www-old.aad.gov.au/goingsouth/tourism/News/2002/default.asp (accessed January 2007).

ASOC (2001) Antarctic tourism. *IP 40, Antarctic Treaty Consultative Meeting XXIV,* St Petersburg, Russian Federation, 9–20 July. Antarctic and Southern Ocean Coalition, Washington, DC.

ATCM (2004) *Final Report of the Twenty-Seventh Antarctic Treaty Consultative Meeting, Cape Town, South Africa.* Antarctic Treaty System.

Bastmeijer, C. and Roura, R. (2004) Regulating Antarctic tourism and the precautionary principle. *American Journal of International Law* 98(4), 763–781.

BBC News (2005) *Explorer dies in fall through ice.* BBC website; available at http://news.bbc.co.uk/go/pr/fr//1/england/cambridgeshire/4183295.stm (accessed January 2007).

Bentley, T., Page, S., Meyer D., Chalmers, D. and Laird, I. (2001) How safe is adventure tourism in New Zealand? An exploratory analysis. *Applied Ergonomics* 32, 327–338.

Chiang, E. (2000) Tourism risks from an operational perspective. In: *Proceedings of the Antarctic Tourism Workshop.* Antarctica New Zealand, Christchurch, New Zealand, pp. 25–28.

COMNAP (2002) The interaction between national operators, tourists and tourism operators. *IP 27, Antarctic Treaty Consultative Meeting XXV*, Warsaw, 9–20 September. Council of Managers of National Antarctic Programs, Hobart, Tasmania.

COMNAP (2003) Interaction between national operators, tourists and tourism operators. *IP 37, Antarctic Treaty Consultative Meeting XXVI*, Madrid, 9–20 June. Council of Managers of National Antarctic Programs, Hobart, Tasmania.

COMNAP (2004) Interaction between national Antarctic programs and non-government and tourism operations. *IP ATME#25, Antarctic Treaty Meeting of Experts on Tourism and Non-governmental Activities*, Tromso, Norway, 22–25 March. Council of Managers of National Antarctic Programs, Hobart, Tasmania.

Ewert, A. and Hollenhorst, S. (1989) Testing the adventure model: empirical support for a model of risk recreation participation. *Journal of Leisure Research* 21(2), 124–139.

Hall, C.M. (1992) Adventure, sport and health tourism. In: Weiler, B. and Hall, C.M. (eds) *Special Interest Tourism*. Belhaven Press, London, pp. 141–158.

IAATO (1997) Overview of Antarctic tourism activities: a summary of 1996–1998 and five year projection 1997–2002. *IP 75, Antarctic Treaty Consultative Meeting XXI*, Christchurch, New Zealand, 19–30 May. International Association of Antarctica Tour Operators, Basalt, Colorado.

IAATO (1998) Overview of Antarctic tourism activities. *IP 86, Antarctic Treaty Consultative Meeting XXII*, Tromso, Norway, 25 May–5 June. International Association of Antarctica Tour Operators, Basalt, Colorado.

IAATO (1999) Overview of Antarctic tourism activities. *IP 98, Antarctic Treaty Consultative Meeting XXIII*, Lima, 24 May–4 June. International Association of Antarctica Tour Operators, Basalt, Colorado.

IAATO (2000) Overview of Antarctic tourism. *IP 33, Special Antarctic Treaty Consultative Meeting XII*, The Hague, The Netherlands, 11–15 September. International Association of Antarctica Tour Operators, Basalt, Colorado.

IAATO (2001) Overview of Antarctic tourism. *IP 73, Antarctic Treaty Consultative Meeting XXIV*, St Petersburg, Russian Federation, 9–20 July. International Association of Antarctica Tour Operators, Basalt, Colorado.

IAATO (2002) Guidelines for tourist operations in Antarctica. *IP 72, Antarctic Treaty Consultative Meeting XXV*, Warsaw, 9–20 September. International Association of Antarctica Tour Operators, Basalt, Colorado.

IAATO (2003a) IAATO overview of tourism. *IP 71, Antarctic Treaty Consultative Meeting XXVI*, Madrid, 9–20 June. International Association of Antarctica Tour Operators, Basalt, Colorado.

IAATO (2003b) Report of the International Association of Antarctica Tour Operators (IAATO) 2002–2003. *IP 78, Antarctic Treaty Consultative Meeting XXVI*, Madrid, 9–20 June. International Association of Antarctica Tour Operators, Basalt, Colorado.

IAATO (2003c) Adventure tourism in Antarctica. *IP 96, Antarctic Treaty Consultative Meeting XXVI*, Madrid, 9–20 June. International Association of Antarctica Tour Operators, Basalt, Colorado.

IAATO (2004a) Overview summarizing the terms of reference. *ATME#12, Antarctic Treaty Meeting of Experts on Tourism and Non-governmental Activities*, Tromso, Norway, 22–25 March. International Association of Antarctica Tour Operators, Basalt, Colorado.

IAATO (2004b) IAATO overview of Antarctic tourism 2003–2004 Antarctic season. *IP 63, Antarctic Treaty Consultative Meeting XXVII*, Cape Town, South Africa, 24 May–4 June. International Association of Antarctica Tour Operators, Basalt, Colorado.

IAATO (2005a) IAATO overview of Antarctic tourism 2004–2005 Antarctic season. *IP 95, Antarctic Treaty Consultative Meeting XXVIII*, Stockholm, 6–17 June. International Association of Antarctica Tour Operators, Basalt, Colorado.

IAATO (2005b) Report of the International Association of Antarctic Tour Operators 2004–2005. *IP 95, Antarctic Treaty Consultative Meeting XXVIII*, Stockholm, 6–17 June. International Association of Antarctica Tour Operators, Basalt, Colorado.

Murray, C. and Jabour, J. (2004) Independent expeditions and Antarctic tourism policy. *Polar Record* 40(215), 309–317.

New Zealand (2004) An analysis of the existing legal framework for the management of tourism and non-governmental activities in Antarctica: issues, some proposals and comments. *ATME#07, Antarctic Treaty Meeting of Experts on Tourism and Non-governmental Activities*, Tromso, Norway, 22–25 March.

New Zealand and United States (2004) Observations on Jon Johanson's South Pole flight. *ATME#26, Antarctic Treaty Meeting of Experts on Tourism and Non-governmental Activities*, Tromso, Norway, 22–25 March.

Outside Online Dispatches (1998) *Tragedy. A pole too far. Three skydivers die in Antarctica, leaving the world to ask, "Why?"* Outside Online website; available at http://outside.away.com/outside/magazine/0298/9802disppole.html (accessed January 2007).

Priest, S. (2001) The semantics of adventure programming. In: Miles, J. and Priest, S. (eds) *Adventure Programming*. Venture Publishing, State College, Pennsylvania, pp. 111–114.

Richardson, M. (2000) Regulating tourism in the Antarctic: issues of environment and jurisdiction. In: Vidas, D. (ed.) *Implementing the Environmental Protection Regime for the Antarctic*. Kluwer Academic Publishers, Dordrecht, The Netherlands, pp. 71–90.

Swarbrooke, J., Beard, C., Leckie, S. and Pomfret, G. (2003) *Adventure Tourism. The New Frontier*. Butterworth-Heinemann, Oxford, UK.

Swithinbank, C. (1989) Non-government aircraft in the Antarctic 1988/89. *Polar Record* 24(154), 252.

Swithinbank, C. (1990) Non-government aircraft in the Antarctic 1989/90. *Polar Record* 26(159), 316.

Swithinbank, C. (1992a) Non-government aircraft in the Antarctic 1990/91. *Polar Record* 28(164), 66–67.

Swithinbank, C. (1992b) Non-government aircraft in the Antarctic 1991/92. *Polar Record* 28(166), 232.

Swithinbank, C. (1993) Non-government aircraft in the Antarctic 1992/93. *Polar Record* 29(170), 244–245.

Swithinbank, C. (1994) Non-governmental aircraft in the Antarctic 1993/94. *Polar Record* 30(174), 221.

Swithinbank, C. (1995) Non-governmental aircraft in the Antarctic 1994/95. *Polar Record* 31(178), 346.

Swithinbank, C. (1996) Non-governmental aircraft in the Antarctic 1995/96. *Polar Record* 32(183), 355–356.

Swithinbank, C. (1997) Non-governmental aviation in Antarctica 1996/97. *Polar Record* 33(187), 340–341.

Swithinbank, C. (1998a) *Forty Years on Ice. A Lifetime of Exploration and Research in the Polar Regions*. The Book Guild Ltd, Lewes, UK.

Swithinbank, C. (1998b) Non-governmental aviation in Antarctica 1997/98. *Polar Record* 34(190), 249.

Swithinbank, C. (1999) Non-governmental aviation in Antarctica 1998/99. *Polar Record* 36(196), 51–52.

Swithinbank, C. (2000) Non-governmental aviation in Antarctica 1999/2000. *Polar Record* 36(198), 249–250.

The poles.com (2006) *Antarctica: ALCI prices and flight schedule from Cape Town*. the poles.com website; available at http://www.thepoles.com/news.php?id=14115 (accessed-January 2007).

United Kingdom (2004a) The regulation of adventure tourism. *ATME#08, Antarctic Treaty Meeting of Experts on Tourism and Non-governmental Activities*, Tromso, Norway, 22–25 March.

United Kingdom (2004b) Managing adventure tourism: the need for enhanced co-operation amongst parties. *ATME#09, Antarctic Treaty Meeting of Experts on Tourism and Non-governmental Activities*, Tromso, Norway, 22–25 March.

United States (2004) US policy on private expeditions to Antarctica and current US framework for regulation of Antarctic tourism. *ATME#05, Antarctic Treaty Meeting of Experts on Tourism and Non-governmental Activities*, Tromso, Norway, 22–25 March.

Uruguay (2005) Visitor programme to the 'Artigas' Antarctic Scientific Base. *Antarctic Treaty Consultative Meeting XXVIII*, Stockholm, 6–17 June.

Weber, K. (2001) Outdoor adventure tourism. A review of research approaches. *Annals of Tourism Research* 28(2), 360–377.

11 Antarctic Scenic Overflights

THOMAS BAUER

School of Hotel and Tourism Management, The Hong Kong Polytechnic University, Hung Hom, Kowloon, Hong Kong SAR, China

Introduction

Ship-based tourism to Antarctica, in particular to the Antarctic Peninsula, has increased significantly during the past decade. Setting foot on the continent and being able to experience the Antarctic environment at first hand is the preferred method of seeing Antarctica for many tourists. Because of the requirements on time and money, however, the dream of stepping on Antarctica comes true for only a small proportion of those who would like to gain an Antarctic experience. It is perhaps against this background that sightseeing overflights of parts of the continent were started in the first place. In the author's experience, overflights of Antarctica are certainly a worthwhile experience. They provide a completely different perspective from that gained during voyages and Zodiac landings. In the mid-1990s the author participated in several overflights and also carried out survey work among the passengers that investigated their motivations. Some of these observations and findings are reported in this chapter.

Early Overflights: the Background

Headland (1994: 275) traces the beginnings of commercial Antarctic tourism back to 1956 when the first commercial tourist flight by Linea Aerea Nacional (Chilean National airline) took 66 passengers on an overflight of the South Shetland Islands and the Trinity Peninsula. In 1977 the concept of seeing Antarctica from above was re-ignited by the Australian entrepreneur Dick Smith, who began a series of charter flights which took passengers from Australia to Antarctica and back in one day without undertaking a landing (Burke, 1994). By creating this new tourism experience he established what can be described as arguably the most unique day-excursion on the planet – a 'visit' to another continent without setting foot on it. This claim to fame will be challenged only when Sir Richard Branson's Virgin Galactic spaceliner starts operating day-flights into space in a few years' time, which will allow paying passengers to gain an astronaut's view of our planet.

Overflights of Antarctica can also be seen as the continuation of other Antarctic aviation milestones, such as the 1929 return flight to the South Pole by Richard Byrd, Bernt Balchen and crew, and the 1963 first non-stop flight from Africa across Antarctica to New Zealand by Richard Dickerson, William Kurlak and crew. Of course neither of these flights was conducted for tourism purposes, but all were of a pioneering nature.

While visiting Antarctica by sea allows passengers to feel the Antarctic environment, the overflight is a completely different experience. Spending nearly 12 h inside an aircraft in flight without landing at a new destination is not everybody's idea of having a good time. Those who have not overflown might argue that a similar experience to sitting in a plane and looking down on Antarctica could be gained by watching a previously recorded video of such a flight, or by spending time in an IMAX theatre that is showing an Antarctic movie. But they are wrong, and Antarctic overflights are a great experience.

Sightseeing flights in other parts of the world are of much shorter duration. Flights in light aircraft over some scenic sights like the Grand Canyon or the Great Barrier Reef rarely exceed 1 or 2 h. Flights that take off regularly from Kathmandu, Nepal, to offer passengers a view of Mount Everest, are also less than 2 h long. One operator, Buddha Air, sells T-shirts that say 'I did not climb Mt Everest ... but I touched it with my heart', and they proudly declare their Everest flights as 'The best mountain flights in the world'. However, they have not seen Antarctica from above. Many of these sightseeing flights are heavily weather-dependent and are often operated on an *ad hoc* basis, using small to medium-sized aircraft.

Many passengers who travel on scheduled air services between Europe and the US West Coast get good views of the Greenlandic ice cap, and flights from North America to Asia fly close to the Geographic North Pole, but most passengers care little for the scenic beauty that can be observed outside their windows opting instead to close the blinds and to sleep or watch movies.

The Arctic region does have short sightseeing flights, for example helicopter and fixed-wing flights over glaciers in Greenland, Alaska and Canada, but does not have any dedicated long-haul ones. The relative ease with which Arctic destinations can be accessed by road, sea and scheduled and chartered air services is one explanation for the absence of longer sightseeing flights in the North. The geography of the Arctic is for the most part arguably less spectacular from the air than the Antarctic. Antarctica is difficult to access and hence a one-day flight can be seen as an easy way to get an impression of this remote southernmost continent. People are happy just to catch a glimpse of Antarctica, even if it is through the window of a wide-bodied jet aircraft.

As noted above, the first Antarctic sightseeing overflight took place in December 1956 when a Douglas DC6B of LAN Chile overflew the South

Shetland Islands and Trinity Peninsula. In 1977 a Pan American Airways Boeing 707 aircraft flew from the USA to London, Cape Town and Auckland, crossing both geographical poles in the process. The first series of sightseeing flights from Australia took place on 13 February and 16 March 1977, and flew over Macquarie Island and the South Magnetic Pole as well as parts of Victoria Land, Oates Land and George V Land. Qantas aircrafts were used for the flights. They were organized by former *Australian Geographic* publisher Dick Smith, who has been quoted as saying that he started them so he could go back to the office on Monday morning and tell people that he had been to the South Pole (*Adelaide Advertiser*, 1977; quoted in Reich, 1979: 33).

The series of 30 Qantas and ten Air New Zealand flights that had carried some 10,000 passengers concluded on 16 February 1980, when the last Qantas plane returned from Antarctica. The end of the first series of overflights came as a result of the crash of Air New Zealand flight TE 901 in Antarctica on 28 November 1979 (commonly referred to as the Mount Erebus disaster) that claimed the lives of all 237 passengers and 20 crew (for a full account of this tragic event and its aftermath see Mahon, 1984).

Resumption of Overflights from Australia

Over 10 years ago, in 1994, the Melbourne-based travel entrepreneur Mr. Phil Asker of Croydon Travel showed great courage in resuming Antarctic day-flights and the company has conducted them successfully ever since. Overflights departing from Australia are domestic flights – the longest in the world – and no passport is required and no duty-free shopping is available. Taking advantage of the near 24-h daylight conditions that prevail in Antarctica during the austral summer the series of six charter flights resumed from Australia on New Year's Eve 1994 when 356 passengers celebrated the occasion aboard Qantas flight 2601. The other five flights took place between 7 January and 18 February 1995. Economy seats sold for Aus$1199, business for Aus$1699 and first for Aus$2099. Demand for seats was strong and 2134 passengers and 151 crew saw Antarctica from above during this first season. The flights continued in 1995/6 with ten departures and air fares ranged from Aus$999 to Aus$2799. A total of 3301 passengers and 224 crew were carried on the ten flights that took place during the 1996/7 season (Asker, personal communication, 1997). Between 31 December 1994 and 27 January 1996 a total of 8393 passengers and 577 crew were carried. Another ten flights took place during the 1997/8 season, ten during 1998/9 and ten during 1999/2000. During the 2001/2 season Croydon Travel carried 2082 passengers on six flights and during the 2002/3 season three flights with 1072 passengers took place. During the 2003/4 season Croydon Travel operated six Antarctic day sightseeing flights carrying a total of 2148 passengers, during the 2004/5 season a total of 2030 passengers were carried, in 2005/6 there were two flights with 715 passengers (IAATO, 2006) and a further two flights were scheduled to take place during 2006/7

(Croydon Travel, personal communication, December 2006). Prior to the start of the 2006/7 season Croydon Travel had operated a total of 76 flights and had carried 27,857 passengers since commencing operations during the 1994/5 season. The decrease in the number of flights scheduled during recent years may be seen as an indication that after a decade the demand for such flights is levelling off and that most Antarctic 'fans' in Melbourne and Sydney have by now satisfied their curiosity of seeing the continent from the safety and convenience of a large commercial aircraft. According to information obtained from Croydon Travel (personal communication, 2006), the shortage of available Qantas aircraft is another explanation for the decline in available flights. Prices for the 2006/7 series of flights range from Aus$899 for a centre seat to Aus$5199 in first class.

The economics of these flights is also interesting. At a conservative estimate of an average ticket price of Aus$1500 the series of flights would have generated gross revenue of nearly Aus$45 million, a considerable boost to the Australian economy and to the bottom line of the organizer.

Flight Details

Between 2 and 3 h is commonly spent flying over the continent with the remaining 8–9 h of flight being spent en route to and back from Antarctica. In addition most flights spend approximately an hour flying over the pack ice. There is enough time to give all passengers the chance to make their way to a window and to admire the spectacular views of snow, ice, mountains and glaciers, and take pictures and video footage.

The flight paths vary and nearly 20 different routes are programmed. A sample route is south from Melbourne over Tasmania, to the South Magnetic Pole at 64.7°S and 138.7°E, and on to the French base of Dumont d'Urville. From there the route is either to the east, flying over Commonwealth Bay, Cape Denison, the Mertz and Ninnis glaciers, Cape Hudson, the now-closed Russian base of Leningradskaya, Cape Adare, Cape Hallett and Cape Washington. An alternative route covers the area west of Dumont d'Urville and includes parts of the coastline of Wilkes Land as far as Australia's Casey Station (Fig. 11.1). Because of the unpredictability of Antarctic weather conditions the organizer of the flights always reserves the right to deviate from a planned flight itinerary and no refund is given to passengers in the case where clouds obstruct the view of the continent. To this author's knowledge the situation where passengers did not see any part of the Antarctic during their flight has not arisen yet.

During the time above Antarctica and the pack ice, the Boeing 747-400 aircraft maintain a minimum height above ground of 2000 feet or 10,000 feet above sea level. This provides excellent viewing conditions (Fig. 11.2) and ensures the safety of the aircraft.

As per standard procedures under the Madrid Protocol the operator has to seek permission from the Australian Antarctic Division prior to each series of flights to show that they have no more than a transitory impact on the

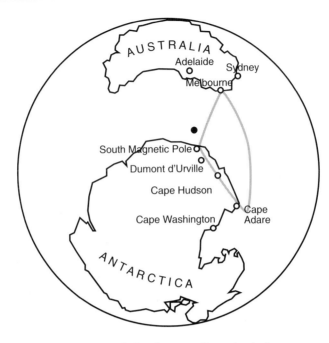

Fig. 11.1. Air routes of Antarctic overflights from southern Australia.

Fig. 11.2. Overflight passengers' view of mountains, glaciers and sea ice along the Borchgrevink Coast, Victoria Land, Antarctica. (Photo: T. Bauer.)

environment. Given that these flights have now been operating successfully for a decade, it is clear that this activity has no unacceptably high environmental impacts.

Overflights from South America

LAN Chile and Aerovias DAP carry out overflights to the Antarctic Peninsula area from Punta Arenas in southern Chile during the period from November to March. Unlike the flights from Australia that are basically serving the domestic [Australian] market as a stand-alone tourism activity, flights from South America are offered as an optional excursion for cruise-ship passengers. On its website (www.iaato.org) the International Association of Antarctica Tour Operators reports that during the 2001/2 season 220 passengers were carried on eight LAN Chile flights. During the 2002/3 season 17 flights carried 480 passengers; during the 2003/4 season 679 tourists observed the Antarctic Peninsula from above; during the 2004/5 season 462 passengers were carried; and during 2005/6 a total of 450 passengers flew south. These flights are used as optional extra excursions for cruise-ship passengers but, because of the much diminished importance of Punta Arenas as an Antarctic gateway, passenger numbers are very low compared with the passenger numbers carried on flights from Australia. It is likely that passenger numbers would be significantly higher if these flights operated from the thriving Antarctic gateway port of Ushuaia.

Antarctic Overflights Observed

The author participated in several of the early overflights from Australia as a researcher and commentator; the following describes his first-hand experiences with the flights. During the flight several Antarctic experts shared their knowledge with the passengers. After crossing Tasmania the plane headed for the South Magnetic Pole and the Antarctic continent was reached at Dumont d'Urville. Due to heavy cloud cover no Antarctic features could be seen. Likewise, Mawson's Hut at Commonwealth Bay could not be seen because of the clouds. Passengers began to be a little restless thinking, no doubt, that they might not be able to see Antarctica at all. Soon after, near Cape Hudson, the clouds disappeared and the Antarctic continent showed itself in all its beauty. From an altitude of 5000 m the viewing range was nearly 400 km and the scenery observed included glaciers, the Transantarctic Mountains, snowfields and, in the distance, the polar ice cap. Geographic features observed during the 3.5 h above the continent included Cape Hudson, Mount Melbourne and Mount Minto. Mounts Erebus and Terror could be seen in the distance. At Cape Adare the huts erected by Borchgrevink could clearly be seen. In addition, sea ice and large trapped tabular icebergs could be viewed. The viewing experience over Antarctica can only be described as spectacular. The written word is inadequate to describe the Antarctic scenery observed from an aircraft and the many photographs taken during the flight will only do marginally better to provide people who have not travelled on the flight with an impression of the sheer magnificence of this the last great wilderness area of the planet.

Qantas uses Boeing 747-400 wide-bodied jet aircraft under the command of highly qualified captains. The aircraft are fuelled to capacity, carrying 170 t of aviation jet fuel for the round trip to Antarctica. On one of the flights observed the captain advised that the aircraft had burned 120 t of fuel during the 11.5-h flight and after the 9000-km flight had a reserve of some 50 t upon landing in Melbourne. The extra safety margin is necessary should the plane have to be diverted because of bad weather.

Safety Issues

In light of the Mount Erebus disaster the safety of Antarctic overflights was called into question prior to the resumption of flights, but such concerns have now been put aside: it is clear that these flights are no more dangerous than other scheduled air services between Australia/New Zealand and Argentina which fly over the remote southern region of the South Pacific Ocean. In the case of an emergency landing in the Southern Ocean there would be no way of rescuing people – the nearest airport is thousands of kilometres away. It follows that Antarctic overflights carried out using the equipment and skills of Qantas, one of the world's most reliable airlines, do not present unacceptable risks to the safety of passengers and crew.

During the flights passengers were introduced to the normal aviation safety procedures, including the use of life vests, and were shown the directions of the emergency exits. The aircraft also carried flotation body suits, but passengers were not told about them. In any case they would be of very limited use should a plane have to make an emergency landing in the freezing Antarctic water.

Passenger Survey

The author carried out passenger surveys on flights that took place between 1994 and 1997. Questionnaires were distributed during the flights and a total of 1761 fully useable responses were received.

The age distribution of passengers was as follows: under 18 years of age (1%), 18–25 (7%), 26–35 (8%), 36–45 (13%), 46–55 (21%), 56–60 (14%), 61–70 (19%), 71–80 (13%) and over 80 years (4%). Antarctic overflights are particularly attractive to more mature people, with 71% of passengers surveyed being over 46 years of age.

The gender distribution was interesting in that there were more female passengers than male passengers; on cruises this tends to be the opposite. As far as occupations were concerned, professionals dominated the sample.

During the 1994/5 and 1995/6 flights passengers were asked what motivated them to undertake the flight. As reported by Bauer (2001: 173), the main motivations for undertaking the flight among the sample of 484 passengers (649 responses with multiple responses possible) included: unique trip to a unique region (23%); always wanted to go to Antarctica (12%); just wanted to see Antarctica (12%); fascinated by Antarctica (10%); an experience (10%); it was a present (8%).

During these two seasons passengers were also asked what images they held of Antarctica as a tourist destination. The main images were: ice features (32%); positive attributes such as beauty, pristine, exciting and peace (16%); negative attributes such as cold, windy, hostile and remote (13%); and fauna (8%). Historic attributes such as the early explorers and historic sites accounted for only 4% and scientific stations for 1% of the total of 1133 responses (multiple responses).

The main expectations that passengers had of the overflights included: good views (18%); learn more about Antarctica (13%); pleasure (12%); scenery (11%); memories (10%); an experience (10%). The 2% who expected to view wildlife were obviously sadly disappointed by the experience. While a substantial number of passengers stated that they were likely to visit Antarctica by sea within 3 years of undertaking the overflight, this appears to have been more wishful thinking than reality.

Conclusions

There is no doubt that sightseeing flights over Antarctica have been a resounding success. They have provided tourists with a unique experience and have contributed to the better understanding of Antarctic issues by a large number of people who, in the absence of these flights, would not have been able to get exposure to the continent and to the many issues that surround it. It can be anticipated that the series of flights will continue indefinitely but that unless new market segments, in particular larger groups of international tourists, can be encouraged to take these flights demand will remain at present levels.

References

Bauer, T. (2001) Tourism in the Antarctic: Opportunities, Constraints and Future Prospects. Haworth Hospitality Press, New York.

Burke, D. (1994) Moments of Terror: The Story of Antarctic Aviation. New South Wales University Press, Sydney, New South Wales.

Croydon Travel (2005) Antarctica sightseeing flights. Croyden Travel, Croyden, Victoria; available at http://www.antarcticaflights.com.au (accessed 3 August 2005).

Headland, R. K. (1994) Historical development of Antarctic tourism. Annals of Tourism Research 21(2), 269–280.

IAATO (2006) IAATO overview of Antarctic tourism 2005–2006 Antarctic season. *IP86, Antarctic Treaty Consultative Meeting XXIX*, Edinburgh, 12–23 June. International Association of Antarctica Tour Operators, Basalt, Colorado.

Mahon, P. (1984) *Verdict on Erebus*. Collins, Auckland, New Zealand.

Reich, R.J. (1979) Tourism in the Antarctic: its present impact and future development. Thesis for the Diploma in Polar Studies, Scott Polar Research Institute, Cambridge University, Cambridge, UK.

12 Antarctic Tourism: What are the Limits?

DENISE LANDAU[1] AND JOHN SPLETTSTOESSER[2]

[1]*International Association of Antarctica Tour Operators, PO Box 2178, Basalt, CO 81621, USA*; [2]*PO Box 515, Waconia, MN 55387, USA*

Introduction

This chapter reviews briefly the history of Antarctic tourism, the development and major functions of the International Association of Antarctica Tour Operators (IAATO), current trends in the Antarctic tourism industry, the industry's viewpoint on inherent limitations, and the need to look towards the future. Numerous changes or trends in the tourism industry and traditional ship-borne activities are discussed in response to an ever-evolving and competitive nature of tour operators and their clientele. These innovations include kayaking, diving, skiing, mountaineering, overflights without landings, cruising without landing, high-risk adventure tourism, Russian icebreaker and helicopter operations, and others. The response of IAATO to these changes in concert and liaison with Antarctic Treaty parties shows a viable means of dealing with management challenges.

Most tourism activities are conducted by ship in the Antarctic Peninsula area, with a small amount in the Ross Sea and East Antarctica. Land-based tourism occurs in three regions of Antarctica. The interior of the continent has two primary locations operating out of temporary camps, one in the southern Ellsworth Mountains and one in Dronning Maud Land. Clientele are flown there, from South America and South Africa, respectively, for adventures in skiing and mountaineering, and visits to remote emperor penguin colonies. One- and two-day fly-in trips take place from South America to the Antarctic Peninsula, King George Island, with minimal stays ashore.

Overflights without landings have been conducted since 1995 from Australia, with flights on Qantas aircraft for sightseeing purposes over the northern Transantarctic Mountains and nearby coasts. Some 1000 to 2000 tourists have been passengers annually on these popular flights. Overflights are also conducted from South America to the Antarctic Peninsula. Nine such flights were made in the 2005/6 austral summer, carrying some 450 tourists.

The Development of IAATO

Tourism in Antarctica essentially began in the mid-1960s, when the tour operator Lars-Eric Lindblad recognized the potential market for transporting adventurous and educationally minded tourists to remote parts of the world by small expedition ship. His foresight was prescient. Combining education and travel with a particular focus on the preservation of the environment and its wildlife proved to be a magical combination, and one that has since created thousands of 'ambassadors' for wilderness destinations like Antarctica. The 'Lindblad model' can now be found throughout the world and is used by numerous vessels.

In 1991 the increase in numbers of operators and concern for the environment resulted in the formation of IAATO by seven companies, in order to act as a single voice in concerns of tourism and to advocate, promote and practise environmentally responsible private-sector travel to Antarctica. A website (www.iaato.org) was formed for the benefit of members and also for the interest of the public. Website contents include information for members preparing for a coming season in Antarctica, Information Papers tabled at Antarctic Treaty Consultative Meetings (ATCMs) and tourism statistics. IAATO's participation in ATCMs began by invitation from the Treaty parties in 1994. Numerous documents on the members-only section of the website detail all the operational procedures and resource links.

By 2006 IAATO included some 80 members operating in 14 countries (Argentina, Australia, Belgium, Canada, Chile, France, Germany, Italy, Netherlands, New Zealand, Norway, Sweden, UK, USA) plus the Falkland Islands (Islas Malvinas). Most members are ship operators, working as business competitors, but members also include two land-based operators, one fly/cruise operator, one company that conducts scenic overflights without making landings, several travel companies who charter ships, one helicopter operator, agencies that cater for adventure travel, shipping agencies, government tourism offices and conservation groups. Many of the operators also offer trips in the Arctic, Amazon and other remote regions of the world where they apply the same guiding principles that they adopt in the Antarctic.

It is important to note that IAATO has no legal authority to keep companies or tour operators from establishing business in Antarctica. What IAATO can do is communicate effectively with each of its members, provide environmental, operational and logistical information, offer assistance in contingency planning and search and rescue, and help with emergency medical evacuations to avoid both impacts on scientific stations and the need for rescue assistance from nearby countries.

IAATO currently includes almost all of the operators working in Antarctica, an indication that almost all choose to qualify for membership of an environmentally sensitive professional association that addresses concerns in such a special part of the world. Antarctic tourism continues to grow gradually, and new operators joining IAATO learn quickly from established best practices, with access to a wealth of knowledge and experience.

IAATO holds annual meetings, and committees may meet throughout the year. Memoranda regarding operational, political and environmental operating strategies are regularly sent to members. Communication amongst members has been an integral part of the running of the organization, and most certainly plays a part in the industry's ability to self-regulate and create self-imposed limits on tourist activities. IAATO representatives participate also in the ATCMs, inter-sessional working groups, specialized meetings on Antarctic-related issues, and other projects or scientific studies as needed. IAATO presents Information Papers to ATCMs that discuss details of operator statistics (numbers of cruises, staff, crew and passengers, and their nationalities, etc.) in its annual reports and other papers (IAATO, 2001, 2006a,b).

Categories of Membership

Up to June 2001, operators of ships carrying more than 400 passengers to Antarctica were barred from membership of IAATO. As more and larger ships came on the scene, after some 5 years of annual debating, members agreed to develop new categories of membership that would encompass this aspect of the industry's changing nature. This decision brought big-ship operators into IAATO, and placed the organization in a stronger position to establish additional limitations if needed. Present levels of tourism create less than minor or transitory impacts, but none of a significant nature, according to present observations. However, the industry is changing, and IAATO is braced for the 21st century, with infrastructure in place that will allow it to evolve further.

There are currently seven categories of membership:

1. Organizers of expedition ships or yachts that carry less than 200 passengers. They are required to limit to 100 the number of passengers ashore at any time.
2. Organizers of vessels carrying 200 to 500 passengers who are intending to land passengers. Stringent restrictions on landing activities of time and place could apply. The limit of 100 passengers ashore at one site at one time also applies.
3. Organizers of cruise ships making no landings (cruise only). Cruise ships carrying more than 500 passengers are not permitted to make landings.
4. Organizers of land-based operations.
5. Organizers of overflight operations.
6. Organizers of air/cruise operations.
7. Companies in support of Antarctic tourism.

The above categories, depending on organizer interests and types of activities, can be grouped into any of the following major types of membership.

1. *Full Members* are experienced organizers who operate travel programmes to the Antarctic and who: (i) pledge to abide by IAATO Bylaws; (ii) agree to the above-mentioned categories and to not have more than 100 passengers ashore at any one site at the same time; (iii) maintain a staff-to-passenger ratio of 1:20 ashore; and (iv) have been formally accepted by two-thirds of the Full Members in good standing after review and fulfilling any other requirements.

2. *Provisional Members* are organizers that operate travel programmes to the Antarctic that are requesting full membership in IAATO. In addition to the requirements for full membership (above), they also agree to carry an observer on board on a voyage in their first season in Antarctica.

3. *Probationary Members* are current or past full or provisional members who have not fully complied with IAATO Bylaws or who otherwise are not in good standing as decided by a two-thirds vote of the full members. Probationary members agree to the requirements for provisional members and, after a satisfactory operating season with an observer, may be reinstated into provisional or full membership.

4. *Associate Members* are other organizations and individuals interested in or promoting travel to the Antarctic that wish to support IAATO objectives and whose application has been formally accepted by two-thirds of the standing members.

Tourism Trends

In the late 1950s Chile and Argentina took more than 500 fare-paying passengers to the South Shetland Islands aboard naval transports that were otherwise used for servicing their research stations. In 1966 Lars-Eric Lindblad led the first traveller's expedition to Antarctica. During the 1980s the tour industry gradually became aware that Antarctica represented an attractive and feasible destination. Prominent tourist attractions included magnificent scenery, unique wildlife (e.g. penguins, seals, whales), the pristine environment, solitude and wilderness, avoidance of crowded tourist areas, a sense of adventure and an opportunity to visit 'the last continent'.

Up until 1985/6 tourists per year numbered 1000 to 2000, followed by a gradual increase as more operators entered the market. Most visited by ship, but in 1977 the industry expanded to include 'flight-seeing' (overflying the continent without landing). Both Qantas and Air New Zealand operated aircraft annually until the crash of an Air New Zealand DC10 on Mount Erebus on 28 November 1979. Tourist flights terminated shortly after, but resumed in the 1994/5 season and have continued annually since then. Other industry offerings presently include land-based adventure tourism (mountain climbing, skiing, catering for private expeditions and wildlife photography). In the 1985/6 season tourists for the first time outnumbered national scientific and logistical support personnel in the area covered by the Antarctic Treaty, though they account for less than 1% of the total time spent ashore by human visitors to Antarctica (Headland, 1994). The collapse of the Soviet Union resulted in Russian vessels becoming available for Antarctic tourism charters, including icebreakers with on-board helicopters, beginning in the 1992/3 austral summer.

No records of tourist numbers and activities were collected during the industry's early years, but from 1992 an important function of IAATO has been to collect accurate statistics on behalf of the US National Science Foundation's Office of Polar Programs. Presented each year to the annual ATCM, these are summarized in Table 12.1. Tourist numbers increased almost continuously between 1992/3 and 1999/2000 (the millennium year), then dropped slightly in 2000/1, but have since continued to rise, as have numbers of operators, ships and voyages.

Simultaneously tourist activities have evolved and diversified. Operators now offer not only the traditional transport by Zodiac and short walks ashore, but also kayaking, camping, scuba diving, skiing and mountain climbing.

Current trends show a continuing gradual increase in tourism, most likely linear in nature. The IAATO website, which is updated as information becomes available, shows current and future statistics and trends.

Numbers of passengers on overflights (i.e. not landing in Antarctica) are not included in IAATO's tourism statistics, though these activities are noted in IAATO's annual reports to the ATCM (IAATO, 2006a). Comprehensive

Table 12.1. Numbers of operators, ships, voyages and passengers involved in Antarctic tourism, 1992–2007. Data are from reports of the International Association of Antarctica Tour Operators to Antarctic Treaty Consultative Meetings 1992–2006. Small numbers of yachts reported in seasons with entries marked * are not included in totals. Air overflights resumed in 1995, following a period of inactivity for several years (n/a, data not available). Most figures represent passenger totals from an Australian-based company; in later years, the figures include flights from South America to the Antarctic Peninsula. Figures for 2006/7 are based on estimates received from operators as of mid-2006.

Year	Operator/ charterer	Ships	Voyages	Tourists landing	Tourists on cruise-only vessels, no landings	Overflights
1992/3	10	12	59	6,704	0	n/a
1993/4	9	11	65	7,957	0	n/a
1994/5	9	14	93	8,098	0	n/a
1995/6	10	15	113	9,312	0	n/a
1996/7	11	13	104	7,322	0	n/a
1997/8	12	13*	92*	9,473	0	3,146
1998/9	13	15*	116	9,857	0	3,127
1999/2000	17	22	153	16,430	936	3,412
2000/1	15*	32	131	12,109	0	2,041
2001/2	19*	37	117	11,429	2,029	2,082
2002/3	26	47	136	13,263	2,424	1,552
2003/4	31	51	180	19,369	4,949	2,827
2004/5	35	52	207	22,294	5,027	2,030
2005/6	47	44	249	25,167	4,632	1,165
2006/7	45	50	278	27,575	7,500	1,600

overviews of the tourism industry can be found in yearly updates entitled 'Overview of Tourism' on the IAATO website (www.iaato.org).

Response to Challenges and Advances

As the tourism industry changed from the traditional ship-borne operations that were developed by Lars-Eric Lindblad in the 1970s, additional operators realized the market opportunities in tourism in Antarctica and the numbers of ships and companies increased, as did a variety of innovations in tourism. The need for an organization such as IAATO became apparent, and IAATO developed in 1991 as discussed above. A delegation was invited to the following ATCM, and IAATO has been represented at every subsequent ATCM, becoming a strong voice in policy decisions by the Treaty parties in the ensuing years (Splettstoesser, 2000).

Selected highlights of IAATO'S management of tourism in Antarctica in the 1990s and 2000s are listed below, some of which are continuing activities:

- Successful management of nearly all forms of Antarctic tourism by establishing a series of mechanisms for all operators to work successfully together, even though all operators are competitors of one another.
- Developed an online ship-scheduling programme whereby all companies input their schedules in order to ensure there is only one ship at one site at one time. This allows for the management of specific sites.
- Developed an online database of information to assist in emergency contingency planning and to be able to access specifications related to all vessels.
- IAATO designed and implemented the first ever comprehensive tourism database which includes companies, ships, site visits, and nationalities of passengers, staff and crew. IAATO took over the compilation of tourism statistics previously coordinated by the US National Science Foundation.
- In 2003 compiled 12 years of information from experienced field staff to produce IAATO's Site Guidelines, in order to quantify parameters of locations visited by tourists to establish time limits and numbers of passengers for environmentally safe visits. Development of site selection criteria, activity guidelines and standard operating procedures to ensure compliance of tourism standards.
- As of 2000, IAATO implemented 'Boot Washing and Clothing Decontamination' procedures as a measure towards preventing alien species from colonizing Antarctica.
- Development of Accreditation Scheme to illustrate that IAATO has credibility to conduct environmentally safe tourism in Antarctica based on its successful record of the same under self-imposed guidelines, its numerous guidelines for wildlife presence, and adherence to strict procedures in its bylaws.

- Improvement of navigational charts by sending International Hydrographic Organization–Hydrographic Committee on Antarctica (IHO-HCA) new sounding data resulting from tourism cruises.
- Development of Treaty Recommendation XVIII-1, Guidance for Tour Operators and Visitors, in concert with Treaty parties and incorporation into IAATO guidelines (Splettstoesser and Folks, 1994).
- Raised substantial funds to support scientific and conservation projects.
- Outreach programme and IAATO ships of opportunity – an offering by IAATO to Treaty parties to avail them of transport of science and support personnel and their gear to stations and field sites.
- Russian icebreakers and other Russian vessels, becoming available for tourism charters during the 1990s, provided means for icebreaker and helicopter activities in ice-infested areas such as Weddell Sea and Ross Sea, reaching emperor penguin colonies, and flights to Taylor Valley. IAATO developed specific guidelines for environmentally safe operation with helicopters, and worked with the US and New Zealand governments to ensure that visits to Taylor Valley did not interfere with ongoing science programmes.
- Circumnavigation of the Antarctic continent by a Russian icebreaker in 1996/7 (the tenth ever) resulted in a census of significant wildlife by IAATO-member staff, published in scientific journals (Gavrilo, 1997; Todd *et al.*, 1999; Splettstoesser *et al.*, 2000; IAATO, 2002).
- Successful outreach by IAATO to increasing yacht traffic in Antarctic tourism.
- Site Inventory begun by Oceanites in 1994 to identify locations visited by tourists that might be sensitive to over-visitation, and to provide census data for penguin breeding colonies (Naveen, 1997, 2003; Naveen *et al.*, 2001). IAATO members provided the transport for this project, which is still in operation today.
- IAATO involvement in assisting Treaty parties with management plans for selected areas (e.g. Dry Valleys, Deception Island, historic huts in Ross Sea area).
- Assistance by IAATO staff in publishing observations on Antarctic wildlife in refereed journals; thus recording new discoveries of emperor penguin colonies, for example, plus distribution of Ross seals and other wildlife.
- Bibliography and library of articles on Antarctic tourism (in preparation), an effort to provide a library of reference material on the subject – more than 100 items to date.

What are the Limits?

Analysing tourism trends and considering possible limiting factors is a further task undertaken by IAATO. While it is difficult to predict future tourist numbers, IAATO considers it likely that, if current trends continue, by 2010/11 numbers of passenger landings will have increased to about 35,000, based on linear growth of about 2000 per year, and of passengers on cruise-only ships to 10,000. The latter figure is difficult to estimate because some of the ships conducting cruise-only itineraries are not annual visitors. Clients flown to the interior for adventure tourism are not expected to total more than 250 per year. One- and two-day trips provide opportunities for some 800 tourists to visit Antarctica for short periods of time.

Despite the almost constant growth since the industry began, Antarctic tourism in world terms remains a relatively small niche market. In some other parts of the world, examining limits and 'sustainability' of tourism can be relatively straightforward, based on well-established prediction models. A comparable example can be mentioned of tourists visiting the Galapagos Islands, a popular location, where upwards of 85,000 tourists visit annually and, under strict management practices, produce virtually no impact on the islands or the wildlife. In considering possible limits, to be imposed on grounds either of market saturation or of ecological impacts on so truly unique an environment, numbers alone can be meaningless, especially where – as in this case – impacts per voyage or passenger are apparently low. The exceptional remoteness, extreme conditions, brevity of season and general difficulties of operations combine to present an extraordinary challenge towards maintaining and planning future operations. Furthermore, given the current economic climate, and the effects of the events of 11 September 2001, international travel and tourism generally are facing an uncertain future.

IAATO has established good cooperation between tour operators, Antarctic Treaty parties, the Council of Managers of National Antarctic Programs (COMNAP), the Scientific Committee on Antarctic Research (SCAR), the IHO-HCA, service providers in ports of departure, and other interested parties. The desirability to continue and further develop this cooperation is necessary given the international nature of Antarctic tourism and the ultimate protection of the continent.

In our analysis we consider three types of limits: (i) industry limitations – economic, psychological, logistical and operational; (ii) regulatory limitations established by governments (for example, by Antarctic Treaty parties), domestic legislation, maritime and aviation regulations; and (iii) limitations imposed by the environment.

Industry limitations

Economic limitations
- Demand: are there enough potential tourists to keep ships and aircraft at capacity?
- International politics (wars, stock market trends, etc.) and how they affect the psyche of the travelling market.
- Dollar exchange rates (most cruise prices are in US$).
- Limited ability to sell cruise/land trips in South America or Asian Pacific itineraries on board cruise ships.

- Size of market: is the Antarctic market sufficient economically to provide profit for tour companies in the off-season northern market?
- General costs of operating a travel/tourist company.
- Extraordinary costs of operating aircraft and ships in Antarctica, particularly rising fuel costs.
- Geographical competition – exciting ecotourism destinations worldwide.
- Very costly range of cruise prices and land-based adventure trips (an average price for a cruise is US$400–800 per day).
- Costs of special insurance for emergency medical evacuation.

Psychological limitations
- Cold-weather destination.
- Lack of full family interest or participation.
- Financial (US$/foreign currency exchange).
- Remoteness (distance to reach port of departure, as most tourists come from the northern hemisphere).
- Lack of instant communication and costs of communicating home (expense of Inmarsat and limited e-mail access).
- Inability to repatriate quickly to meet family or business emergency.
- Need to purchase cold-weather clothing.
- Preferred cabin or sleeping arrangements.
- Requirement of single cabin: reluctance to share with strangers.
- Need for physical ability.
- Sailing rough waters in Drake Passage and the Southern Ocean.
- Apprehension of expedition travel and preference for luxury vacations.
- Lack of vacation time.

Logistical limitations
- Lack of ability to choose exact itinerary or duration of each landing.
- Lack of ability to book at last moment.
- Flexibility and time needed to wait out bad weather for land-based adventures.
- Austral summer departures only.
- Uncertainty of commercial airlines and schedule changes.
- Lack of hotels or permanent tourist infrastructure in Antarctica. Will scientific stations become future hotels to supplement the costs of research programmes?

Operational limitations
- Lack of capital to start airlines and/or construct dependable runways.
- Costs of operating air service (i.e. availability of appropriate aircraft, adequate runways, fuel caches, etc.).
- Operational costs of expedition ships and travel companies' ability to operate vessels year-round.
- Current lack of low-cost ships.
- Lack of airstrips available for tourist operations (i.e. not used solely for government operations).

- Lack of suitable facilities and support services in port departure cities (i.e. fuel, water, food, waste disposal, recycling and personnel available in Australia, New Zealand, Argentina, Chile, Falkland Islands, South Africa).
- Cost of search and rescue and transport for adventure tourism and the need for emergency backup.

Regulatory limitations

As tourism has increased, so has the level of concern among Antarctic Treaty parties as to how tourism may be effectively addressed or regulated. Due to the international nature of Antarctic governance, numerous challenges are presented. Tourism is currently being regulated by the following mechanisms:

- *Regulations imposed under the Antarctic Treaty System*, including the Antarctic Treaty, Agreed Measures for the Conservation of Antarctic Fauna and Flora, Convention for the Conservation of Antarctic Seals, Convention on the Conservation of Antarctic Marine Living Resources, the 1991 Environmental Protocol, other decisions, measures and recommendations, and the Antarctic Conservation Act of 1978.
- *Self-regulatory mechanisms* imposed by the industry itself, including IAATO Bylaws, Guidelines for Visitors, Guidelines for Tour Operators, guidelines on wildlife watching, camping, kayaking and scuba diving, helicopter operations, certification of Zodiac drivers, and other regulations established by IAATO.
- *Domestic law* or national legislation governing tourist companies, for example the US Antarctic Conservation Act of 1978 (Public Law 95-541) as amended by the Antarctic Science, Tourism and Conservation Act of 1996.
- *Port regulations* set forth by gateway ports and other departure cities.
- *Shipping regulations*: international shipping standards [e.g. International Safety Management (ISM), International Convention for the Prevention of Pollution From Ships (MARPOL), Safety of Life at Sea (SOLAS)], insurance companies, etc.
- *Aircraft regulations*: international and domestic aircraft regulations and legal and regulatory issues relating to constructing runways and permanent infrastructure.

Environmental limitations

To determine if limits should be placed on tourism it is essential to have a clear understanding of the diversification of tourism activities, and whether each of them, singularly and scientifically, is or is not likely to have more than a minor or transitory impact on the specified landing sites, or on the marine or atmospheric environments. To date all the evidence suggests that tourists have no significant negative effects on the ecology of the Antarctic environment (Fraser and Patterson, 1997; Cobley and Shears, 1999; IAATO,

2001). Additional long-term, multi-year baseline data are necessary in order to achieve scientific results that might indicate whether tourism is harmful or benign. It is important to note that all current Antarctic tour operators file yearly environmental impact assessments to their national authorities.

Both IAATO and the scientific community will need to develop methods to measure the biological diversity at the site location, relative to the distributional ranges of the species present, and identify indicator species, robustness of the species, availability of open space, general topography, novelty of the site, ice and weather conditions, availability of safe anchoring or holding sites, acoustic characteristics, location of comparable sites nearby, meteorological information, amount of krill and other food sources in close vicinity to ensure that wildlife can have sufficient resources, snow pack, etc. IAATO is actively involved in encouraging scientific studies that contribute to our understanding of the Antarctic ecosystem. Until that time, the tour industry will continue to operate under the precautionary approach, proceeding with extreme care in the planning of its activities and ensuring that disruptions and potential environmental damage do not occur.

Why Set Limits?

Looking to the future, we can only assume that Antarctic tourism will continue to increase, and as an industry it is the responsibility of IAATO to ensure that sufficient limitations are established by operators to protect the biological, physical, spiritual and aesthetic qualities of Antarctica.

Our 5-year forecast of growth to 2011 is a 'guesstimate' based mainly on trends of cruise vessels looking for new itineraries and more exotic destinations, and possibilities of more travel to the continent by air. Such trends beg the question of the bearing capacity of the continent and its attractions for tourists. It remains to be seen whether our projection is accurate, and whether increasing numbers translate to potential environmental damage that could occur only through long-term visits by humans.

IAATO is dedicated to the development of a more detailed and tactical approach, in order to attempt to manage the industry, minimize environmental impacts and create the highest possible standards for operators and visitors. Tourism in Antarctica is a legitimate activity. As Lars-Eric Lindblad once said, 'You can't protect what you don't know'. It is without doubt one of the greatest adventures and wilderness experiences currently available. With proper precautions and safeguards in place, future generations will be able to enjoy all that Antarctica offers forever.

Acknowledgements

We acknowledge the work of IAATO's Environmental Operations Manager, Dr Kim Crosbie; members of IAATO's various committees who volunteered their time to accomplish needed projects and make decisions on behalf of the membership; and Wisteria IT and Zenn for computer IT design and implementation. Antarctica is protected because of the dedication of IAATO members and the talented field staff who take great care in ensuring they are operating at the highest possible standards.

References

Cobley, N.D. and Shears, J.R. (1999) Breeding performance of gentoo penguins (*Pygoscelis papua*) at a colony exposed to high levels of human disturbance. *Polar Biology* 21, 355–360.

Fraser, W.R. and Patterson, D.L. (1997) Human disturbance and long-term changes in Adélie penguin populations: a natural experiment at Palmer Station, Antarctic Peninsula. In: Battaglia, B., Valencia, J. and Walton, D.W.H. (eds) *Antarctic Communities: Species, Structure and Survival*. Cambridge University Press, Cambridge, UK, pp. 445–452.

Gavrilo, M. (1997) Antarctic circumnavigation in austral summer 1996/97. Preliminary results of sea bird and mammal survey. *Korean Journal of Polar Research* 8(1/2), 105–111.

Headland, R.K. (1994) Historical development of Antarctic tourism. *Annals of Tourism Research* 21(2), 268–280.

IAATO (2001) Issues relating to cumulative environmental impacts of tourist activities. *IP 52, Antarctic Treaty Consultative Meeting XXIV*, St Petersburg, Russian Federation, 9–20 July. International Association of Antarctica Tour Operators, Basalt, Colorado.

IAATO (2002) Bibliography of publications by staff/naturalists/lecturers involved in tour activities in Antarctica, 1991–2001. *IP 71, Antarctic Treaty Consultative Meeting XXV*, Warsaw, 9–20 September. International Association of Antarctica Tour Operators, Basalt, Colorado.

IAATO (2006a) IAATO overview of Antarctic tourism 2005–2006 season. *IP 86, Antarctic Treaty Consultative Meeting XXIX*, Edinburgh, 12–23 June. International Association of Antarctica Tour Operators, Basalt, Colorado.

IAATO (2006b) Report of the International Association of Antarctica Tour Operators 2005–2006. *IP 90, Antarctic Treaty Consultative Meeting XXIX*, Edinburgh, 12–23 June. International Association of Antarctica Tour Operators, Basalt, Colorado.

Naveen, R. (1997) *The Oceanites Site Guide to the Antarctic Peninsula*. Oceanites, Inc., Chevy Chase, Maryland.

Naveen, R. (2003) *Compendium of Antarctic Peninsula Visitor Sites: A Report to the United States Environmental Protection Agency*, 2nd edn. Ron Naveen Oceanites, Inc., Chevy Chase, Maryland (also available on CD).

Naveen, R., Forrest, S.C., Dagit, R.G., Blight, L.K., Trivelpiece, W.Z. and Trivelpiece, S.G. (2001) Zodiac landings by tourist ships in the Antarctic Peninsula Region, 1989–99. *Polar Record* 37(201), 121–132.

Splettstoesser, J. (2000) IAATO's stewardship of the Antarctic environment: a history of tour operators' concern for a vulnerable part of the world. *International Journal of Tourism Research* 2(1), 47–55.

Splettstoesser, J. and Folks, M.C. (1994) Environmental guidelines for tourism in Antarctica. *Annals of Tourism Research* 21(2), 231–244.

Splettstoesser, J., Gavrilo, M., Field, Carmen, Field, Conrad, Harrison, P., Messick, M., Oxford, P. and Todd, F.S. (2000) Notes on Antarctic wildlife: Ross seals *Ommatophoca rossii* and emperor penguins *Aptenodytes forsteri*. *New Zealand Journal of Zoology* 27(2), 137–142.

Todd, F.S., Splettstoesser, J.F., Ledingham, R. and Gavrilo, M. (1999) Observations in some emperor penguin *Aptenodytes forsteri* colonies in East Antarctica. *Emu* 99(2), 142–145.

13 Antarctic Tourism Research: the First Half-Century

BERNARD STONEHOUSE[1] AND KIM CROSBIE[2]

[1] Scott Polar Research Institute, University of Cambridge, Lensfield Road, Cambridge CB2 1ER, UK; [2] International Association of Antarctica Tour Operators, PO Box 2178, Basalt, CO 81621, USA

Introduction

Commercial tourism to the Antarctic region began during the late 1950s, contemporaneously with the International Geophysical Year (IGY, 1957/8). IGY publicity identified Antarctica as an area of outstanding scientific interest: early tourists reported a world of photogenic mountains, ice and fascinating wildlife. From the early 1960s the more accessible points of the continent and neighbouring islands were visited each summer by increasing numbers of cruise ships and aircraft, accelerating from the 1990s to the present. This chapter surveys some of the literature of research on the industry, in particular accounts of fieldwork and discussion concerned with its management during the second half of the 20th century.

That it is not an exhaustive survey will be apparent to all who work in the field. Antarctic tourism spans many disciplines from geopolitics to physiology: its studies appear in a wide range of publications and guises, some of which we have probably missed. There are gaps; we have omitted, for example, discussion papers tabled at Antarctic Treaty Consultative Meetings (ATCMs) and similar venues, which may well have been important in determining policies and can readily be found on government websites. Nor have we listed more than a sample of the publications that have been generated within the industry itself; we understand that John Splettstoesser is currently compiling such a list. We provide only a representative sample of an array of broadly scattered, cross-disciplinary research contributions, which we hope will be helpful both to established workers and to new students entering the field.

The Industry

In the recent summer season of 2005/6 over 27,500 tourists landed in the Antarctic region, mainly on the Antarctic Peninsula and islands of the Scotia

Arc in the South American sector. Though a small total compared with popular tourist venues elsewhere, the number marked a doubling and redoubling in 14 years (Chapter 12, this volume) with commensurate growth in the numbers of ships, voyages and air passages that brought them. For accounts and reviews of Antarctic tourism as a whole see Galimberti (1991), Headland (1994), Stonehouse and Crosbie (1995), Bauer (2001) and Chapters 9 and 12, this volume. Scientists, on-board lecturers, travel writers, artists and photographers have provided guidebooks and illustrated tourist-friendly literature: see for example Sage and Hosking (1982), Chester (1993), Soper (1994), Chester and Oetzel (1995), Lucas (1996), Naveen (1997), Rubin (2000) and Stonehouse (2000, 2005). Carr and Carr (1998) and Poncet and Crosbie (2005) provide similar coverage for South Georgia; Fraser (1986) for New Zealand's sub-Antarctic islands.

The modest start of Antarctic tourism, and its slow growth in comparison with the spread of scientific stations during the 1960s and 1970s, failed to presage its more recent rapid growth and development. The advent of a new industry, on a continent dedicated by the Antarctic Treaty to the pursuit of science, was at first viewed with little more than mild apprehension by scientists and administrators, who feared mainly for its possible adverse effects on research. Little investigation into its nature, effects and potential was undertaken until the 1990s, when enhanced growth, coupled with demands for better conservation and management of Antarctica, stimulated both discussion and field research.

Early Studies: the 1960s to 1990

Starting with overflights and ship-borne cruises to the South Shetland Islands and Antarctic Peninsula in December 1956–1958, early Antarctic tourism attracted news media attention but little serious investigation. *The Explorer's Journal* in 1957 noted the 'first commercial flight to Antarctica', by a Pan American Boeing Stratocruiser chartered to carry naval personnel from Christchurch, New Zealand to McMurdo Station. A reliable source of information on early voyages and flights is *Antarctic*, the quarterly news bulletin of the New Zealand Antarctic Society; see for example *Antarctic* (1968: 51–54), reporting the first visit of a cruise ship to McMurdo Sound. The US Library of Congress's Antarctic Bibliography introduced 'Tourism' as an index heading in 1968, to accommodate a note in the *United States Antarctic Journal* on tourist cruises by the Argentine naval transport *Lapataia* (McDonald, 1967). Of the few scientists and support staff who encountered ship-borne tourists on beaches in the Peninsula area, some commented on their presence, but none saw them as material for research.

Tourism was discussed briefly during the second biological conference of the Scientific Committee for Antarctic Research (SCAR) in 1968, reported in Holdgate (1970: 951–952). The New Zealand biologist George Knox (a later president of SCAR) remarked on the scientific community's negative attitude to tourism, and urged a more positive approach. In his view the industry was bound to expand:

> we must have adequate information available for the tourists and not just rely on their party leaders. We want SCAR to tell people what their responsibilities are towards conservation and science in the Antarctic zone.

SCAR indeed produced a visitor's guide to Antarctica some 14 years later (SCAR, 1982). The US biologist Joel Hedgepeth, commenting on complaints by American tourists over their hostile reception at McMurdo Station during recent cruises to the Ross Sea region, raised a more telling point:

> One of the most important things about Antarctic tourism is that it allows the scientists to show the non-scientists on whom they depend for money how valuable the work they are doing is. Only well-off and potentially influential people are likely at the present time to tour such far distant parts, and I think it is a good plan to give them well prepared brochures and politeness, and show them rather more than the McMurdo dump.

Again this message took time to penetrate: several years elapsed before station visits could be booked in advance, and a warm welcome offered to visiting tax-payers.

E.A. Bauer noted the advent of Antarctic tourism in two semi-popular accounts (Bauer, 1974a,b). In the first detailed study of Antarctic tourism, R.J. Reich (later Codling) outlined the development of the industry (Reich, 1979, 1980), chronicling 80 cruises and over 40 commercial overflights between 1958 and 1980. A later paper (Codling, 1982) described and evaluated sea-borne tourism, with notes on her own cruise-ship voyage to the Antarctic Peninsula. In still-later papers Codling (1995, 2002) reviewed respectively early 20th century efforts to establish tourist visits to the continent and concepts of wilderness and aesthetic values applied to the region. Reich reported on relevant ATCMs, of which the 4th in 1966 recognized that 'the effects of tourist activities may prejudice the conduct of scientific research, conservation of flora and fauna and the operation of Antarctic stations', and the 8th in 1975 acknowledged that tourism was 'a natural development in the Area' that required regulation.

Reich listed only two research-based studies: the first considering tourism as a possible economic development in Antarctica (Potter, 1970), the second describing management of a colony of Adélie penguins at Cape Royds, McMurdo Sound, that had come under stress from an early and non-commercial form of tourism (Thomson, 1977). Before commercial tours first came to the area, Taylor (1962) and Stonehouse (1963, 1965a,b,c, 1967) had shown that insensitive helicopter-borne visits by official guests of the US and New Zealand governments had significantly reduced the colony. Thomson outlined protective measures taken jointly by the two governments – perhaps the earliest record of research-based action to protect an Antarctic wildlife resource from recreational visitors.

Literature of the 1980s, a period of slow and unspectacular growth, began with accounts of the industry's first major disaster – the 1979 crash of an overflying Air New Zealand DC10 aircraft on the flanks of Mount Erebus, McMurdo Sound (Chippindale, 1980; Mahon, 1981, 1984; Pyne, 1986), which effectively terminated East Antarctica overflights for 15 years. Headland and Keage (1985) noted tourist activities on King George Island, South Shetland Islands, and Boswall (1986) recorded overflights to that area. British and other authorities with Antarctic responsibility produced visitor's guides, generally admonitory and intended mainly for scientists and support staff (e.g. SCAR, 1982, 1984). Increasing numbers of articles by travel writers appeared in popular journals and newspapers, which began to run advertisements for Antarctic cruises. Adams and Lockley (1982), Zuckerman (1985) and Snyder and Shackleton (1986) produced well-illustrated books of Antarctic cruises.

Scientists too continued to agonize over possible effects of ill-regulated tourism. Sage (1985: 362) for example wrote:

> Anxiety over tourism arises for a number of reasons. Firstly, the landing of large parties of tourists who cannot be properly supervised is likely to have important and lasting effects on the local flora and fauna. Secondly, the visits to scientific stations, although often welcomed by the scientists, cause disruption of work and social stresses. And thirdly, accidents associated with the cruise ships or aircraft often require the use of scarce personnel, time or fuel from one or more of the Treaty countries.

These points caricatured rather than represented an industry that, after 15 relatively blameless years, had established itself as fully responsible to both the natural and human environments. International lawyers and politicians too were concerned, in particular with legal implications of Antarctic tourism, including weaknesses in management arising from lack of sovereign authority (Auburn, 1982; Nicholson, 1985). Triggs (1986: 184), for example, was concerned that, on the issue of non-Treaty nationals entering Antarctica as tourists, Treaty consultative parties could:

> merely recommend to their governments that ... they use their best endeavours to ensure that all those who enter the Antarctic Treaty Area ... are aware of the Statement of Accepted Practices and the relevant provisions of the Antarctic Treaty.

Thankfully, non-Treaty nationals behaved as well as those of signatory governments, though to Boczek (1988: 465) controls based on such recommendations 'lacked conceptual rigor', and Auburn (1982: 280) had earlier found them 'too vague to be enforceable'.

A study group including prominent Antarctic-oriented scientists and diplomats, considering in 1985 the question of 'Antarctica, the next decade', produced a wide-sweeping, book-length survey of the Antarctic Treaty and its implications (Parsons, 1987: 73). Taking into account science, mineral and living resources, military potential and other considerations, they devoted just two token paragraphs to tourism. In their view the industry would grow, but control was at that time almost non-existent. They added uneasily:

there is nothing in the current system to stop a land-based tourist operation being set up in a vulnerable habitat, nor even to require that there be some form of assessment of the impacts of such a development ... It is an uncomfortable thought that such activities could in theory go ahead unregulated. Indeed on the Fildes Peninsula, the Chilean station at Teniente Marsh contains a tourist hotel served by the large airstrip. This development has never been discussed by parties other than the Chilean authorities.

The hotel (Chileans still prefer 'hostel') would soon be far exceeded in size by several scientific stations catering for large numbers of scientists and support staff, all erected without benefit of international discussion. Two years later a far-thinking parliamentary standing committee of the Government of Australia considered seriously a proposal for combining research and tourism facilities at Davis Station, involving an all-weather airstrip and hotel facilities for scientists, support staff and some 16,000 visitors per year (Commonwealth of Australia, 1989; Hall, 1992; Enzenbacher, 1995a,b). The concept was rejected, but has recently resurfaced in a similar scheme currently under consideration for Casey Station (Chapter 8, this volume).

It fell to four naturalists who had served on Antarctic cruise ships to publish the first practical code of conduct designed for tourists and tour operators (Naveen et al., 1989). Antarctic Treaty parties, in their view, had not fashioned 'sufficiently specific guidelines to govern tourism and other Antarctic visits'. Their simple code, relevant for all visitors including tourists, was based on:

> basic conservation principles, the ethics underlying the Antarctic Treaty's Agreed measures for the conservation of Antarctic Fauna and Flora, prevailing international conservation treaties, and the authors' collective experience as expedition leaders and naturalists in the field.

As such it formed a refreshing contrast to the complexities of official statements, which were of little help in managing parties of tourists on Antarctic beaches (Stonehouse, 1990). The same four authors in the following year produced *Wild Ice: Antarctic Journeys* (Naveen et al., 1990), a popular account of Antarctica that cannot have failed to stimulate interest among intending travellers.

Conservation and the Protocol

Several publications of the 1980s reflected increasing concern, from individuals and international environmental interests, that the Treaty powers were allowing Antarctica and the surrounding ocean to be ruined by commercial development. Barnes's (1982) *Let's save Antarctica!*, an appeal to the public to save the area from further exploitation, especially from uncontrolled commercial fishing and mining, contained little reference to tourism. The author's main concerns, reflecting the conservationist views of the Antarctic and Southern Ocean Coalition (ASOC), were that the newly introduced Convention for the Conservation of Antarctic Marine Living

Resources (CCAMLR) would not adequately control harvesting of marine species, and that the 14 Treaty nations were in secret planning a convention that would permit minerals exploitation on the continent. In the presence of these greater and more immediate threats, neither conservationists nor ATCMs found time to consider tourism.

The Convention on the Regulation of Antarctic Mineral Resource Activities (CRAMRA) was indeed presented to the Treaty in 1988, but rejected on various grounds, not least that it provided insufficient guarantees of environmental protection to satisfy a groundswell of public opinion. Following its demise, conservation became the most pressing concern for discussion at ATCMs. The 15th ATCM of 1989, notable for its emphasis on environmental matters, gave rise to an Antarctic Treaty Special Consultative Meeting (1990) to consider conservation. From this in turn arose the comprehensive 1991 Protocol on Environmental Protection, the Treaty instrument which has since regulated all human activities in Antarctica, including tourism (Heap, 1994).

At the subsequent 16th ATCM (1991), the question was raised as to whether a Protocol that applies impartially to scientists and all other visitors would be adequate to deal with tourism, or whether a special annex was required to cover it. At the 17th ATCM (1992, first of the annual meetings), a draft annex on tourism was presented, discussed and discarded: tourism, the delegates decided, could be dealt with like all other visitor activities under existing Protocol regulations. Critics continued to point out that such even-handedness failed to recognize fundamental differences in the ways that scientific research and tourism are conducted – one mainly from permanent static stations, the other almost entirely by transient landings from ships. Once established, the principle has remained.

Research During the 1990s

Soon after the Protocol's appearance, the International Union for the Conservation of Nature and Natural Resources (IUCN) produced a report on Antarctic conservation (IUCN, 1991). Ill-timed, it had originally been intended for tabling at the 11th Special Antarctic Treaty Consultative Meeting, but was delayed (Holdgate, personal communication) and thus not considered in the Protocol deliberations. The report singled out tourism as a matter for special concern, recommending among other issues a comprehensive review of the industry, development of tour management guidelines, proactive planning for Areas of Special Tourist Interest, followed by careful monitoring of subsequent impacts; and controlling choice of tourist destinations.

These recommendations indicated possibilities for both field and desk research that were adopted independently by several groups throughout the Treaty nations. Among early starters was the Polar Ecology and Management Group of the Scott Polar Research Institute, University of Cambridge, UK. Led by Bernard Stonehouse (who had recently revisited Antarctica aboard a cruise ship), the group's Project Antarctic Conservation (PAC) had already completed a study of the heavily impacted environment of King George Island, South Shetland Islands (Harris, 1991a,b; Harris and Meadows, 1992) and begun research on tourism (Christensen, 1990; Enzenbacher, 1991; Davis, 1992). PAC and similar groups produced a substantial increase in research studies, here summarized under three broad headings: (i) growth, development, logistics and the future; (ii) landing operations, impacts and remedies; and (iii) management, regulation and tourist views.

Growth, development, logistics and the future

Reich's (1980) record of early cruises and flights was followed by a gap of almost a decade, though many, perhaps all, voyages up to 1989 are reported by Headland (1989, 1994); see also Enzenbacher (1993a,b). From 1989/90 the US National Science Foundation (NSF) became responsible for reporting data on Antarctic tourism, to be gathered annually by the International Association of Antarctica Tour Operators (IAATO). Records thereafter are with few exceptions complete, and available on the IAATO website (www.iaato.org). We are unaware of any definitive listing of cruise ships and voyages from the start of the industry. One assembled from existing information would show a tendency towards increasing tonnage and passenger capacity. The first ocean liner carrying more than 900 passengers entered the industry in the millennium year (Stonehouse and Brigham, 2000); one with capacity for over 3000 joined in 2007.

The activities of non-government aircraft in Antarctica between 1988 and 2001, many of them involving tourism, appear in a series of papers by Swithinbank. The first (Swithinbank, 1988: 313–316) reported the first airline set up specifically to carry tourists to Antarctica. Two later papers (Swithinbank, 1993a: 103–110, 1997a: 243–244) dealt respectively with developments in air-borne tourism and the opening of an air route between Cape Town and Dronning Maud Land, Antarctica. The remaining notes (Swithinbank, 1989, 1990, 1992a,b, 1993b, 1994, 1995, 1996, 1997b, 1998, 2000, 2001) were annual activity reports. Swithinbank's examination of wind-cleared blue-ice landing grounds on the continental ice cap opened much of the interior of Antarctica to air-borne research and tourism (Mellor and Swithinbank, 1989). Further details of overflights appear in Headland and Keage (1995) and in Bauer (2001 and Chapter 11, this volume). For accounts of 'adventure tourism' which is largely serviced by flights, see Lamiers *et al.* (Chapter 10, this volume).

Bauer (1991) like many others predicted continuing growth for the industry; Jones (1998) sought precision in applying traditional models developed for warmer venues, but reported problems rather than solutions. Landau and Splettstoesser (Chapter 12, this volume), with many years' experience in polar tourism, tentatively predict continuing growth.

Landing operations, impacts and remedies

The 1990s' surge in ship-borne tourism focused attention on management of visitors at landing sites. PAC observers evaluated the visitor guidelines issued by the recently formed International Association of Antarctica Tourist Operators (IAATO, 1991), finding them more practical and effective for controlling passengers at landings than guidelines prepared outside the industry. With slight modifications these were later adopted in 1994 by the 18th ATCM as Recommendation XXVIII-1. Codling's (1982) early description and analysis of landing operations, confirmed in detail by Stonehouse and Crosbie (1995), established the 'Lindblad pattern' as the norm to which all responsible cruise operators aspired. A landing site database developed by NSF and IAATO from 1989/90 (see above), subsequently taken over solely by IAATO, has been updated annually since then; this records some 230 sites, of which over half are currently used, and is available on the IAATO website.

PAC's three research sites were chosen for their attractiveness to the industry, guaranteeing visits that could be monitored throughout each season (Stonehouse 1992, 1993). From 1992 PAC also published detailed assessments of 12 popular, heavily used landing sites, highlighting visitor attractions and points of environmental sensitivity, and suggesting ways of minimizing impacts (Stonehouse, 1995). Little official interest in the guidelines was shown at that time, but the assessments were circulated to and used by IAATO, cruise operators and naturalists.

PAC fieldwork began in 1991/2, in association with scientists of Instituto Antártico Argentino, with a survey of tourist impacts on Halfmoon Island, South Shetland Islands (Stonehouse, 1992). In the following three seasons studies centred on a small field station accommodating six observers on Cuverville Island, Danco Coast of Antarctic Peninsula. A second station of similar size operated in 1993/4 at Hannah Point, Livingston Island, South Shetland Islands. Cuverville Island, selected as a site that had previously received few visitors, was surveyed topographically (Crosbie, 1998) and botanically (De Leeuw and Aptroot, 1998) and breeding bird populations were catalogued. Part of the landing beach was segregated as a no-visit control area, maintained with the full cooperation of most visiting parties. All ship visits were recorded, and ways in which visiting parties made use of the island's resources were examined. From these studies emerged an overview of the site and operations and, later, key management parameters for assessing visitor impacts and designs for effective measures for long-term monitoring of this and other vulnerable landing sites (Crosbie, 1998).

The colony of gentoo penguins, the main visitor attraction of the site, was studied in detail. Breeding success was assessed in visited and unvisited areas of their colony, and techniques were devised for studying changes in heartbeat of incubating birds as indicators of stress, without handling or otherwise traumatizing the subjects (Nimon, 1992, 1995, 1997; Nimon *et al.*, 1994, 1995, 1996; Nimon and Stonehouse, 1995). During the single-season study at Hannah Point, a popular visitor site unusually rich in wildlife, a team made similar overall surveys of flora and fauna to identify particular risks arising from tourist visits (Davis, 1998). Particular attention was paid to the large breeding populations of giant petrels and to non-breeding elephant seals (Cruwys and Davis, 1994, 1995, 1996a,b).

Acero and Aguirre (1994) and Davis B. (1999) surveyed other sites and made management recommendations. During the late 1990s Stonehouse reviewed tourist management at the Polish research station Henryk Arctowski (Admiralty Bay, South Shetland Islands), a station which welcomed tourist visits but sought to provide diversions from the living quarters and laboratories (Ciaputa and Salwicka, 1997). Management plans including trails and walks were drawn up for both station staff and visitors (Stonehouse, 1999).

Studies on interactions between visitors and birds, some involving physiological measurements, were made by other teams elsewhere in Antarctica. As in Nimon's research, emphasis lay on monitoring changes in heartbeat and general deportment of penguins (Culik *et al.*, 1990; Young, 1990; Wilson *et al.*, 1991; Woehler *et al.*, 1994; Regel and Pütz, 1995; Giese, 1996, 1998; Chupin, 1997). Emslie (1997) reviewed human impacts on bird populations in general. From Palmer Station, Anvers Island, Fraser and Patterson (1997) and Patterson *et al.* (2001) reported on longer-term studies of colonies of Adélie penguins, some subject to tourist visits, others retained as controls. Their evidence indicated population changes due to shifts in sea ice distribution, rather than to human visitation. Cobley and Shears (1999) assessed the breeding performance of gentoo penguins at a heavily visited site (Port Lockroy), concluding that disturbance from tourism was unlikely to have been a major determinant of breeding populations. Pfeiffer and Peter (2004) reviewed giant petrel and skua behaviour in the presence of visitors on Penguin Island (South Shetland Island). Salwicka and Stonehouse (2000) similarly monitored changes in respiration rates and heartbeat of seals resting on beaches close to Henryk Arctowski Station, concluding that resting seals relaxed into an energy-saving diving mode, which only noisy visitors were likely to disturb.

Since 1994 all sites in regular use in the Antarctic Peninsula region have been surveyed and inventoried by 'Oceanites' – ship-borne teams operating under the direction of Ron Naveen (1996, 1997), who has also analysed NSF records of landings during the decade starting 1989/90 (Naveen *et al.*, 2000, 2001). These observations form a basis for management plans recently developed (2006) for the Maritime Antarctic's 12 most-popular landing sites.

The sinking of *ARA Bahia Paraiso* off Palmer Station in January 1989 (an Argentine naval vessel carrying 81 tourists) released oil that severely damaged neighbouring shores and biota (Kennicutt, 1990; Kennicutt and Sweet, 1992; Bruchhausen, 1996). The incident raised important questions of responsibility and financial liability. Remediation of damage to the over-visited Antarctic sites was investigated by Campbell *et al.* (1998), who discussed the appearance of foot-worn paths on dry ground – in mild form a common phenomenon on well-trampled sites, often remedied naturally by winter solifluction. Bölter and Stonehouse (2002: 403), reviewing human impacts on Antarctic coastal soils and possibilities for damage remediation, recommended natural process and patience: 'Recovery that, in temperate regions, might be expected within a human life span may take two or three times longer in Antarctica.' More general remediation issues are discussed by Bertram and Stonehouse (Chapter 17, this volume).

Management, guidelines, legislation and tourist views

Enzenbacher (1991) assessed conflict and cooperation in Antarctic tourism management policies, providing reviews of the industry's growth and development (Enzenbacher, 1992a,b,c, 1993a,b,c, 1994a,b). She contributed also a case study of tourist visits to Faraday Station, Argentine Islands (Enzenbacher, 1994c) and an analysis of how the Antarctic Treaty System was seeking to regulate tourism during this period (Enzenbacher, 1995a,b). Waugh (1994) applied Geographical Information Systems techniques to tourism management in the Antarctic and sub-Antarctic, with particular reference to Campbell Island.

Splettstoesser and Folks (1994), Johnston and Hall (1995) and Johnston (1997, 1998) debated the adequacy of guidelines as management tools; Riffenburgh (1998) contrasted their effectiveness in managing well-briefed tourist groups with their inadequacy in controlling ill-briefed and unsupervised scientists and support staff. The requirement of tour operators to monitor for environmental changes at landing sites as a condition of their permitting under the Protocol was reviewed by Minbashian (1997) and Crosbie (1998), who suggested strategies based respectively on biological and ecological integrity; *see also* Bertram (2005) and Bertram and Stonehouse (Chapter 17, this volume).

Antarctic tourism management concepts were discussed in symposia on ecotourism (Stonehouse, 1994a), tourism in relation to Antarctica's protected areas (Stonehouse, 1994b), sustainable tourism (Stonehouse *et al.*, 1995), and management of tourism in both polar regions (Enzenbacher, 1995a; Girard, 1995; Stonehouse and Crosbie, 1995; Stonehouse, 1996, 1998). Continuing legal and geopolitical problems arising from tourism were aired by Beck (1994), who postulated a 'political nightmare' which might result from a shipping-based accident in, for example:

> an area claimed by both Argentina and the United Kingdom involving a Liberian registered vessel, with an Italian captain, a Filipino crew, and a multinational tour group organized by a US travel agent jointly with other agents in Australia, Britain, Japan, and the United Kingdom.

Beck concluded that there would be no easy answer. Chaturvedi (1996: 217–220) summarized in geopolitical terms the development of Antarctic Treaty policies in relation to tourism, and Dodds (1997: 130–131) analysed in particular Chile's motivation in actively promoting tourism in its southernmost territory.

Following the passing in 1996 of the US Antarctic Science, Tourism, and Conservation Act (Public Law 104-227), the US NSF's Office of Polar Programs, State Department and Environmental Protection Agency became jointly responsible for establishing environmental regulations to implement the

1991 Protocol. The proposal of a Final Interim Rule to ensure that Antarctic tourism would actively support environmental protection caused a flurry of data assembly, research activity and the solicitation of diverse public opinions, culminating in the February 2001 Draft Environmental Impact Statement (EIS) for the Proposed Rule on Environmental Assessments of Nongovernmental Activities in Antarctica, and the Final EIS published in August of the same year. On public record and readily available for scrutiny, both documents contain valuable records of factual information and US public perceptions of the period.

Tourists themselves give abundant evidence of satisfaction with their cruises and flights, but appear seldom to have been invited to comment in structured surveys. Davis (1995a,b) conducted shipboard surveys in the Maritime Antarctic, Cessford and Dingwall (1998) conducted similar enquiries among visitors to the Ross Sea region, and Bauer (2001) made extensive analyses of motivations, expectations and images of Antarctica perceived by tourists both before and after visits. Medical practice aboard cruise ships was the subject of a PAC-hosted Cambridge conference in 1995 (Levinson and Ger, 1998).

Conclusions

Research on Antarctic tourism mirrors almost precisely the growth of the industry – slow to develop during the first three decades, accelerating during the 1990s, and since then burgeoning. Much has been written, mainly by independent rather than government-sponsored researchers, to explore and explain the industry, in particular to foster understanding between the various participants: the entrepreneurs, ship owners, planners, cruise directors, master mariners, aviation crews and staff who operate in the field; the diplomats, conservationists and civil servants who determine policies; and the tourists who are the consumers. Much remains to be explored and examined. As the industry continues to grow and diversify, so will opportunities for continuing research into an interdisciplinary field involving aspects of environmental sciences, international relations and law, industrial develop-ment, management and human behaviour.

References

Acero, J.M. and Aguirre, C.A. (1994) A monitoring research plan for tourism in Antarctica. *Annals of Tourism Research* 21(2), 295–302.

Adams, R. and Lockley, R. (1982) *Voyage Through the Antarctic*. Lane, London.

Antarctic (1968) First tourists arrive. *Antarctic* 5(1), 54–57.

Auburn, F.M. (1982) *Antarctic Law and Politics*. Indiana University Press, Bloomington, Indiana.

Barnes, J.N. (1982) *Let's Save Antarctica!* Greenhouse Publications, Richmond, Victoria.

Bauer, E.A. (1974a) Tourism comes to Antarctica. *Oceanus* 7(1), 12–19.

Bauer, E.A. (1974b) The race is on for Antarctica. *International Wildlife* 4(6), 42–46.

Bauer, T.G. (1991) The future of commercial tourism in Antarctica. *Annals of Tourism Research* 21(2), 410–413.

Bauer, T. (2001) *Tourism in the Antarctic: Opportunities, Constraints and Future Prospects.* Haworth Hospitality Press, New York, New York.

Beck, P.L. (1994) Managing Antarctic tourism: a front-burner issue. *Annals of Tourism Research* 21(2), 375–386.

Bertram, E. (2005) Tourists, gateway ports and the regulation of shipborne tourism in wilderness regions: the case of Antarctica. PhD thesis, University of Cambridge, Cambridge, UK.

Boczek, B.A. (1988) The legal status of visitors, including tourists, and non-governmental expeditions in Antarctica. In: Wulfrum, R. (ed.) *Antarctic Challenge III: Conflicting Interests, Cooperation, Environmental Protection, Economic Development.* Duncker and Humblot, Berlin, pp. 455–490.

Bölter, M. and Stonehouse, B. (2002) Uses, preservation and protection of Antarctic coastal regions. In: Beyer, L. and Bölter, M. (eds) *Geoecology of Ice-free Coastal Landscapes* [Special Issue]. *Ecological Studies* 154, 393–407.

Boswall, J. (1986) Airborne tourism 1982–84: a recent Antarctic development. *Polar Record* 43(123), 187–191.

Bruchhausen, P. (1996) The grounding and sinking of the *ARA Bahia Paraiso.* In: Giraud, L. (ed.) *Ecotourism in the Polar Regions.* A pas de Loup, Paris.

Campbell, I.B., Claridge, G.G.C. and Balks, M.R. (1998) Short- and long-term impacts of human disturbance on snow-free surfaces in Antarctica. *Polar Record* 34(188), 15–24.

Carr, T. and Carr, P. (1998) *Antarctic Oasis; Under the Spell of South Georgia.* W.W. Norton Company, New York.

Cessford, G. and Dingwall, P.R. (1998) Research on shipborne tourism in the Ross Sea region and the New Zealand sub-Antarctic islands. *Polar Record* 34(189), 99–106.

Chaturvedi, S. (1996) *The Polar Regions: A Political Geography.* John Wiley & Sons, Chichester, UK.

Chester, S.R. (1993) *Antarctic Birds and Seals.* Wandering Albatross, San Mateo, California.

Chester, S.R. and Oetzel, J. (1995) *South to Antarctica: A Handbook for Antarctic Travelers.* Wandering Albatross, San Mateo, California.

Chippindale, R. (1980) *Aircraft Accident Report No. 79-139.* Government Printer, Wellington.

Christensen, T.R. (1990) Tourism in polar environments with special reference to Greenland and Antarctica. MPhil thesis, University of Cambridge, Cambridge, UK.

Chupin, I. (1997) Human impact and breeding success on southern giant petrel *Macronectes giganteus* on King George Island (South Shetland Islands). *Korean Journal of Polar Research* 8, 113–116.

Ciaputa, P. and Salwicka, K. (1997) Tourism at Antarctic Arctowski Station 1991–1997: policies for better management. *Polish Polar Research* 18(3–4), 227–239.

Cobley, N.D. and Shears, J.R. (1999) Breeding performance of gentoo penguins (*Pygoscelis papua*) at a colony exposed to high levels of human disturbance. *Polar Biology* 21, 355–360.

Codling, R.J. (1982) Sea-borne tourism in the Antarctic: an evaluation. *Polar Record* 21(130), 3–9.

Codling, R.J. (1995) The precursors of tourism in the Antarctic. In: Hall, C.M. and Johnson, C.E. (eds) *Polar Tourism: Tourism in the Arctic and Antarctic Regions.* John Wiley & Sons, Chichester, UK, pp. 167–177.

Codling, R.J. (2001) Wilderness and aesthetic values in the Antarctic. *Polar Record* 37(203), 337–352.

Commonwealth of Australia (1989) *Tourism in the Antarctic. Report of the House of Representatives Standing Committee on Environment, Recreation and the Arts.* Australian Government Publishing Service, Canberra.

Crosbie, P.K. (1998) Monitoring and management of tourist landing sites in the Maritime Antarctic. PhD thesis, University of Cambridge, Cambridge, UK.

Cruwys, E. and Davis, P.B. (1994) Southern elephant seal numbers during moult on Livingston Island, South Shetland Islands. *Polar Record* 30(175), 313–314.

Cruwys, E. and Davis, P.B. (1995) The effect of local weather conditions on the behaviour of moulting elephant seals *Mirounga leonina* (L). *Polar Record* 31(179), 427–430.

Cruwys, E. and Davis, P.B. (1996a) Moulting juvenile male Southern elephant seals *Mirounga leonina* (L) at Hannah Point, Walker Bay, Livingston Island, South Shetland Islands. *Polish Polar Research* 14(3), 1–5.

Cruwys, E. and Davis, P.B. (1996b) Moulting southern elephant seals, *Mirounga leonina* (L) and local weather conditions on Livingston Island, South Shetlands. *Polish Polar Research* 15(3–4), 133–144.

Culik, B., Adelung, D. and Woakes, A.J. (1990) The effect of disturbance on the heart rate and behaviour of Adélie penguins (*Pygoscelis adeliae*) during the breeding season. In: Kerry, K.R. and Hempel, G. (eds) *Antarctic Ecosystems: Ecological Change and Conservation.* Springer-Verlag, Berlin, pp. 177–182.

Davis, B. (1999) Management requirements for tourist landing sites in the Maritime Antarctic, and a model plan for Deception Island, South Shetland Islands. MPhil thesis, University of Cambridge, Cambridge, UK.

Davis, P.B. (1992) Planning for a changing environment: administration and management of South Georgia. MPhil thesis, University of Cambridge, Cambridge, UK.

Davis, P.B. (1995a) Wilderness visitor management and Antarctic tourism. PhD thesis, University of Cambridge, Cambridge, UK.

Davis, P.B. (1995b) Antarctic visitor behaviour: are guidelines enough? *Polar Record* 31(178), 327–334.

Davis, P.B. (1998) Understanding visitor use in Antarctica: the need for site criteria. *Polar Record* 34(188), 45–52.

De Leeuw, C. and Aptroot, A. (1998) The lichen and bryophyte vegetation of Cuverville Island, Antarctica. *Nova Hedwigia* 67(3–4), 469–480.

Dodds, K. (1997) *Geopolitics in Antarctica: Views from the Southern Oceanic Rim.* John Wiley & Sons, Chichester, UK.

Emslie, S.L. (1997) Natural and human-induced impacts to seabird productivity and conservation in Antarctica: a review and perspectives. In: De Poorter, M. and Dalziell, J.C. (eds) *Cumulative Impacts in Antarctica: Minimization and Management.* World Conservation Union, Washington, DC.

Enzenbacher, D.J. (1991) A policy for Antarctic tourism: conflict or cooperation? MPhil thesis, University of Cambridge, Cambridge, UK.

Enzenbacher, D.J. (1992a) Tourism in Antarctica: numbers and trends. *Polar Record* 28(164), 17–22.

Enzenbacher, D.J. (1992b) Tourism in Antarctica: an overview. In: Kempf, C. and Girard, L. (eds) *Tourism in Polar Areas: Proceedings of the First International Symposium in Colmar, France, 21–23 April 1992.* Conseil General du Haut Rhin, Colmar, France.

Enzenbacher, D.J. (1992c) Antarctic tourism and environmental concerns. *Marine Pollution Bulletin* 25(9–12), 258–265.

Enzenbacher, D.J. (1993a) Tourists in Antarctica: numbers and trends. *Tourism Management* 14(2), 142–146.

Enzenbacher, D.J. (1993b) Antarctic tourism: 1991/92 season activity. *Polar Record* 29(170), 240–242.

Enzenbacher, D.J. (1994a) Antarctic tourism: an overview of 1992/93 season activity, recent developments and emerging issues. *Polar Record* 30(173), 105–116.

Enzenbacher, D.J. (1994b) NSF and Antarctic tour operators' meetings. *Annals of Tourism Research* 21(2), 424–427.

Enzenbacher, D.J. (1994c) Tourism at Faraday Station: an Antarctic case study. *Annals of Tourism Research* 21(2), 303–317.

Enzenbacher, D.J. (1995a) The regulation of Antarctic tourism. In: Hall, C.M. and Johnston, M.E. (eds) *Polar Tourism: Tourism in the Arctic and Antarctic Regions*. John Wiley & Sons, Chichester, UK, pp. 179–215.

Enzenbacher, D.J. (1995b) The management of Antarctic tourism: environmental issues, the adequacy of current regulations and policy options within the Antarctic Treaty System. PhD thesis, University of Cambridge, Cambridge UK.

The Explorer's Journal (1957) First commercial flight to Antarctica. *The Explorer's Journal* 35(1), 18–19.

Fraser, C. (1986) *Beyond the Roaring Forties: New Zealand's Subantarctic Islands*. Government Printing Office, Wellington, New Zealand.

Fraser, W.R. and Patterson, D.L. (1997) Human disturbance and long-term changes in Adélie penguin populations: a natural experiment at Palmer Station, Antarctica. In: Battaglia, B., Valencia, J. and Walton, D.W.H. (eds) *Antarctic Communities: Species, Structure and Survival*. Cambridge University Press, Cambridge, UK, pp. 445–452.

Galimberti, D. (1991) *Antarctica: An Introductory Guide*. Zagier and Urruty Publications, Miami Beach, Florida.

Giese, M. (1996) Effects of human activity on Adélie penguin *Pygoscelis adeliae* breeding success. *Biological Conservation* 75, 157–164.

Giese, M. (1998) Guidelines for people approaching breeding groups of Adélie penguins. *Polar Record* 34(191), 287–292.

Girard, L. (1995) Le tourism dans les régions polaires: du rêve à la réalité. *Espaces* 132(Mars–Avril), 39–44.

Hall, C.M. (1992) Tourism in Antarctica: activities, impacts and management. *Journal of Travel Research* 30(4), 2–9.

Harris, C.M. (1991a) Environmental effects of human activities on King George Island, South Shetland Islands, Antarctica. *Polar Record* 27(162), 193–204.

Harris, C.M. (1991b) Environmental management on King George Island, South Shetland Islands, Antarctica. *Polar Record* 27(163), 313–324.

Harris, C.M. and Meadows, J. (1992) Environmental management in Antarctica: instruments and institutions. *Marine Pollution Bulletin* 25(9–12), 239–249.

Headland, R.K. (1989) *Chronological List of Antarctic Expeditions and Related Historical Events*. Cambridge University Press, Cambridge, UK.

Headland, R.K. (1994) Historical development of Antarctic tourism. *Annals of Tourism Research* 21(2), 269–280.

Headland, R.K. and Keage, P. (1985) Activities on the King George island group, South Shetland Islands, Antarctica. *Polar Record* 22(140), 475–484.

Headland, R.K. and Keage, P.L. (1995) Antarctic tourist day-flights. *Polar Record* 31(178), 347.

Holdgate, M.W. (ed.) (1970) *Antarctic Ecology*, two volumes. Academic Press, London.

Heap, J. (ed.) (1994) *Handbook of the Antarctic Treaty System*, 8th edn. US Department of State, Washington, DC.

IAATO (1991) Antarctica tour operators form association. *Press release* 29 August. International Association of Antarctica Tour Operators, Basalt, Colorado.

IUCN (1991) *A Strategy for Antarctic Conservation*. International Union for the Conservation of Nature and Natural Resources, Gland, Switzerland and Cambridge, UK.

Johnston, M.E. (1997) Polar tourism regulation strategies: controlling visitors through codes of conduct and legislation. *Polar Record* 33(184), 11–18.

Johnston, M.E. (1998) Evaluating the effectiveness of visitor regulation strategies for polar tourism. *Polar Record* 34(188), 25–30.

Johnston, M.E. and Hall, C.M. (1995) Visitor management and the future of tourism in polar regions. In: Hall, C.M. and Johnston, M.E. (eds) *Polar Tourism: Tourism in the Arctic and Antarctic Regions*. John Wiley & Sons, Chichester, UK, pp. 297–313.

Jones, C.S. (1998) Predictive tourism models: are they suitable in the polar environment? *Polar Record* 34(190), 197–202.

Kennicutt, M.C. II (1990) Oil spillage in Antarctica. *Environmental Science and Technology* 24(5), 620–624.

Kennicutt, M.C. and Sweet, S.T. (1992) Hydrocarbon contamination on the Antarctic Peninsula: the Bahia Paraiso – two years after the spill. *Marine Pollution Bulletin* 25(9–12), 303–306.

Levinson, J.M. and Ger, E. (eds) (1998) *Safe Passage Questioned: Medical Care and Safety for the Polar Tourist*. Cornell Maritime Press, Centreville, Maryland.

Lucas, M. (1996) *Antarctica*. New Holland Publishers, London.

McDonald, E.A. (1967) Antarctic tourism in 1967. *United States Antarctic Journal* 2(3), 82–83.

Mahon, P.T. (1981) *Report of the Royal Commission to Enquire into the Crash on Mount Erebus, Antarctica, of a DC10 Aircraft operated by Air New Zealand Ltd*. Government Printer, Wellington.

Mahon, P. (1984) *Verdict on Erebus*. Collins, Auckland, New Zealand.

Mellor, M. and Swithinbank, C. (1989) *Airfields on Antarctic Glacier Ice*. CRREL Report 89-21. National Science Foundation, Washington, DC.

Minbashian, J. (1997) Biological integrity: an approach to monitoring human disturbance in the Antarctic Peninsula region. MPhil thesis, University of Cambridge, Cambridge, UK.

Naveen, R. (1996) Human activity and disturbance: building an Antarctic site inventory. In: Ross, R., Hofman, R.E. and Quetin, L. (eds) *Foundations for Ecosystem Research in the Western Antarctic Peninsula Region*. American Geophysical Union, Washington, DC, pp. 389–400.

Naveen, R. (1997) *The Oceanites Site Guide to the Antarctic Peninsula*. Oceanites, Chevy Chase, Maryland.

Naveen, R., de Roy, T., Jones, M. and Monteith, C. (1989) Antarctic traveller's code. *Antarctic Century* 4(July–October).

Naveen, R., Monteath, C., de Roy, T. and Jones, M. (1990) *Wild Ice: Antarctic Journeys*. Smithsonian Institution, Washington, DC.

Naveen, R., Forrest, S.C., Dagit, R.G., Blight, L.K., Trivelpiece, W.Z. and Trivelpiece, S.G. (2000) Censuses of penguin, blue-eyed shag, and southern giant petrel populations in the Antarctic Peninsula region, 1994–2000. *Polar Record* 36(199), 323–334.

Naveen, R., Forrest, S.C., Dagit, R.G., Blight, L.K., Trivelpiece, W.Z. and Trivelpiece, S.G. (2001) Zodiac landings by tourist ships in the Antarctic peninsula region, 1989–99. *Polar Record* 37(201), 121–132.

Nicholson, I.E. (1985) Antarctic tourism – the need for a legal regime? In: Wulfrum, R. (ed.) *Antarctic Challenge*. Duncker and Humblot, Berlin, pp. 191–203. Reprinted in Jørgensen-Dahl, A. and Østreng, W. (eds) (1986) *The Antarctic Treaty System in World Politics*. Macmillan, Basingstoke, UK, pp. 415–427.

Nimon, A.J. (1992) Human–animal interaction in the Antarctic: an animal behaviour approach to human disturbance of penguin colonies. MPhil thesis, University of Cambridge, Cambridge UK.

Nimon, A.J. (1995) Responses of gentoo penguins (*Pygoscelis papua*) to humans in the Antarctic. *International Society for Anthrozoology Newsletter* 9, 7–8.

Nimon, A.J. (1997) Gentoo penguin responses to humans. PhD thesis, University of Cambridge, Cambridge UK.

Nimon, A.J. and Stonehouse, B. (1995) Penguin responses to humans in Antarctica: some issues and problems in determining disturbance caused by visitors. In: Dann, P., Norman, I. and Reilly, P. (eds) *The Penguins: Ecology and Management. Proceedings of the 2nd International Conference on Penguins*. Surrey Beatty, Sydney, New South Wales, pp. 420–439.

Nimon, A.J., Oxenham, R.K.C., Schroter, R.C. and Stonehouse, B. (1994) Measurement of resting heart rate and respiration in undisturbed and unrestrained incubating Gentoo penguins (*Pygoscelis papua*). *Journal of Physiology, London* 481, 57–58P.

Nimon, A.J., Schroter, R.C. and Stonehouse, B. (1995) Heart rate of disturbed penguins. *Nature* 374(6521), 415.

Nimon, A.J., Schroter, R.C. and Oxenham, R.K.C. (1996) Artificial eggs: measuring heart rate and effects of disturbance in nesting penguins. *Physiology and Behaviour* 60(3), 1019–1022.

Parsons, A. (1987) *Antarctica: The Next Decade*. Cambridge University Press, Cambridge, UK.

Patterson, D.L., Easter-Pilcher, A.L. and Fraser, W.R. (2001) The effects of human activity and environmental variability on long-term changes in Adèlie penguin populations at Palmer Station, Antarctica. In: Huiskes, A.H.L., Gieskes, W.W.C., Rozema, J., Schorno, R.M.L., van der Vies, S.M. and Wolff, W.J. (eds) *Antarctic Biology in a Global Context*. Backhuys, Leiden, The Netherlands, pp. 301–307.

Pfeiffer, S. and Peter, H.-U. (2004) Ecological studies toward the management of an Antarctic landing site (Penguin Island, South Shetland Islands). *Polar Record* 40(215), 345–353.

Poncet, S. and Crosbie, K. (2005) *A Visitor's Guide to South Georgia*. Wildguides Ltd, Old Basing, UK.

Potter, N. (1970) The Antarctic: any economic future? *Science and Public Affairs* 26(10), 94–99.

Pyne, J. (1986) *The Ice: A Journey to Antarctica*. Arlington Books, London.

Regel, J. and Pütz, K. (1995) Effect of human disturbance on body temperature and energy expenditure in penguins. *Polar Biology* 18, 246–253.

Reich, R.J. (1979) Tourism in the Antarctic: its present impacts and future developments. Dissertation for the Diploma in Polar Studies, Scott Polar Research Institute, University of Cambridge, Cambridge, UK.

Reich, R.J. (1980) The development of Antarctic tourism. *Polar Record* 20(126), 203–214.

Riffenburgh, B. (1998) Impacts on the Antarctic environment: tourism vs government programmes. *Polar Record* 34(190), 193–196.

Rubin, J. (2000) *Antarctica*. Lonely Planet Publications, Melbourne, Victoria.

Sage, B. (1985) Conservation and exploitation. In: Bonner, W.N. and Walton, D.W.H. (eds) *Key Environments: Antarctica*. Pergamon Press, Oxford, UK, pp. 351–369.

Sage, B. and Hosking, E. (1982) *Antarctic Wildlife*. Croom Helm, London.

Salwicka, K. and Stonehouse, B. (2000) Visual monitoring of heartbeat and respiration in Antarctic seals. *Polish Polar Research* 21(3), 189–197.

SCAR (1982) *A Visitor's Introduction to the Antarctic and Its Environment*. Scientific Committee on Antarctic Research, Cambridge, UK.

SCAR (1984) *A Visitor's Introduction to the Antarctic and Its Environment*. British Antarctic Survey/Scientific Committee on Antarctic Research, Cambridge, UK.

Snyder, J. and Shackleton, K. (1986) *Ship in the Wilderness: Voyages of MS Lindblad Explorer Through the Last Wild Places on Earth*. Dent and Sons, London.

Soper, T. (1994) *Antarctica, A Guide to the Wildlife*. Bradt Publications, Chalfont St Peter, UK.

Splettstoesser, J. and Folks, M.C. (1994) Environmental guidelines for tourism in Antarctica. *Annals of Tourism Research* 21(2), 231–244.

Stonehouse, B. (1963) Observations on Adelie penguins *Pygoscelis adeliae* at Cape Royds, Antarctica. *Proceedings of the 13th International Ornithological Congress*. Louisiana State University/American Ornithologists' Union, Baton Rouge, Florida, pp. 766–779.

Stonehouse, B. (1965a) Too many tourists in Antarctica? *Animals* (17), 12–16.

Stonehouse, B. (1965b) Counting Antarctic animals. *New Scientist* 25, 273–276.

Stonehouse, B. (1965c) Animal conservation in Antarctica. *New Zealand Science Review* 23(1), 3–7.

Stonehouse, B. (1967) Penguins in high latitudes. *Tuatara* 15(3), 129–132.

Stonehouse, B. (1990) A traveller's code for Antarctic visitors. *Polar Record* 26(156), 56–58.

Stonehouse, B. (1992) Monitoring shipborne visitors in Antarctica: a preliminary field study. *Polar Record* 28(166), 213–218.

Stonehouse, B. (1993) Shipborne tourism in Antarctica: Project Antarctic Conservation studies 1992/93. *Polar Record* 29(171), 330–332.

Stonehouse, B. (1994a) Ecotourism in Antarctica. In: Cater, E.A. and Lowman, G.A. (eds) *Ecotourism – A Sustainable Option?* John Wiley & Sons, Chichester, UK, pp. 195–212.

Stonehouse, B. (1994b) Tourism and protected areas. In: Lewis Smith, R.I., Walton, D.W.H. and Dingwall, P.R. (eds) *Improving the Antarctic Protected Areas*. International Union for the Conservation of Nature and Natural Resources, Cambridge, UK, pp. 76–83.

Stonehouse, B. (1995) *Management Recommendations for Visitor Sites in the Antarctic Region*. Scott Polar Research Institute, Cambridge, UK.

Stonehouse, B. (1996) Arctic and Antarctic tourism: can one learn from the other? In: Lange, M.A. (ed.) *Proceedings of the Arctic Opportunities Conference, Rovaniemi, Finland, September 1994*. Arctic Centre Reports 22. University of Lapland, Rovaniemi, Finland, pp. 347–356.

Stonehouse, B. (1998) Polar shipborne tourism: do guidelines and codes of conduct work? In: Humphreys, B.H., Pedersen, Å.Ø., Prokosch, P.P. and Stonehouse, B. (eds) Linking Tourism and Conservation in the Arctic. Proceedings from Workshops in 20–22 January 1996 and 7–10 March 1997 in Longyearbyen, Svalbard. *Norsk Polarinstitutt Meddelelser* 159, 49–58.

Stonehouse, B. (1999) Antarctic shipborne tourism: facilitation and research at Arctowski Station, King George Island. *Polish Polar Research* 20(1), 65–75.

Stonehouse, B. (2000) *The Last Continent: Discovering Antarctica*. SCP Books, Burgh Castle, UK.

Stonehouse, B. (2005) *Antarctica from South America*. SCP Books, Burgh Castle, UK.

Stonehouse, B. and Brigham, L. (2000) The cruise of *MS Rotterdam* in Antarctic waters, January 2000. *Polar Record* 36(199), 347–349.

Stonehouse, B. and Crosbie, P.K. (1995) Tourist impacts and management in the Antarctic Peninsula area. In: Hall, C.M. and Johnson, C.E. (eds) *Polar Tourism: Tourism in the Arctic and Antarctic Regions.* John Wiley & Sons, Chichester, UK, pp. 217–233.

Stonehouse, B., Crosbie, P.K. and Girard, L. (1995) Sustainable tourism in the Arctic and Antarctic. *Insula: International Journal of Island Affairs* 4(1), 24–31.

Swithinbank, C.W. (1988) Antarctic Airways: Antarctica's first commercial airline. *Polar Record* 24(151), 313–316.

Swithinbank, C.W. (1989) Non-governmental aircraft in the Antarctic 1988/89. *Polar Record* 25(154), 254.

Swithinbank, C.W. (1990) Non-governmental aircraft in the Antarctic 1989–90. *Polar Record* 26(159), 316.

Swithinbank, C.W. (1992a) Non-government aircraft in the Antarctic 1990–91. *Polar Record* 28(164), 66.

Swithinbank, C.W. (1992b) Non-government aircraft in the Antarctic 1991–92. *Polar Record* 28(166), 232.

Swithinbank, C.W. (1993a) Airborne tourism in the Antarctic. *Polar Record* 29(169), 103–110.

Swithinbank, C.W. (1993b) Non-government aircraft in the Antarctic 1992–93. *Polar Record* 29(170), 244–245.

Swithinbank, C.W. (1994) Non-government aircraft in the Antarctic 1993–94. *Polar Record* 30(174), 221.

Swithinbank, C.W. (1995) Non-government aircraft in the Antarctic 1994–95. *Polar Record* 31(178), 346.

Swithinbank, C.W. (1996) Non-government aircraft in the Antarctic 1995–96. *Polar Record* 32(183), 355–356.

Swithinbank, C.W. (1997a) New intercontinental air route: Cape Town to Antarctica. *Polar Record* 33(186), 243–244.

Swithinbank, C.W. (1997b) Non-government aircraft in the Antarctic 1996–97. *Polar Record* 33(187), 341.

Swithinbank, C.W. (1998) Non-government aviation in Antarctica 1997/98. *Polar Record* 34(190), 249.

Swithinbank, C.W. (2000) Non-government aviation in Antarctica 1998/99. *Polar Record* 36(196), 51.

Swithinbank, C.W. (2001) Non-government aviation in Antarctica 1999/2000. *Polar Record* 36(198), 249.

Taylor, R.H. (1962) The Adelie penguin *Pygoscelis adeliae* at Cape Royds. *Ibis* 104, 176–204.

Thomson, R.B. (1977) Effects of human disturbance on an Adelie penguin rookery and measures of control. In: Llano, G.A. (ed.) *Adaptations within Antarctic Ecosystems: Proceedings of the Third SCAR Symposium on Antarctic Biology.* Smithsonian Institution, Washington, DC, pp. 1177–1180.

Triggs, G.D. (1986) *International Law and Australian Sovereignty in Antarctica.* Legal Book Pty, Ltd, Sydney, New South Wales.

Waugh, S.M. (1994) Monitoring and management of Antarctic and sub-Antarctic tourist sites: a GIS case study. MPhil thesis, University of Cambridge, Cambridge, UK.

Woehler, E.J., Penney, R.L., Creet, S.M. and Burton, H.R. (1994) Impacts of human visitors on breeding success and long-term stability in Adélie penguins at Casey, Antarctica. *Polar Biology* 14, 269–274.

Wilson, R.P., Culik, B.M., Danefield, R. and Adelung, D. (1991) People in Antarctica: how much do penguins care? *Polar Biology* 11, 363–370.

Young, E.C. (1990) Long-term stability and human impact in Antarctic skuas and Adélie penguins. In: Kerry, K.R. and Hempel, G. (eds) *Antarctic Ecosystems: Ecological Change and Conservation.* Springer-Verlag, Berlin, pp. 231–236.

Zuckerman, A. (1985) *A Voyage to Adventure: Antarctica.* Eye on the World, Los Angeles, California.

IV Managing the New Realities: Introduction

JOHN M. SNYDER AND BERNARD STONEHOUSE

Prospects for Polar Tourism ends with a section of four chapters, gathered under the heading 'Managing the New Realities', that are concerned primarily with management – the management of a boisterous industry in a range of remarkable and sensitive wilderness areas of the world. In Chapter 14 John Snyder introduces some of the principles involved in managing polar wilderness, in particular wilderness that is open to public usage. He points out that this is not a new departure: the USA especially has well over a century's experience of managing wilderness for recreational purposes. While polar regions provide special problems, those problems are neither insuperable nor even particularly difficult, once the underlying principles have been grasped. His conclusion – that polar tourism depends largely on tour operators and the tourists themselves for conservation its resources, but still requires firm management objectives to be set outside the industry – applies equally to both polar regions.

In Chapter 15 John Snyder and Bernard Stonehouse review tourism on South Georgia, a fringe Antarctic island notable for its scenic beauty, abundant wildlife, an unusual legacy of whaling stations (left derelict since their closure in the 1960s) and an enterprising government that is interested to develop it – with due caution – as a tourist attraction. The island's many resources require a special approach to management, for which these authors propose a five-phase multiple resource management strategy.

In Chapter 16 Phillip Tracey, of the Australian Government Antarctic Division, draws attention to the 23 isolated islands and island groups of the southern oceans, all of which are claimed by sovereign states and managed more or less for their wilderness values. Within the past few years all – even the most remote and inaccessible – have been visited by ship-borne tourists, a fact that those who claim responsibility for them must now take into account while considering their further management. Tracey outlines the various ways in which governments have responded to the threat (or promise) of tourism, illustrating points of relevance to the broader issues of management of polar and sub-polar wilderness areas.

In Chapter 17 Esther Bertram and Bernard Stonehouse return to the question of Antarctica – a wilderness area that, unique among its kind, is managed under an international treaty rather than a sovereign government. The merits of the Antarctic Treaty as an experiment in peacekeeping and international accord are beyond dispute. Whether it is equally successful in managing tourism – the lively and burgeoning industry that has appeared on its doorstep – is the subject of this chapter. The authors propose relatively simple solutions to complexities that Treaty delegates have generated.

14 Managing Polar Tourism: Issues and Approaches

Strategic Studies, Inc., 1789 E. Otero Avenue, Centennial, CO 80122, USA

Introduction

> We have met the enemy and he is us.
>
> (Walt Kelly, *Pogo*)

Responsible stewardship of a region, defined as the conservation of its natural and cultural resources, requires the creation and implementation of management techniques. In polar regions as elsewhere, a primary management goal is to protect resources from loss or damage. Issues regarding the tourist's 'appropriate or allowable uses' of the polar regions require careful consideration.

Fortunately, the opportunity exists to learn from more than a century of polar tourism experiences. Historical patterns of visitor behaviour, industry practices, jurisdictional responses, economic strategies, and expressions of community opinions and expectations provide valuable knowledge upon which tourism management plans can not only be produced, but also evaluated. The purposes of this chapter are to identify critical issues that characterize the polar regions' management settings and briefly discuss the prominent approaches now employed to respond to these issues.

Managing Wilderness

Wilderness is defined as places where human presence and development are virtually absent or not readily apparent. The polar regions contain the world's largest expanses of land and marine wilderness. The continent of Antarctica is entirely a wilderness land mass. North America's and Europe's largest wilderness regions are located in the Arctic. Comprehensive sets of marine charts and hydrographic information for the Southern and Arctic oceans are not yet complete. These enormous land and marine wilderness regions provide permanent habitat for highly adapted indigenous wildlife species and seasonal habitat for immense populations of migratory wildlife. The Arctic is the homeland of native peoples who have practised cultural traditions for millennia. The Antarctic contains important artefacts of the history of exploration and scientific discovery. For most of their existence, these vast wilderness regions, their people and their historical artefacts have been protected by their remoteness and climatic conditions. But since the end of World War II, tourism has ventured steadily into all of these once inaccessible regions.

Wilderness regions are exceedingly difficult to manage. Comprehensive inventories of their natural and cultural resources are time-consuming and expensive to obtain. A competent knowledge of their dynamic ecological systems requires long-term investigations that are equally complicated and costly to accomplish. Establishing mutually acceptable methods to facilitate stakeholder participation in wilderness planning and management are difficult to implement, and frequently contentious (Wright, 2001).

Governments throughout the world have responded to wilderness management challenges in a variety of ways (Loomis, 1993). The range of wilderness management techniques generally extends from comparatively open approaches that combine resource inventories and environmental assessments with public participatory processes. One example of this type of wilderness management approach is called Limits of Acceptable Change (Lucas, 1985; Stankey and McCool, 1985). At the other end of the management spectrum are laws that prohibit or severely restrict public access to wilderness areas. The goal of these stringent measures is to achieve preservation by preventing human entry. But it must also be recognized that at some point in time all wilderness areas, even Antarctica, have experienced various types and intensities of human activity. The wilderness management paradox is to determine when, where and how people will be allowed to use these areas without destroying their natural character. In other words, what human activity should be permitted in an area universally defined as having no human presence?

The philosophical challenges of wilderness management are matched by equally impressive operational constraints. The absence of development may be essential for sustaining the wilderness character of a region, but it is also a major obstacle in the performance of environmental and tourism management responsibilities. The lack of infrastructure, such as transportation systems, telecommunications and support facilities, directly impacts critical management functions. Specifically, essential operational functions such as resource monitoring, scientific research, security patrolling, visitor safety and emergency response capabilities are all affected by the lack of infrastructure. All of these management duties need to be performed to ensure wilderness resource protection.

Jurisdictional Authority and Enforcement Capability

International treaties and agreements, national laws and regulations, territorial jurisdictions and sanctioned traditional resource uses form a complex network of laws, regulations and customs that define the allowable uses of the Arctic's resources. Ostensibly, all human activity, wildlife management, land uses and maritime activities in the Arctic are subject to, and governed by, this network of treaties, laws and customs. In reality, however, the effectiveness of these legal mechanisms is entirely dependent upon the enforcement capabilities of each jurisdiction.

The sparsely populated Arctic region is governed by eight sovereign nations who strenuously endeavour to manage their vast territorial lands and waters with very limited enforcement capabilities. To a certain extent, their efforts are reinforced by a collection of international treaties that support the conservation of Arctic resources, especially fisheries and wildlife populations. But overall, their planning and management capacities are restricted by extremely limited human and financial resources. The allocation of these scarce resources in the Arctic results from difficult decision-making processes in which human needs, such as health care, education, housing and transport, compete with wildlife management, cultural preservation and environmental conservation.

One resource management technique increasingly used by Arctic jurisdictions for environmental conservation is the legal designation of protected areas. These areas include National Parks, Nature Reserves and Wildlife Refuges, National Forests, Wilderness Areas, Marine Sanctuaries and World Heritage Sites. Important tourism management implications result from these designations. In a positive sense, the establishment of a protected area provides jurisdictional authority for its perpetual conservation, prohibits or minimizes destructive uses, and focuses world attention on its environmental significance. But it must also be noted that protected areas immediately attain high visibility as tourist attractions and this requires management commitments to sustain their integrity. The establishment of a protected area does not, by itself, provide resource protection from tourist activity. The creation of this type of premier tourist attraction also requires the creation of management practices best suited to its conservation. The Arctic's recognition of this need is evidenced by the several national park systems, and other government agencies, now working on behalf of protected areas.

International treaties play an important role in the management of polar resources. Their scope ranges from the protection and management of a single wildlife species, such as the polar bear (*Ursus maritimus*), to the conservation of an entire continent, Antarctica. Several international commissions are tasked with the responsibility of monitoring the diverse wildlife species that reside or migrate to polar regions, particularly fishery stocks and marine mammals. Other international consortia comprised of government, research institutions and non-governmental organizations (NGOs) cooperatively monitor and evaluate the environmental conditions of the polar regions' oceanographic, atmospheric and land resources. National park and other types of protected area management agencies in the Arctic have a history of positive international cooperation. Although the law enforcement capabilities of the treaty parties are frequently stretched thin, all of these international endeavours represent valuable precedents for conserving the resources of the polar regions.

To date, the only international treaty that addresses the management of polar tourism is the Antarctic Treaty. In 1991, the Antarctic Treaty parties adopted the Protocol on Environmental Protection to the Antarctic Treaty which sets out principles, procedures and obligations for the comprehensive protection of Antarctica. The purpose of the protocol is to protect the Antarctic environment and its scientific and aesthetic values. In 1994 the parties adopted Recommendation XVIII-I: Guidance for Visitors to the Antarctic, which defines both acceptable visitor behaviour and tourism operations in terms of 'best management practices'. Numerous forms of visitor behaviour are prescribed. Tourist admonitions range from environmental and heritage protection to personal safety. Respect for Protected Areas and scientific research is emphasized. Techniques for the protection of wildlife and the conservation of environmental conditions are explicitly enumerated, and appropriate methods for conducting tourism operations are advocated.

Although Antarctic tourism management issues are specifically agreed upon by the international community, enforcement remains a challenge. The Antarctic has no permanent inhabitants to either monitor conditions or enforce regulations. There is no tax revenue base to pay for resource management. And the international treaty itself necessarily involves the use of a protracted, consensus-based decision-making process to influence any form of tourism management. The enforcement situation is further complicated by the fact that tourist ships and aircraft are registered in countries that have demonstrated either little inclination or capacity to prosecute alleged wrong-doing in the distant southern continent. Collectively, these circumstances represent extraordinary difficulties for either creating or adequately enforcing resource and visitor management regimes in Antarctica.

Because the Treaty parties are unable to commit substantial law enforcement resources to Antarctica, adherence to the guidelines is the operational duty of the tour operators. These duties have been assumed and responsibly performed by the International Association of Antarctica Tour Operators (IAATO). It must also be recognized, however, that IAATO has no jurisdictional authority and can only report alleged violations, rather than prosecute them. IAATO 'enforces' the guidelines by means of educational techniques designed to promote ethical behaviour and resource conservation among its passengers, staffs and crews. IAATO also strongly advocates the employment of scientists, polar experts and well-trained cruise directors qualified to monitor Antarctica's environmental conditions. To date, these personnel have contributed to both environmental protection and visitor safety, but admittedly client satisfaction is their primary responsibility.

Operational difficulties also extend to the severe shortage of tourism support facilities in the Antarctic. The only facility and infrastructure goal

TITLE

Contents

- Geographic/auto rdbc
- Climate change in the world
- impacts
- who observes?
- any other problems?
- future?
- Class activity

Love

Love overcom 8 hate

Love has no colour

Lot free to on'wahen h o.

all islove

Adam Lambert

developments permitted in Antarctica are associated with multi-national scientific stations. Given the complete absence of any other form of human settlement, the scientific stations have become surrogate tourism support facilities. They derive this unsolicited role from the facts that they possess the only emergency response capabilities on the continent, store the only provisions, possess transport and communication resources, and offer the only human shelter to be found in the Antarctic. Given the scarcity of these essential resources, and their obvious scientific attraction, the stations have become reluctant partners in the Antarctic tourism business.

Determining Environmental Cause and Effect – Who Did What?

The environmental conditions of the polar regions are extremely dynamic and, to a considerable extent, not fully understood. They are simultaneously experiencing significant climatic, plant succession, marine and wildlife behavioural changes. Wildlife populations that seasonally migrate to the polar regions from other regions of the world are impacted by changing oceanographic conditions, environmental pollution, hunting and fishing pressures, and habitat transformations that are well beyond the jurisdictional boundaries and management control of polar resource agencies. The cumulative impacts of these complex environmental changes are a colossal challenge to comprehend, much less competently manage. When polar tourism is mixed with this collection of dynamic, naturally occurring events, then a reliable understanding of these interdependencies is further complicated (Watson *et al.*, 1998).

Polar tourism must be conducted in a responsible manner and some form of tourism management is often suggested for conserving the resources of the polar regions. One of the supreme difficulties in accomplishing this objective is to accurately attribute environmental cause and effect and then manage tourism accordingly. In other words, when an environmental condition changes was it caused by a natural event, was it altered by tourism activity, or by some combination of both? The capacity to accurately monitor tourism, revise tourism management plans, alter visitor activities and behaviour, and implement appropriate environmental conservation measures depends on the answers to these questions. Realistically, the ability to assess and manage tourism impacts in the polar regions is vitally dependent on a competent understanding of these relationships (Cater and Lowman, 1994; Mieczkowski, 1995). The creation and modification of tourism management plans requires that knowledge.

Tourism Management from a Cultural Perspective

Numerous cultural traditions pervade the Arctic, many of them derived from centuries of habitation by indigenous peoples. Others are the culmination of settlement patterns, resource uses, economic systems and social customs evolved from empire building and the desires of sovereign nations. The heritage and cultural traditions of Arctic societies are simultaneously tourist attractions and sensitive tourism management issues.

A fundamental Arctic tourism management goal is to achieve balance between the public, commercial display of cultural features and the preservation of cultural integrity. Each cultural group has its own tolerance level for visitation and willingness to share its resources. Individual Arctic societies define acceptable visitor behaviour in terms of the types, magnitude, geographic location and seasons of resource use. Consequently, tourism management in the Arctic is subject to a set of 'host conditions' in which cultural preservation, economic necessity and scarce natural resources are continuously debated. Perhaps the foremost topic among these intense discussions is the willingness to share cultural traditions and art forms. The conservation of ancestral homelands, scarce resources and vulnerability to irreversible change are all significant tourism management issues facing Arctic indigenous societies (Dressler *et al.*, 2001). But the context within which all of these discussions occur is the certain knowledge that Arctic societies have very limited economic development alternatives. Given the fact that virtually all of these alternatives involve natural resource exploitation, Arctic societies are keenly aware of their vulnerability to change.

In the Antarctic and sub-polar islands of the southern hemisphere the 'allowable and acceptable' visitor uses of heritage sites and resources is an especially difficult task. Internationally significant heritage resources associated with polar discovery, scientific inquiry, historical economic development and human settlement are located throughout the southern polar region. When human use or settlement was abandoned, many of the sites started to deteriorate. Their abandonment, the absence of a permanent population to perform conservation and severe weather led to further deterioration. In an effort to preserve a record of these sites, several national governments, universities and historic preservation groups have made substantial efforts to produce accurate inventories and maps. Occasionally they have been able to implement environmental remediation and preservation programmes to conserve historic facilities (Basberg and Rossnes, 1993; South Georgia Association, 2005; Chapter 15, this volume).

The sub-polar islands possess sovereignty status and therefore have jurisdictional authority to implement heritage conservation programmes (Chapter 16, this volume). Various resource conservation activities have taken place, but the scarcity of financial resources and the remoteness of the sites are ever-present impediments to this effort. Antarctica's lack of clear jurisdictional authority and insufficient law enforcement result in major heritage resource management problems.

Tourism management of the Antarctic and sub-polar islands' heritage resources is an intriguing challenge. Many of the heritage sites are key attractions because of their historical significance and public notoriety. Still others are sought because of family and cultural ties to the pioneering settlers who once worked and inhabited the region's remote sites. And in a few instances, the existence of historic sealing camps was discovered by tourists who arrived to pursue wildlife viewing experiences. Tourism management responses to these demands for visitation have ranged from prohibited entry to cordial invitations. For example, despite the popularity of Sir Ernest Shackleton's exploits, tourist access is denied to Stromness, South Georgia. Meanwhile, visitor access to Grytviken, South Georgia (where 'the Boss' is traditionally toasted at his graveside) and Antarctic heritage sites in the Ross Sea are permitted with strict, but hospitable controls. In contrast, the refurbished historic structure at Port Lockroy, in the Antarctic Peninsula Region, actively promotes tourism to share its colourful heritage and to commercially support its postal concession (Hughes, 2000; Blanchette et al., 2002).

Managing Tourist Behaviour and Numbers

Certainly the most enduring criticism of tourism, and its most intractable problem, is tourist behaviour. Since tourism's inception, tourists have been universally criticized for 'inappropriate' behaviour, cultural insensitivities and even their appearance. When tourist activities result in resource damage, then condemnations and tough responses are well-deserved. In other instances, criticisms of visitor behaviour are more accurately described as expressions of opinion rather than proof of harm. An abundance of negative impacts are attributed to 'inappropriate' visitor behaviour and efforts are made to hold both the tourist and the tourism industry accountable. Governments, private industries and NGOs endeavour to control tourism and influence tourist behaviour. But ironically, in the face of constant derision and increasing regulatory constraints, tourism has evolved to become the world's largest industry. Growing numbers of international destinations fiercely compete for the wealth it generates, and tourists show no signs of limiting their pursuit of leisure activities.

Throughout the ages a considerable amount of criticism regarding tourist behaviour has been motivated by pure elitism. The earliest forms of touring were exclusively reserved to the highest strata of society. The introduction of mass tourism in the mid-1800s broke down social barriers, progressively diminished exclusivity, and began a process of increased entry to a world of leisure previously unknown and unattainable to most people. For many scholars, tourism exemplified society's evolution towards greater democratization, and it held promise for improving cultural tolerance and educational attainment (Loschberg, 1979; Feifer, 1986; Hibbert, 1987).

The issue of exclusivity raises a fundamental question concerning tourism management: should the social and economic barriers to tourism attractions be lowered or reinforced? For some, improved access to tourist opportunities is evidence of a process of social and economic equality. Others are motivated to preserve exclusivity. Regardless of how that question is answered, the legacy of tourist elitism remains strong. By the start of the 20th century, one Edwardian wit observed: 'It's funny isn't it, how every traveller is a tourist, except oneself', and as the 21st century commences, the gated recreational communities, 'members only' resorts and exorbitant prices for 'deluxe tours' effectively succeed in preserving this attitude.

The evolution of tourism also proves that increased access inevitably leads to increased numbers of tourists. That fact is a prominent and justifiable tourism management concern. In addition to efforts to influence visitor behaviour, tourism management is directly confronted with the need to respond to growing numbers of tourists, their expanding geographical distribution, longer seasons of use, greater duration of stay, and their pursuit of increasingly diverse recreational activities. The tourism manager is forced to address issues such as the exposure of environmental resources to additional risks, the economic dependencies of local communities on tourism, threats to privacy, more cultural contacts, and growing demands for infrastructure. From the tourist's viewpoint, the issue of numbers equates to perceptions of congestion, and this directly affects the quality of their leisure experience. Tourist perceptions of congestion are especially important in locations strongly promoted as wilderness. Obviously, as the number and distribution of tourists steadily increase, impacts resulting from 'inappropriate' behaviour are magnified.

Polar tourism shares this peculiar history and collection of tourism management dilemmas. Remarkably, the first promoters of the polar tourism industry were among its harshest critics. In the 1880s Ms Eliza Ruhamah Scidmore used her prestigious position as Honorary Associate Editor of the *National Geographic Magazine*, her access to prominent politicians and publishers, and her journalism credentials to extol, promote and profit from Arctic tourism. The personal, professional and monetary benefits she derived from these promotional efforts did not, however, prevent her from deriding the behaviour of her fellow tourists. A representative critique, following rhapsodic passages about the pleasure of polar attractions, stated:

> A small iceberg, drifted to shore, was the point of attack for the amateur photographers, and the Indian children marveled with open eyes at the 'long legged gun' that was pointed at the young men, who posed on perilous and picturesque points of the berg.
>
> (Scidmore, 1885)

Ms Scidmore was hardly alone in rebuking polar tourists for behaviour perceived to be 'inappropriate'. Famous contemporaries of Scidmore made comments about polar tourism that mirror today's cruise-ship experiences and tourist behaviour. One of the first tourists to write about Arctic tourism was Mrs Septima Collis, the wife of a famous general. Mrs Collis's description (1890) of a shore excursion in Glacier Bay, Alaska is reminiscent of today's polar tourism experiences:

> The next sensation in store for the tourist is the climb to the top of the glacier. All the rowboats were lowered, and about a dozen passengers in each, armed with alpenstocks (climbing sticks), were ferried in successive groups from the ship to the eastern beach, a distance of perhaps half a mile, instructions being

given to keep sharp lookout for falling icebergs. And here your trouble commences unless you are well advised.

Upon regaining her stateroom aboard the cruise ship, Mrs Collis summarily chastised the behaviour of her companions as recklessly dangerous and demeaning the beauty of the site. She recorded that: 'The thermometer was 70 degrees in the shade, and the Kodak fiends were at work everywhere preserving as best they could the counterfeit presentments of each other.' Whatever hazards, either real or imagined, the polar tourist experienced ashore were quickly relieved by the pleasantries aboard the cruise ship. In the late 1800s polar tourism experiences aboard Arctic steamships included first-class accommodations, sumptuous meals and beverages, on-board string orchestras, a barbershop and a dark room for photographers. During this era, Rudyard Kipling's comment about cruise-ship travel was: 'for sheer comfort, not to say padded sloth, the life was unequalled' (Kipling, 1920). Today, the fleets of cruise ships travelling throughout the polar regions compete with one another to happily replicate the luxurious settings that Kipling so aptly characterized.

Legitimate concerns about visitor behaviour in the polar regions persist to this day and the impacts associated with tourism are now vital topics. Reconciling the benefits and costs of polar tourism is a complex task. Host communities in the Arctic attempt to reconcile the economic benefits they obtain from tourism with the social and cultural intrusion they bear. Decisions concerning alternative types of economic development often favour tourism and the jobs and income it generates in preference to other, more extractive and less sustainable, resource development activities. But it is fully recognized that tourism causes both consumptive and non-consumptive resource uses. Natural resource management in the polar regions, especially wildlife and fisheries management, seeks to achieve difficult balances between a variety of conflicting uses. In all instances, the tourism management dilemma is nothing less than influencing the number, distribution and behaviour of tourists to achieve tolerant social practices, economic benefits and sustainable resource conservation.

Fortunately for the polar regions, polar tourists often exhibit behaviour that is generally consistent with sound tourism management principles. Specifically, polar tourists actively seek and are willing to pay premium prices for 'pristine' environmental conditions. They support a wide range of wildlife management programmes. They admire cultural tourism activities and constitute an important market for indigenous products. And they contribute time, talent and money to numerous resource conservation organizations. These positive traits are so well known to polar jurisdictions and the tourist industry that they are actively exploited as integral parts of tourism development strategies.

Self-regulation in Managing Polar Tourism

Self-regulation is now the dominant form of tourism management actively employed throughout the polar regions. Although a multitude of conservation laws, treaties and regulations exist, the functional reality is that the overwhelming majority of polar lands and seas are almost entirely reliant upon the tour industry and the tourists themselves to regulate their own operations and behaviour in order to safeguard polar resources. The most powerful methods currently available for managing polar tourism are the tour industry's respect for laws and customs, and the tourist's willingness to exercise self-restraint. The significance of these facts cannot be overstated. This remarkable situation is a direct consequence of the complexity of jurisdictional authorities, scarcity of law enforcement resources, social and economic pressures, and wilderness immensity previously described.

Self-regulation of tourist operations and behaviour is the only management technique currently available for conserving the resources of the entire continent of Antarctica. Further, the Antarctic's offshore region has no fleet of coastguard vessels to ensure the conservation of its marine resources either. The existence of Treaty protocols and annexes intended to conserve Antarctic resources by controlling 'non-Governmental' activities, i.e. tourism, are fully acknowledged by the Treaty parties and sincere efforts to comply with them are actively pursued. But actual, on-site tourism management is dependent on the following facts: (i) the tourist industry must instruct their clients about appropriate behaviour without alienating them; (ii) voluntary coordination among cruise and tour directors is the only means of managing site visits; and (iii) responses to emergency situations are completely dependent on voluntary efforts and the skills of those who, fortuitously, are willing to respond. Noticeably absent from this approach to tourism management is the ability to control the number of tourists who want to visit Antarctica.

The tourism companies that operate in Antarctica have endeavoured, both individually and collectively, to create and implement visitor guidelines that will conserve Antarctic resources. For example, individual companies such as Abercrombie and Kent distribute a flyer to each of their passengers entitled 'Guidelines for Responsible Tourism' and the several members of IAATO collectively endeavour to comply with Recommendation XVIII-I: Guidance for Visitors to the Antarctic. To date, the tour industry's maritime, shore excursion and aviation personnel have been well qualified to conduct Antarctic tours, and their practice of influencing tourist behaviour has been beneficial. But the inherent environmental conditions of Antarctica, its remoteness and, most importantly, the growing number of tourists will test the limits of this approach to tourism management. Recognizing that tourism is now the single largest and fastest-growing human activity in Antarctica, it is indeed significant that the management of tourism for a whole continent is entirely reliant on self-regulation. Anticipating that law enforcement, resource conservation and infrastructure resources will not soon be available to cope with growing numbers of Antarctic tourists, it is imperative to improve upon techniques for verifiably influencing positive visitor behaviour.

The Arctic, in sharp contrast with Antarctica, has a variety of jurisdictional authorities and has instituted several tourism management techniques to conserve its resources. But the immensity of the region and the scarcity of monitoring, enforcement and emergency response resources can impair the effectiveness of these management tools. Consequently there is, again, a significant dependency upon self-regulation to achieve tourism management in the Arctic. Especially revealing examples of this situation exists within the Arctic's most intensely managed environments, its legally established Protected Areas.

Environmental merit, aesthetic quality and political strengths are needed to legally establish a protected area. The Arctic's profusion of national parks, wildlife refuges, national forests, national marine sanctuaries, wild and scenic rivers, world heritage sites and officially designated wilderness preserves evidences great concern for the conservation of polar resources. Among all tourist attractions located throughout the Arctic, these are the sites that garner the most management attention and resources. At the time of their designation, a government agency is assigned management responsibility, including tourism management if that is permitted by the enabling legislation. From that time onwards, the government agency must compete with all other government agencies and programmes to secure its budget. Agencies responsible for the management of protected areas have confronted this reality for many years. In some instances, such as Canada, Sweden and the USA, the competition to obtain sufficient funds for protected area management has persisted for more than a century.

With a long history of protected area management in the face of scarce financial and human resources, government agencies have devised a variety of innovative tourism management techniques. In the Arctic, most of these techniques involve some form of self-regulation, because of the wilderness features that characterize many of the protected areas. The most prominent self-regulatory management techniques deserve mention. Before departure the tourists are requested to read all pertinent regulations, obtain maps, essential supplies and equipment, and relevant guidebooks. Given a scarcity of agency personnel or lack of funds for visitor centres, this information is frequently provided at entry kiosks, or by means of signage. Obviously, selecting the foreign languages to be used will directly impact the effectiveness of this information. The backcountry tourist is then requested to identify their route of travel, frequently by means of trailhead registrations. They are also requested to identify their intermediate campsites and destinations by means of backcountry permit registrations. They are instructed what clothing and equipment to possess because of the absence of any sources en route. And they are advised that their personal safety is their personal responsibility. In locations where search and rescue services are available, emergency instructions, such as communication frequencies and weather radio broadcasts, are provided with the assumption that the tourist has compatible communication equipment. Persons travelling by boat or kayak are required to possess proper vessel licenses and to have adequate safety and navigational equipment on board, but there are rarely any coastguard vessels to conduct inspections.

All of these self-regulatory techniques are enforced by means of periodic patrolling by authorized resource agency personnel. One method for accomplishing this is to locate backcountry rangers at remote sites that are seasonally popular tourist attractions. Patrol boats and aircraft are utilized by resource agencies when money for their purchase and maintenance is available. Search and rescue and emergency response capabilities vary widely in the Arctic's protected areas. Their availability is generally dependent on agency resources, proximity to communities, terrain conditions and the severity of the weather.

Parks Canada, the world's oldest national park system, has the most experience with federally sanctioned and financed tourism management. Since its first park was established in 1885, Parks Canada has exercised management control over a rapidly expanding domain. As of 2005 there were 40 National Parks, 146 National Historic Sites and two National Marine Conservation Areas located throughout the Provincial, Territorial and Nunavut regions of Canada. Many of these locations utilize the self-regulatory management techniques previously described in combination with professional park ranger supervision. This is especially true in the Canadian Arctic.

A glimpse of the resource management commitment needed to accomplish Parks Canada's management functions is revealed in its Annual Reports and Corporate Plans. From 2000 to 2004 the sites managed by Parks Canada contributed Can$1.2 billion annually to the Canadian economy, and provided 38,000 full-time equivalent jobs. The Parks Canada budget dedicated Can$155.5 million to Visitor Services (Parks Canada, 2004). Significantly, Can$99.1 million of that amount was salaries for visitor management personnel. Substantial additional monies, measured in the hundreds of millions of dollars, were allocated for park maintenance and expansion. The enormous cost of visitor management is a stark reality that even a century of financial commitment, professional experience and institutional support cannot offset. Clearly, if Parks Canada replaced its self-regulatory techniques with more direct tourist supervision, then management costs would grow substantially.

The newest sovereign entity to assume responsibility for protected area tourism management is the Russian Federation. Since its birth in 1991, the Federation has been engaged in the huge legal task of completely transforming its property ownership system. Given the immensity of that task, virtually all national and local attention has understandably been dedicated to legal designations, rather than resource management systems. Progress is being made to legally establish a diversity of protected areas that will, among several other uses, be tourist attractions.

The legal categories of the Federation's protected areas range from nationally and internationally designated UNESCO World Heritage Sites, such as Kamchatka Volcanoes, to legislation designed to protect indigenous 'traditional rights', such as the Federation's Law on Territories of Traditional Nature Use. The goal of this 2001 law is the 'defense of the age-old environment and traditional way of life of aboriginal peoples, and the maintenance of biodiversity' (Russia, 2001). The resource protections that the UNESCO Kamchatka and Law on Territories envisioned have yet to be reinforced with effective resource management support. The deadly poaching of Kamchatka bears to supply illegal markets endangers resource conservation efforts in that protected area, and there are no monies available to implement a resource conservation programme on behalf of the Law on Territories. The implication of these circumstances for tourism management in the world's largest Arctic land mass is significant. While the Russian Federation ambitiously seeks to attract tourists, it will be seriously challenged to provide even basic tourism management support because of severe financial constraints (Whelan, 2004). By default, it is probable that polar tourist behaviour in that immense Arctic region will continue to be largely self-regulating for the next several years.

The Arctic's tourism management opportunities and constraints are obviously well-known to its people, governments, conservation organizations and the tour companies that operate there. In response to a strong desire to influence the growth and development of polar tourism, a meeting of affected stakeholders was facilitated in 1995 by the World Wildlife Fund Arctic International Programme (WWF Arctic, 1996). The result of that collaborative effort was the publication in 1996 of *Linking Tourism and Conservation in the Arctic*, which contains ten principles for Arctic tourism and associated codes of conduct for both tour operators and tourists. The participatory process and its product, as described by WWF Arctic, involved:

> Representatives from local communities, governments, different sectors of the tourism industry, conservation organizations and scientific institutions used their experience to create these guidelines for arctic tourism. The principles were also adopted into Codes of Conduct for tourism business and tourists which contain more specific information on what to consider when doing business or traveling in the Arctic.

The WWF Arctic's consensus-based Principles and Codes of Conduct offer a responsible and well-intentioned approach to polar tourism management. It must also be noted that the realization of these Principles and Codes is primarily dependent on self-regulation. The ten principles constitute a very appropriate collection of resource conservation goals, the essential starting point for resource management. Environmental conservation, cultural integrity, economic benefits, visitor safety and respect for polar resources are comprehensively included in the Principles for Arctic Tourism. The ten principles are to be realized by ten Codes of Conduct for Tour Operators in the Arctic, and ten Codes of Conduct for Arctic Tourists. The direct relationships between the conservation goals and tourism practices by both operators and tourists are clearly described and represent a strength of this approach.

All of the Principles and Codes are intended to be achieved by means of a comprehensive advocacy programme. The stakeholders collectively promote the Principles and Codes through information dissemination, consumer education and personnel training. To date, they have been diligent in their efforts. Ultimately, however, the effectiveness of their endeavours, the conservation of Arctic resources and 'appropriate' visitor behaviour rely upon self-regulation by the tour operators and tourists themselves.

Licensed Guides

One of the most successful management techniques for conserving Arctic resources and directly influencing lawful visitor behaviour is guide licensing. Wildlife managers realized many decades ago that an effective way to ensure regulatory compliance with conservation laws was to require anglers and hunters to employ licensed guides. Guide licensing programmes were established by wildlife management agencies throughout the Arctic. The programmes instruct specialized knowledge of environmental conditions, resource laws and regulations, survival skills and emergency response skills. Based on demonstrable competency, guide licenses are issued and in most instances refresher courses are required to sustain both educational knowledge and practical skills. Guide licensing in the Arctic has expanded well beyond angling and hunting. The pursuit of Arctic recreation activities such as mountaineering, rafting, kayaking and wildlife photography frequently require licensed guides. Licensing requirements very considerably among Arctic jurisdictions and responsible resource agencies, but this tourist management technique has proved effective.

Summary and Conclusions

Appropriate visitor behaviour, jurisdictional cooperation, cultural tolerance and, of course, environmental conservation are the primary objectives of tourism management in the polar regions. When placing these objectives within a 'real world' context, it is also evident that economic feasibility, from both a public and private sector perspective, and maintaining tourist safety and satisfaction must also be included as crucial elements of tourism management policies, plans and programmes. The scarcity of law enforcement resources adds another dimension that requires careful consideration when contemplating alternative approaches to tourism management in the polar regions. Because all of these factors are important, it is essential to thoughtfully integrate each of them into a comprehensive approach to tourism management.

In contrast with other regions of the world, tourism management in the polar regions depends heavily on either the private sector, i.e. the tour operators and licensed guides, or the tourists themselves to conserve its resources. Given this remarkable situation, it seems advisable to focus on ways in which the tourist experience can be designed and delivered so that the polar resources are best protected.

One prospect for effective tourism management is to accurately define and evaluate tourism experiences in terms of the ways in which the host environment and communities are willing and able to provide these experiences. This approach emphasizes the need to identify the full range of potentially 'allowable' tourist activities in terms of the ways they will actually be delivered, and then apply sustainability criteria to determine if they are consistent with resource management objectives. For example, tourism management issues surrounding wildlife viewing in the polar regions are most dependent on how, when and where the actual experience should or can be delivered. Thus, offshore wildlife viewing with a spotting scope is an entirely different experience from backcountry wildlife viewing and, correspondingly, the management techniques required to perform environmental conservation, visitor safety and the prevention of cultural intrusion are exceedingly different. Recognizing that self-regulation represents the major tourism management tool currently available to polar regions, then methods for defining and delivering tourist experiences seem particularly relevant and useful line of inquiry.

References

Basberg, B.L. and Rossnes, G. (1993) *Dokumentasjon av hvalfangststasjonen Grytviken, SydGeorgia*. En rapport fra prosjektet Hvalfangstminneregistrering på Syd Georgia, NARE publikasjon nr. 128. Interiøroppmålinger, Trondheim/Oslo, Norway.

Blanchette, R.A., Geld, B.W. and Farrell, R.L. (2002) Defibration of wood in the expedition huts of Antarctica: an unusual deterioration process occurring in the polar environment. *Polar Record* 38(207), 313–322.

Cater, E. and Lowman, G. (1994) *Ecotourism, A Sustainable Option?* John Wiley & Sons, Chichester, UK.

Collis, S.M. (1890) *A Woman's Trip to Alaska, Being an Account of a Voyage through the Inland Seas of the Sitkan Archipelago in 1890*. Cassell Publishing Company, New York.

Dressler, W.H., Berkes, F. and Mathias, H. (2001) Beluga hunters in a mixed economy: managing the impacts of nature-based tourism in the Canadian western Arctic. *Polar Record* 37(200), 35–48.

Feifer, M. (1986) *Tourism in History from Imperial Rome to the Present*. Stein and Day, New York.

Hibbert, C. (1987) *The Grand Tour*. Thames Methuen, London.

Hughes, J. (2000) Ten myths about the preservation of historic sites in Antarctica and some implications for Mawson's huts at Cape Denison. *Polar Record* 36(197), 117–130.

Kipling, R. (1920) *Letters of Travel*. Macmillan, London.

Loomis, J.B. (1993) *Integrated Public Lands Management Principles and Applications to National Forests, Parks, Wildlife Refuges, and BLM Lands*. Columbia University Press, New York.

Loschberg, W. (1979) *History of Travel*. Edition Leipzig, Leipzig, Germany.

Lucas, R.C. (1985) *Visitor Characteristics, Attitudes, and Use Patterns in the Bob Marshall Wilderness Complex, 1970–82.* Research Paper INT-RP 345. US Department of Agriculture Forest Service, Intermountain Research Station, Ogden, Utah.

Parks Canada (2004) *Parks Canada Agency Corporate Plan 2003/04–2007/08.* Parks Canada, Ottawa.

Mieczkowski, Z. (1995) *Environmental Issues of Tourism and Recreation.* University Press of America, Lanham, Maryland.

Russia (2001) *O territoriyakh tradisionnogo prirodopolzovaniya korennykh malochislennykh narodov Severa Sibiri I Dal'nego Vostoko Rossiyskoy Federatsii [On Territories of Traditional Nature Use of the Indigenously Small Peoples of the North, Siberia, and the Far East of the Russian Federation].* Russian Federal Law 49-F3, May.

Scidmore, E.R. (1885) *Journeys in Alaska. Alaska – Its Southern Coast and the Sitkan Archipelago.* D. Lothrop and Co., Boston, Massachusetts.

South Georgia Association (2005) *The Future of South Georgia: A Programme for the Next 10 Years.* Report on a Conference, 18–20 September 2003. Scott Polar Research Institute, Cambridge, UK.

Stankey, G.H. and McCool, S.F. (1985) *Proceedings – Symposium on Recreation Choice Behavior.* General Technical Report INT-GTR 184. US Department of Agriculture Forest Service, Intermountain Research Station, Ogden, Utah.

Watson, A.E., Aplet, G.H. and Hendee, J.C. (1998) *Personal, Societal, and Ecological Values of Wilderness. Proceedings of the Sixth World Congress on Research, Management, and Allocation, Bangalore, India, Vol. 1.* Proceedings RMRS-P-4. US Department of Agriculture Forest Service, Rocky Mountain Research Station, Ogden, Utah.

Whelan, H. (2004) Walking the bear. *Outside Magazine* 29(3), 61–67.

WWF Artic (1996) *Linking Tourism and Conservation in the Arctic.* World Wildlife Fund Arctic Programme, Oslo.

Wright, V. (ed.) (2001) *Defining, Managing, and Monitoring Wilderness Visitor Experiences: An Annotated Reading List.* General Technical Report RM-GTR-79. US Department of Agriculture Forest Service, Rocky Mountain Research Station, Fort Collins, Colorado.

15 Tourism on South Georgia: a Case for Multiple Resource Management

JOHN M. SNYDER[1] AND BERNARD STONEHOUSE[2]

[1]Strategic Studies, Inc., 1789 E. Otero Avenue, Centennial, CO 80122, USA; [2]Scott Polar Research Institute, University of Cambridge, Lensfield Road, Cambridge CB2 1ER, UK

Introduction

South Georgia is a mountainous island, approximately 125 miles long and up to 22 miles wide, in the southern Atlantic Ocean, location 54°30'S, 37°00'W (Fig. 15.1). The island and its surrounding waters provide vital habitats for many species of birds and large populations of marine mammals and fish. Following its discovery in the 17th century, South Georgia became a centre for sealing, and for six decades of the 20th century supported a land-based whaling industry. Currently its main revenues come from licensing fisheries in its territorial waters, but since the 1970s the island's spectacular scenery, abundant wildlife and historic interest have attracted increasing numbers of ship-borne tourists. The responsible government, the Government of South Georgia and the South Sandwich Islands (GSGSSI), based in the Falkland

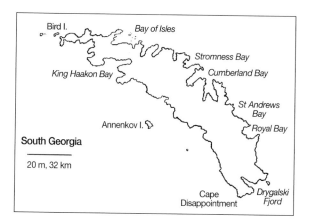

Fig. 15.1. South Georgia.

Islands, is concerned to develop tourism on a sustainable basis. This chapter reviews the current state of tourism on the island and proposes a multiple resource management stratagem for its future development.

South Georgia: Heritage and Natural Resources

South Georgia was first sighted in 1675 from the British merchant ship *Antoine de la Roche*. Captain James Cook in 1775 mapped its northern coast, in the hope that it represented the edge of the sought-after southern continent. Cape Disappointment, at its south-eastern corner, signified the end of that hope, but Cook's published description of an abundance of seals was no disappointment to the sealing and whaling industry in Britain and America. At a time when stocks of northern oceans' fur seals were declining due to commercial hunting pressure, his reports initiated the hunt for fur seals, and later elephant seals, on South Georgia and throughout the southern oceans (Gurney, 1997). Sealers thus became the first to explore the island's many sheltered harbours and beaches, in an industry that lasted over a century. Now-faded remnants of sealing, mostly overgrown campsites, try-pots and graves scattered along the coast (Headland, 1984), form an important part of South Georgia's historical heritage.

More recent heritage sites are the whaling stations established on the island from 1904 onwards. Southern oceans whaling was initiated by the Norwegian Captain Carl Anton Larsen, who, during three exploratory voyages to the South American sector of Antarctica between 1892 and 1904, discovered South Georgia's potential for land-based whaling (Quartermain, 1967; Headland, 1989; Baughman, 1994). During his third voyage, in May 1902, Larsen located a fine harbour, which he called Grytviken ('pot cove', for the sealers' try-pots he found there). In 1904 he established at Grytviken the island's first whaling station (Carr and Carr, 1998; Berg *et al.*, 1999).

Whaling from South Georgia proved immensely successful. Ultimately a further six stations were built: Godthul (1908–1929), Ocean Harbour (1909–1920), Leith Harbour (1909–1965), Husvik (1910–1931, 1945–1960), Stromness (1912–1961, famous as the final destination of Sir Ernest Shackleton's epic journey across the island) and Prince Olav Harbour (1917–1931). For full accounts of whaling from South Georgia and the South American sector of Antarctica see Hart (2001, 2006). The industry's employees came mostly from Norway, the UK, Argentina, the USA and Japan, labouring in dangerous occupations and practising a wide diversity of skills that enabled the stations to operate self-sufficiently. They organized social and athletic associations, minted their own coins, occasionally went to jail, and at Grytviken practised their religion in the most distant church of the Norwegian Lutheran realm. Several died and were buried on the island. The abandoned stations, and in particular the Grytviken whaling museum, are important mementos of this once-flourishing industry.

Grytviken was also a port-of-call and source of supplies for several Antarctic explorers (Fuchs and Hillary, 1958). It was the point of departure for Shackleton's 1914 *Endurance* expedition (Shackleton, 1920), and the place where the explorer died during his *Quest* expedition in 1922. Parties from visiting ships today gather at the granite memorial marking Sir Ernest's grave in the whalers' graveyard, toasting the memory of 'The Boss'. In 1925 the UK's Discovery Committee established a scientific station at nearby King Edward Point. From 1969 to 1982 the British Antarctic Survey (BAS) operated a biological station at the Point, now replaced by a fisheries research station of the GSGSS, while BAS continues seabird and seal studies on Bird Island (Stonehouse, 2006).

South Georgia's natural history is as compelling as its human history; for an illustrated visitor's guide to both see Poncet and Crosbie (2005). The island's interior is heavily glaciated (Fig. 15.2): ice-free coastal plains, headlands and islands supports 31 breeding species of birds, and a further 25 species have been seen in neighbouring waters (Prince and Poncet, 1996) (Fig. 15.3). The nutrient-rich surrounding oceans continue to support stocks of whales and seals. In 1994 the International Whaling Commission designated the Southern Ocean a whale sanctuary, in which commercial whaling is now prohibited. Whaling is prohibited also in the waters around South Georgia. While the recovery of whale populations after hunting is uncertain (McIntosh and Walton, 2000), stocks of fur seals and elephant seals currently thrive (Boyd, 1993; Boyd *et al.*, 1996), to the extent that breeding fur seals dominate many beaches that would otherwise become popular tourist venues (see 'Environment, culture and wildlife' below).

Fig. 15.2. Tidewater glacier in Drygalski Fjord. (Photo: J.M. Snyder.)

Fig. 15.3. Fledgling king penguin chicks examine a visitor on a South Georgia colony. (Photo: B. Stonehouse.)

Tourism: Background and Growth

A few early 20th century visitors to South Georgia, notably scientists and photographers, explored in private yachts or small inshore craft, helped by the presence of whalers. Lars Eric Lindblad, an enterprising cruise operator, was among the first to visit South Georgia as a variant on his newly developing Antarctic Peninsula and Scotia Arc cruises. Spectacular scenery, close encounters with penguins, albatrosses and other seabirds, plus the interest of the recently abandoned whaling stations, gave visitors of the 1970s a very personal acquaintance with the island.

This phase ended abruptly in April 1982 when Argentine forces invaded South Georgia. Although their occupation lasted only three weeks, security and political issues disrupted tourism until the late 1980s, when the island again became a popular destination. The International Association of Antarctica Tour Operators (IAATO) extended its sphere of interest without questioning to include South Georgia, creating and implementing safe operational practices, visitor guidelines, resource conservation and impact mitigation programmes, documentation of tourism activity, protocols for rendering emergency aid, and logistical support to scientists working in the region. For an account of IAATO see Landau and Splettstoesser (Chapter 12, this volume).

Table 15.1 illustrates the growth of tourism during the past 15 seasons at South Georgia, from 11 ships carrying 954 passengers in 1991/2 to 46 ships (plus 15 yachts) carrying an estimated 5200 passengers in 2005/6. For resource management purposes the complete measure of tourism impacts includes not only tourists, but also the staff who accompany them ashore and crew members who may make shore visits. Available data do not make clear how many of these visitors landed. The table indicates a fivefold increase in passengers, most of whom may be assumed to have landed, but a tenfold increase in total visitors, an unknown number of whom, in addition to passengers, may have landed. Management plans now need to respond to the potential impacts and needs of at least 6000 visitors annually landing on South Georgia – a number which currently appears to be increasing at 15–20% per year.

No less important than numbers of ship visits, tourists, staff and crew are:

- The number of landing sites visited.
- The natural and heritage resources located at the landing sites.
- Numbers of passengers landed at the most popular sites.
- The steadily increasing passenger capacity of tourist vessels and yachts.
- Increasing lengths of site visits, especially by yacht passengers.
- The growing diversity of tourist activities, especially adventure tourism.

These all need to be included if statistics are to be meaningful.

Table 15.1. Tourist visits to South Georgia, 1991–2006. (Sources: Harbour Master, Grytviken, South Georgia, personal communication; IAATO website, www.iaato.org; GSGSSI website, www.sgisland.org.)

Season	Ship visits	Passengers	Staff	Crew	Other	Total visitors
1991/2	11					954
1992/3	6					546
1993/4	13					1658
1994/5	18					1753
1995/6	19					1378
1996/7	23					1677
1997/8	25					1789
1998/9	29	2179	295	1555	1	4030
1999/2000	34	2718	345	2014	18	5095
2000/1	27	2100	312	1448	3	3873
2001/2	42 (16)	NA	NA	NA	NA	NA
2002/3	45 (14)	3606		3177		6783
2003/4	42 (NA)	3600		3100		6700
2004/5	40 (18)	3965		NA		NA
2005/6	49 (28)	5436	500	3491	NA	9427

IAATO, International Association of Antarctica Tour Operators; GSGSSI, Government of South Georgia and the South Sandwich Islands; NA, not available. Figures in brackets represent yachts.

The Tourism Experience

South Georgia offers tourists the grandeur of its glaciated mountains, coupled with abundant wildlife and historic authenticity. In fine weather, such popular landing sites as Gold Harbour, Salisbury Plain, Cooper Bay, Saint Andrews Bay, Albatross Island, Prion Island and Whistle Cove in Fortuna Bay offer wildlife viewing and photo opportunities in beautiful locations (Burton, 1997). Whaling stations (Fig. 15.4), sealing camps and graveyards provide the historic perspective. These traditional pursuits are enjoyed by passengers from cruise ships who seldom have time for more than a few brief landings. However, for those on adventure cruises, often in specially chartered yachts, South Georgia also offers more active forms of tourism, including mountaineering, trekking, cross-country skiing, snowshoeing, climbing Mount Paget and other challenging peaks, crossing the island on the route taken by Shackleton's party in 1916, and sea kayaking (including two recent circumnavigations of the island). Adventure tours are often accompanied by film crews who give them further publicity.

The small cruise ships that formerly brought most tourists to the island are gradually being displaced by larger ships, but passenger satisfaction remains constant. Passenger surveys regularly conducted by IAATO tour operators and comments received from visitors at the South Georgia Whaling Museum are generally positive, especially when the weather has been favourable and the island has appeared at its best. Future tourism worldwide is expected to increase as more elderly people, with wealth and leisure time, survive to pursue unique tourism experiences. GSGSSI, IAATO and such interested conservation organizations as the South Georgia Association and

Fig. 15.4. Tourists visit the abandoned Grytviken whaling station. (Photo: J.M. Snyder.)

Norwegian institutions seeking to preserve their cultural heritage, all expect tourism on South Georgia to continue increasing, and most are interested to sustain the public's awareness and interest in the island.

Management Issues

Growing popularity has generated resource management issues and liability concerns for the GSGSSI. The principal challenges include:

- Conserving the island's environmental integrity.
- Selecting techniques for conserving historical and cultural resources.
- Determining that economic benefits from tourism will exceed costs (i.e. be self-funding).
- Sustaining visitor satisfaction while simultaneously providing visitor safety.

These are being addressed by the government, in association with BAS, IAATO, Norwegian whaling heritage institutions, several non-governmental organizations (NGOs) and independent contractors. Scientific and archaeological research has been undertaken to gain understanding of the island's resources, and to examine techniques that can be applied for management purposes. The remainder of this chapter summarizes the salient resource issues and discusses possible management responses.

Environment, culture and wildlife

An environmental inventory of South Georgia, commissioned by the government to establish a benchmark of environmental conditions and management plan (McIntosh and Walton 2000), pointed out that South Georgia is by no means a pristine environment. Sealers and other early visitors introduced rats, which prey upon nesting birds. Whaling left derelict structures, hazardous materials and sunken vessels that produced many forms of environmental contamination and risks to personal safety. Whalers also introduced reindeer which now graze over extensive areas of the island. A primary management objective must be site remediation and habitat restoration.

The plan assigned high priority to a clean-up of the most hazardous sites, 'to protect the health of wildlife, thousands of visitors, and critical land and marine habitats'. The government's response was to prohibit public access to all whaling stations but Grytviken, and limit access at Grytviken to selected safe sites, including the whaling museum (in the manager's restored villa) and graveyard – a regrettable but necessary safety precaution.

The plan also pointed out that:

> The steady disintegration of Grytviken's structures and artifacts represent an irreplaceable loss of world heritage and cultural resources. The Government of South Georgia's efforts to nominate the island for World Heritage Status may be jeopardized if the damage is allowed to continue. Cultural resource management

is required to preserve heritage resources and sustain the island's eligibility for World Heritage Status.

It added further that:

> The cemetery at Grytviken is a consecrated area that preserves the memory of those who perished in this remote part of the Southern Ocean. The shared respect of both the Norwegians and the British for the persons buried at Grytviken's cemetery should contribute to the long-term preservation of this heritage resource.

Subsequent clean-up operations at the whaling stations, undertaken with the cooperation of Norwegian research institutions (strongly supported by their Ministry of Culture), included comprehensive inventories of all the structures, artefacts and cemeteries, documenting how the stations functioned. Experts from the Norwegian Antarctic Research Expedition produced industrial archaeological surveys, architectural drawings and photographic documentation of the facilities and artefacts at all of the stations. Results of their research may be found in Basberg and Nævestad (1990, 1993), Basberg et al. (1996, 1997) and Basberg and Rossnes (1997). Tourist access is currently restricted to a much-tidied and sanitized Grytviken now open again to visitors, with more information than before on what had been an unprepossessing derelict site. Enactment of the South Georgia Museum Trust Ordinance (1992) gave government support to the excellent South Georgia Whaling Museum in the restored manager's house.

Wildlife conservation presented different management challenges, exemplified by tourist visits to nesting albatrosses and other seabirds. Breeding colonies undergo season-to-season changes unrelated to tourism, and marine species risk hazards far remote from their home ground. Numbers of breeding albatrosses decreased as tourist numbers rose, but currently albatross numbers are decreasing throughout the southern hemisphere due to long-line fishing. Thus effects of tourist visits cannot be assumed: they may be considerable or negligible, at their worst representing a further strain of unknown significance on an already stressed population. Landing on beaches en route to the colonies, today's visitors to South Georgia may be confronted by aggressive fur seals that can be very dangerous, particularly when hidden among coastal tussock grass. Though tour guides are charged to take great care of passengers ashore, responsibility may arguably be shared by a government that permits tourist visits.

Environmental management responses to South Georgia's dynamic conditions are being thoughtfully and creatively applied. Research and wildlife monitoring investigations have been conducted at numerous sites, building upon the island's endowment of scientific research. Results are being used to determine optimal numbers of visitors, allowable activities and their season of use. Rat eradication programmes have been introduced, and innovative techniques involving the participation of tourists to monitor the numbers and behaviour of the fur seal population have also been implemented. During the 2003/4 season, for example, BAS asked tourists to estimate the number of breeding fur seals at the sites they visited. According to the South Georgia government, they then used these observations to try and determine the extent to which their habitats are expanding. The role of tourists as conservation partners proved to be a beneficial initiative.

Since the 2002 adoption of the Environmental Management Plan, refinements have been prepared and evaluated, reinforced by scientific research. Prominent among these are recommendations contained in *The Land and Visitor Management Report* (Poncet, 2005), prepared by an ecologist with wide practical experience on the island. Sixty recommendations submitted to the South Georgia government have been carefully reviewed, and will be considered in a revised management plan to be produced in 2007. Discussion of the government's responses may be found at the South Georgia Government website (www.sgisland.org).

In summary, tourist management techniques now employed by the South Georgia Government extend from conventional limitations and prohibited uses to innovative participatory involvement. The use of permits, regulations and operational best practices is supplemented by extensive visitor education and multiple stakeholder involvement in a participatory process. Guidebooks, brochures and maps are now available and distributed to tourists – materials intended to promote appreciation for the island and advocate visitor behaviour that supports resource conservation. South Georgia's efforts to achieve environmental integrity are matched with an equivalent concern for visitor satisfaction and risk minimization.

Economic and financial implications

Prospects for continuing resource management in South Georgia depend on three factors:

- Continued coordination between the government and NGOs, such as the South Georgia Association, to design and implement heritage conservation plans.
- Obtaining the money needed for preservation, restoration, maintenance and enforcement.
- Effective visitor education that will simultaneously promote appropriate behaviour, visitor satisfaction and potential financial support for cultural conservation.

Environmental remediation, habitat restoration and education are costly tasks requiring long-term commitment. The ability of the government to continue depends entirely on financial resources. Ideally, all tourism destinations should pay for essential investments (with a small profit for contingencies) from revenues derived from tour operators and tourists. The South Georgia Government has proved willing to pay for essential environmental remediation and habitat protection, and has invested in such basic infrastructure as refurbishing the Grytviken jetty, constructing public toilets, providing for waste management and subsidizing interpretive publications. It is also keenly aware of other potentially large costs, for example, corporate and personal liability, expressed in its comprehensive environmental management documents (McIntosh and Walton, 2000):

> Corporate vulnerability to personal liability and financial risks are, and will continue to be, direct consequences of knowingly placing their guests at an environmentally hazardous site. The liability waivers signed by guests of the tour industry do not provide sufficient indemnification for the conditions found at Grytviken. The personal hazards that may be present at passenger landing sites in wilderness areas are unpredictable natural events that tour operators cannot reasonably prevent. In sharp contrast, the environmental and structural hazards at Grytviken are exceedingly well known and it would be difficult to establish a legal defense for knowingly placing the guests in harms way. Risk management needs careful consideration.
>
> The Government of South Georgia is vulnerable to financial risk even though it most probably enjoys sovereign immunity from liability suits. The Government legally requires and collects revenues for tourist access to Grytviken. The revenues sources include *The Visitor Ordinance* (1992) that currently exacts a 50 pound per capita fee; *The Customs (Fees) Regulations* (1992) that sets fees for ships and yachts requiring the services of a custom officer for any purposes; and *The Harbour Fees Regulation* (1994) that sets harbor dues for South Georgia and the South Sandwich Islands. In addition to the fees collected from these sources, the Government of South Georgia is proposing that tour operators pay for on-board observers who will enforce government regulations, as well as potential costs for entry permits (BAS, 1999). In summary, the financial vulnerability of the Government of South Georgia will potentially arise from: (1) the loss of revenue from tour operators concerned about their liability exposure; and (2) tour operators who consider the fees a prohibitive cost of operations given the disrepair and dangers of the site. Again, risk management deserves consideration.

Resource management implications in this statement are applicable to all polar tourism destinations where environmental hazards exist.

In addition to mandated fees, tourist-based revenue accrues from sales of stamps, publications and souvenirs in the museum gift shop. Donations are received also from grateful tourists: for example, during the 2005/6 season, passengers of the Lindblad Expeditions ship *National Geographic Endeavour* contributed $28,000 to the museum from the on-board auction of a bronze fur seal pup sculpture, and one passenger individually donated an additional $5000 in memory of his father (http://www.sgisland.org/pages/main/news30.htm). Given that polar regions have few revenue sources, this potential source warrants attention.

Multiple Resource Management

Multiple resource management is a tourism management technique of proven success throughout the world. Providing resource management plans that may be modified over time in response to changing conditions, its principles may readily be applied to tourism in South Georgia, and indeed elsewhere in polar tourism. Figure 15.5 indicates an adaptable five-stage model.

Multiple resource management starts by determining what is to be achieved – defining objectives that ideally have been reached by consensus between managers and clients. As the Cheshire cat advised Alice, if you don't know where you are going, then any road will take you there. As indicated above, reasonable objectives for tourism on South Georgia include:

- Natural environment conservation.
- Cultural preservation.
- Economic and financial feasibility.
- Visitor enjoyment and safety.

These elements can be combined within the model as follows.

Phase I: Identifying the critical elements

The environment (Box 1 in Fig. 15.5) includes the island, the wildlife and the historic artefacts, which need to be identified and evaluated selectively. Only areas where tourism is permitted need initial scrutiny: the rest of the island can be set aside as appropriate reserves, possibly to be opened later to tourism as need arises. This argues strongly for limiting permits initially to areas where environmental factors are already fully evaluated and understood, and closing areas, however attractive, that are insufficiently known.

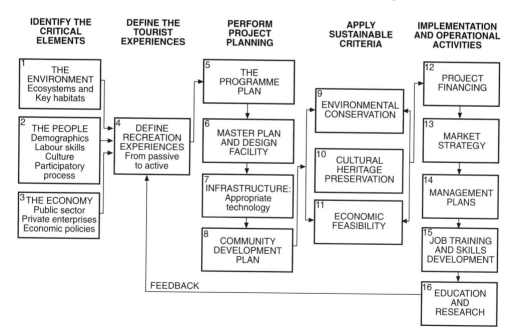

Fig. 15.5. A systems approach to multiple resource management.

The people (Box 2) present little problem on South Georgia: the island has no permanent human population whose interests need to be protected against tourist incursions. There are, however, transient populations of administrators, scientists and technicians at King Edward Point and museum helpers at Grytviken with first-hand knowledge of the island and tourism developments, ecologists who have studied the area intensively over many years, and a constituency of lay-people, including former administrators and scientists, whose interests are expressed through the South Georgia Association. Their knowledge is an asset, and all are interested in the proper regulation of tourism management on the island. The government wisely regards them as stakeholders and gives them opportunities to intervene at every stage. Here as elsewhere, tourism management depends upon partnerships with such communities: their participation adds to visitors' experience by providing a perspective on the local culture that is impossible to gain from outsiders (Cater, 1994; Lea, 1998).

The economy (Box 3) is again simple enough to be well understood. In the absence of local entrepreneurs or investors all ship-side developments in South Georgia tourism depend on a small number of interested tour operators, and all shore-side developments depend on the government. Operators provide the visitors; government provides the infrastructure and management. Both require covering costs and gaining a reasonable return on their investments. Governments in similar situations elsewhere, for example in Svalbard (Bertram, 2007: this volume), provide more services, including policing for law enforcement and rangers, for a very much wider range of tourism activities. This is unlikely to be mirrored in South Georgia until much greater financial returns accrue – a future possibility for an industry in its infancy.

Phase II: Defining the recreational experiences

The preliminary deliberations of Phase I lead directly to a clear exposition in Box 4 of exactly what opportunities for tourist activities are on offer, including where, when and with what levels of supervision. Arising from the fact that tourist visits began before the government saw reason to intervene, tour operators were for several years allowed to do more or less as they pleased, bound only by their own concepts (fortunately very appropriate ones) of environmentally sound behaviour. More active government intervention alters the rules: it is for operators to seek, and for managers to provide, the opportunities and levels of permitting, and to determine the condition within which the activities take place. Box 4, developed cooperatively between operators and government, catalogues the permissible activities and the terms under which they can be undertaken.

Phase III: Project planning

Often overlooked in planning for tourism management is the need for constant cooperation between operators and managers. In a properly run operation their interests are common and objectives identical, and both parties benefit from periodic reviews of the formal plans on which their cooperation depends.

The *programme plan* (Box 5) provides the definitive statement on ways in which the tourism programme will actually be implemented, identifying which resources are involved and the associated costs. Possibly superfluous at this early stage in development on South Georgia, its importance increases as more activities, possibly requiring closer supervision, come on stream.

The *master plan and facility design* (Box 6) is based on the managers' perceptions of appropriate and allowable uses of the various resources available. In South Georgia, the rigorous field and historical research produced by Norwegian and British scholars and institutions have contributed greatly to identifying the probable uses for the island's heritage resources. Based on this knowledge, management techniques may extend from stringent prohibitions against selected tourist uses, locations and seasons to types of visitor use that are specifically encouraged. All resource uses are evaluated by means of monitoring programmes.

The *infrastructure plan* (Box 7) determines what infrastructure is required for the range of activities agreed, and how costs will be allocated or shared between the parties.

The *community development plan* (Box 8), of key importance where native populations are involved, on South Georgia ensures that representatives of the Grytviken and King Edward Point communities are taken fully into consultation on all tourism-related matters.

Phase IV: Applying sustainability criteria

Project planning for tourism on South Georgia as elsewhere must be based on an assumption that sustainability is an important criterion – sufficiently important to warrant rigorous testing. Hence the need for tests involving each of the separate issues.

Environmental conservation and *Cultural heritage preservation* (Boxes 9 and 10) evaluate the tolerances of wildlife and heritage sites to contact with human use. The basis of this evaluation is thorough knowledge of the species, sites or artefacts involved, the usages to which they are subject, management objectives that state clearly what degree of alteration (if any) is tolerable, and monitoring to ensure that those objectives are achieved.

Economic feasibility (Box 11) is based on the fact that sustainable management depends on sustainable income; managers need constantly to be alert to changes in markets, cost structures and other financial variables. Techniques for monitoring operations include market analyses that determine tourism demand and cost–benefit analyses.

Phase V: Implementation and operational activities

Operational activities are perpetually challenged to strike a balance between allowable human uses and the resiliency of environmental and cultural resources to accommodate those uses. All of this must be viewed by tour operators and local government as economically feasible, or there will be no tourism whatsoever. Phase V Boxes 12–16 are reminders of business concerns that are likely to need attention from time to time. *Project financing* (Box 12) requires: (i) knowledge of the likely costs and revenues associated with particular tourism projects; (ii) identification of sources of financing; and (iii) a plan of finance for acquiring both the equity and debt financing required to implement the projects. A *marketing strategy* (Box 13) may be required to publicize South Georgia's attractions or create specific markets. New *management plans* (Box 14) will be required from time to time to respond to ongoing remediation and resource management needs. *Job training and skills development* (Box 15) will be needed to sustain and develop tourism, which benefits from well-trained personnel who provide quality services to visitors. *Education and research* (Box 16) remind operators of a continuing need to discover new facts about South Georgia, its wildlife and heritage, and make them available to the public who are paying for the privilege of sharing them.

Conclusions

South Georgia possesses a rich diversity of natural and heritage resources that appeal to a growing number of tourists. That fact is perfectly clear to the several government agencies, tour companies, academic institutions and scientists who, quite remarkably, mutually agree that these resources require protection. The existence of this consensus is unique in the polar world and the direct benefits to South Georgia have been considerable. Environmental remediation, heritage preservation, codes of visitor conduct, and habitat rehabilitation are some notable examples. All of those endeavours are associated with an ongoing government-sponsored planning process that actively seeks public participation, the most recent of which is the 'Plan for Progress – Managing the Environment 2006–2010'.

Anticipating South Georgia's continued tourism growth, the primary focus of multiple resource management should concentrate on defining allowable tourism experiences, and discovering the best ways for delivering and supporting those experiences. This can be done only by considering tourism activities within the context of resource sustainability, requiring a systems approach that investigates cause-and-effect relationships between tourism and its environmental setting. Using sustainability criteria provides the means for demonstrating that management objectives are being achieved, and whether or not management practices should be modified. The value of correlating tourism management practices with resource management objectives cannot be over-emphasized. In summary, the Multiple Resource Model enhances the excellent research that has already been accomplished by enabling managers to better understand tourism's cause-and-effect relationships within the context of mutually agreed resource management objectives.

References

Basberg, B.L. and Nævestad, D. (1990) Industrial archaeology at Husvik and Stromness, South Georgia. *Norsk Polarinstitutt Meddelelser* 113.

Basberg, B.L. and Nævestad, D. (1993) *Dokumentasjon av hvalfangststasjonen Husvik Harbour, Syd Georgia. Fotojournal.* En rapport fra prosjektet Hvalfangstminneregistrering på Syd Georgia. NARE publikasjon nr. 123. Interiøroppmålinger, Trondheim/Oslo, Norway.

Basberg, B.L. and Rossnes, G. (1997) Industrial archaeology at Grytviken, South Georgia. *Norsk Polarinstitutt Meddelelser* 125.

Basberg, B.L., Nævestad, D. and Rossnes, G. (1996) Industrial archaeology at South Georgia. Methods and results. *Polar Record* 32(180), 51–66.

Basberg, B.L., Nævestad, D., Rossnes, G. and Løkken, G. (1997) Industrial archaeology at Leith Harbour, South Georgia. *Norsk Polarinstitutt Meddelelser* 148.

Baughman, T.H. (1994) *Before the Heroes Came: Antarctica in the 1890's.* University of Nebraska Press, Lincoln, Nebraska.

Berg, K., Devig, T., Johansen, O. and Ulven, H. (1999) *Norwegian Maritime Explorers and Expeditions over the Past Thousand Years.* Index Publishing AS, Oslo.

Boyd, I.L. (1993) Pup production and distribution of breeding Antarctic fur seals at South Georgia. *Antarctic Science* 5, 17–24.

Boyd, I.L., Walker, T.R. and Poncet, J. (1996) Status of southern elephant seals at South Georgia. *Antarctic Science* 8, 237–244.

Burton, R. (1997) *South Georgia.* Government of South Georgia and the South Sandwich Islands, Stanley, Falkland Islands.

Carr, T. and Carr, P. (1998) *Antarctic Oasis: Under the Spell of South Georgia.* W.W. Norton Company, New York.

Cater, E. and Lowman, G. (1994) *Ecotourism: A Sustainable Option?* John Wiley & Sons, New York.

Fuchs, V.E. and Hillary, E. (1958) *The Crossing of Antarctica – The Commonwealth Trans-Antarctic Expedition 1955–1958.* Lenton, Cassell and Co. Ltd, London.

Gurney, A. (1997) *Below the Convergence – Voyages Toward Antarctica 1699–1839.* W.W. Norton and Company, New York.

Hart, I.B. (2001) *PESCA: A History of the Pioneer Modern Whaling Company in the Antarctic.* Aiden Ellis, Salcombe, UK.

Hart, I.B. (2006) *Whaling in the Falkland Islands Dependencies 1904–31.* Pequena, Newton St Margarets, UK.

Headland, R.K. (1984) *The Island of South Georgia.* Cambridge University Press, Cambridge, UK.

Headland, R.K. (1989) *Chronological List of Antarctic Expeditions and Related Historical Events.* Cambridge University Press, Cambridge, UK.

Lea, J. (1998) *Tourism and Development in the Third World.* Routledge, London.

McIntosh, E. and Walton, D.W.H. (2000) *Environmental Management Plan for South Georgia.* British Antarctic Survey, Cambridge, UK.

Poncet, S. (2005) *The Land and Visitor Management Report.* South Georgia Island Official Website; available at http://www.sgisland.org/pages/gov/reports.htm (accessed December 2006).

Poncet, S. and Crosbie, K. (2005) *A Visitor's Guide to South Georgia.* Wildguides Ltd, Old Basing, UK.

Prince, P.A. and Poncet, S. (1996) The breeding and distribution of birds in South Georgia. In: Trathan, P.N., Daunt, F.H.J. and Murphy, E.J. (eds) *South Georgia: An Ecological Atlas.* British Antarctic Survey, Cambridge, UK.

Quartermain, L.B. (1967) *South to the Pole: The Early History of the Ross Sea Sector, Antarctica.* Oxford University Press, London.
Shackleton, E.H. (1920) *South!* Macmillan and Company, New York.
Stonehouse, B. (2006) *Antarctica from South America.* SCP Books, Burgh Castle, UK.

16 Tourism Management on the Southern Oceanic Islands

PHILLIP TRACEY

Australian Government Antarctic Division, Department of Environment and Heritage, Hobart, Tasmania, Australia

Introduction

Around 23 major oceanic islands or island groups lie between the subtropical convergence and the Antarctic continent, spanning warm temperate, cool temperate, sub-Antarctic and maritime Antarctic biogeographic zones (Higham, 1991; Dingwall, 1995; Stonehouse, 2002). Most are without trees or shrubs: some are ice-covered or include remnant glaciers. All have highly endemic flora and fauna, including many rare and endangered species. Integral to southern ocean ecosystems, they provide important breeding grounds for seabirds and seals that depend on the oceans for food, and in consequence are of immense conservation significance. Four are listed as UNESCO World Heritage Sites – Macquarie Island, Gough and Inaccessible Islands, Heard Island and McDonald Islands, and New Zealand's southern islands (traditionally called 'sub-Antarctic' though climatically temperate). Macquarie Island is listed also as a UNESCO Biosphere Reserve.

This chapter outlines how tourism is managed on several of the island groups, selected to illustrate particular points of relevance to broad issues of management of polar and sub-polar wilderness areas. The survey does not include the South Orkney or South Shetland islands, the Balleny Islands, Scott Island or Peter I Øy, all of which lie south of 60°S and are subject to governance under the Antarctic Treaty.

English territorial names used in this chapter follow editorial guidelines, without necessarily reflecting the author's views. Opinions expressed are those of the writer and do not in any way reflect Australian government policy.

Tourism on the Oceanic Islands

The islands discussed here are shown in Fig. 16.1 and listed in Table 16.1. Most were discovered or first charted during the late 18th and early 19th centuries, and subjected immediately to large-scale commercial exploitation of seals and penguins. Over-exploitation led to massive population declines and local extinctions. Harvesting, attempts at settlement (mostly abortive), grazing, fires and damage caused by introduced animal and plant species have substantially degraded many of the islands. Remnants of exploitation, notably sealing camps, graves and castaways' huts from a later period, now have historic and cultural value. Only the Falkland Islands and Tristan da Cunha currently support permanent human populations. Some other islands have permanent scientific or meteorological stations, or receive periodic visits for scientific or management purposes. Few can be regarded as pristine: only Heard Island and some of the New Zealand sub-Antarctic islands remain

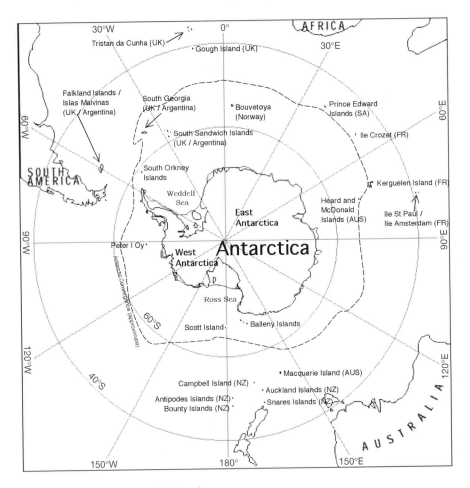

Fig. 16.1. The southern oceanic islands.

substantially free of introduced species. As such they are particularly vulnerable to change:

> experience has shown that plants and animals which have evolved on oceanic islands in the absence of terrestrial mammals are highly vulnerable and sensitive to disturbance. They are readily destroyed, but are virtually impossible to replace.
>
> (Department of Conservation, 1995: 20)

Though cold, wet, cloudy and wind-swept, and by no means typical tourist destinations, these islands support small and specialized expedition cruising industries offering unique and highly valued experiences. Visitors who endure long sea voyages (often in rough conditions), wet landings in small

Table 16.1. Southern oceanic islands. (Sources: Higham, 1991; Dingwall, 1995; Wouters and Hall, 1995a; Headland, 1996; Stonehouse, 2002.)

Island or island group and administration	Biogeographic zone	Area (km²)	Ice cover (%)	Tourist visits?
Atlantic sector				
Tristan da Cunha group (UK)	Warm temperate	111	0	Yes
Gough Island (UK)	Warm temperate	65	0	No
Falkland Islands (UK)	Cool temperate	13,000	0	Yes
South Georgia (UK)	Sub-Antarctic	3,755	57	Yes
Bouvetøya (NOR)	Sub-Antarctic	54	93	No
South Orkney Islands (Antarctic)	Maritime Antarctic	622	85	Yes
South Shetland Islands (Antarctic)	Maritime Antarctic	3,687	80	Yes
South Sandwich Islands (UK)	Maritime Antarctic	310	80	No
Indian Ocean sector				
Ile Saint-Paul (FR)	Warm temperate	7	0	Yes
Ile Amsterdam (FR)	Warm temperate	85	0	Yes
Prince Edward Islands (SA)	Cool temperate	317	1	No
Iles Kerguelen (FR)	Cool temperate	7,215	10	Yes
Iles Crozet (FR)	Cool temperate	325	0	Yes
Heard and McDonald Islands (AUS)	Sub-Antarctic	390	70	Yes
Pacific sector				
Auckland Islands (NZ)	Warm temperate	626	0	Yes
Bounty Islands (NZ)	Warm temperate	1.4	0	Yes
Snares Islands (NZ)	Warm temperate	3.3	0	Yes
Macquarie Island (AUS)	Cool temperate	128	0	Yes
Campbell Island (NZ)	Cool temperate	113	0	Yes
Antipodes Islands (NZ)	Cool temperate	21	0	Yes
Peter 1 Øy (Antarctic)	Maritime Antarctic	157	95	No
Scott Island (Antarctic)	Maritime Antarctic	0.4	0	No
Balleny Islands (Antarctic)	Maritime Antarctic	400	95	Yes

boats, and an absence of amenities on shore, are rewarded with rugged mountain landscapes, volcanoes, fjords and glaciers, unique vegetation, massive congregations of birds and seals, occasional historic sites, and perhaps above all remoteness and wilderness.

The tourist's experience centres on the islands' unique features – remoteness, wilderness qualities, environmental sensitivity, cultural sites, spectacular wildlife – and significant scientific and conservation values. As Rubin (1996: 247) points out: 'the subantarctic and Southern Ocean islands are in many ways more interesting than large sections of the continental coast. Most of them, for instance, have more wildlife than does Antarctica' and Headland (1994a: 270) notes that 'the fauna is more varied, the weather better, and access easier than for the continent, and the scenery is comparable in many instances'.

Visits to most of the islands are ship-borne: only the Falkland Islands and South Shetland Islands have runways and associated infrastructure serving tourism. Tourism is seasonal, and varies according to accessibility: a few of the islands have substantial and regular visits each summer, while others are visited only occasionally, and others again, though open to tourism, receive few or no visits.

On islands where regular visits occur, passengers living aboard small to medium vessels make shore excursions and sightseeing cruises using inflatable rubber boats. There is an educational approach, focusing on nature, with experienced guides ashore and lectures on board (Stonehouse, 1994; Stonehouse and Crosbie, 1995). This form of tourism, known as expedition cruising, involves some discomfort associated with sea voyages and small boat landings. In addition to commercial cruise vessels, private and commercial yachts now visit in increasing numbers.

Visits to the islands fall into three main categories. Most visits to the New Zealand islands, Macquarie Island, South Georgia and the South Shetland, South Orkney and South Sandwich islands occur during cruises to or from Antarctica (Rubin, 1996; Cessford and Dingwall, 1998). A substantial growth in demand at some islands (in particular South Georgia) has mirrored the rapid rise in Antarctic visitation experienced over the past decade or so (British Antarctic Survey, 2006).

The second, much smaller category includes 'occasional visit' tourism to the more remote islands (Headland, 1994b; Rubin, 1996), for example those of the Indian Ocean sector, which take longer to reach and cost more to visit than those on regular Antarctic routes. The third category, 'island-only' tourism, includes specific tours from Australia and New Zealand (Department of Conservation, 1995; Cessford and Dingwall, 1998), for example on small ships carrying fewer than 40 passengers to the New Zealand sub-Antarctic islands. Chartered or private yachts are particularly involved in this traffic (Department of Conservation, 1995; Wouters and Hall, 1995a), especially to islands in the New Zealand sub-Antarctic and Atlantic sectors.

Possible Impacts of Tourism

Several possible impacts of ship-borne tourism on southern oceanic islands are identified as being of concern for management. Risk of introducing pest species is a key concern. Disturbance of wildlife or habitat, physical impacts such as trampling, soil disturbance and track formation, accidental wildfire, impacts on cultural heritage, pollution and improper waste disposal are all potential impacts of tourism (Wouters and Hall, 1995a; Department of Conservation, 1998; Parks and Wildlife Service Tasmania, 2003; Australian Government Antarctic Division, 2005). There is also potential for interference with other uses, in particular science. Impacts may vary according to the form that tourism takes. Smaller ships tend to visit more sites than large ones, and therefore increase risk of pest species introductions, while larger vessels concentrate more people on fewer sites and in smaller areas at these sites (Department of Conservation, 1998). Management to mitigate or avoid impacts is essential: properly managed expedition cruising can be operated to ensure very low levels of environmental impact. Stonehouse and Crosbie (1995) attribute the lack of obvious impacts of Antarctic tourism to the prevalence of this model of cruising.

Links to Antarctic Tourism

Tourism to the southern oceanic islands is closely related to Antarctic tourism, sharing many of the operators, vessels and itineraries. As well as being destinations in their own right, many island groups act as a staging destination to break the long sea voyage to or from Antarctica. There are also, not surprisingly, many similarities in the visitor experience – remoteness, lack of facilities and infrastructure and the 'nature-based' character of the southern oceanic islands – as well as the predominance of expedition cruising (Tracey, 2001).

The visitors themselves are similar. Cessford and Dingwall (1996, 1998), for example, found that visitors to the New Zealand sub-Antarctic islands and Macquarie Island in the 1990s were predominantly older, retired and professional people, many of whom were involved in conservation groups. Thirty per cent had visited polar or sub-polar regions before. Visitors strongly supported most management restrictions on visits, although there was evidence of a demand for close-up access for viewing and photographing wildlife, and overnight stays. This profile is consistent with descriptions of visitors to Antarctica, for example by Davis (1995).

Because of these similarities, many of the management issues are similar to those that apply in Antarctica. There is one stark difference: tourism on most of the islands is regulated under the sovereignty of a country, while tourism in Antarctica is regulated under the international Antarctic Treaty. Individual countries base their legislation on many years' experience of tourism management: the Treaty has no such precedents and is in many ways restricted from following management practices that apply elsewhere. However, tourism management practices on the different southern oceanic island groups, predicated on sovereignty, offer common features that Treaty management might find it expedient to follow.

Tourism Management on Selected Islands

The following case studies outline tourism management practices on five islands and groups, selected to illustrate particular points of conservation management interest: the New Zealand sub-Antarctic islands; Macquarie Island; Heard Island and the McDonald Islands; the Tristan da Cunha group including Gough Island; and South Georgia. Other islands and groups are dealt with more briefly after the case studies.

New Zealand's southern islands

The southern oceanic islands of New Zealand include five widely separated groups – the Snares, Bounty, Antipodes and Auckland islands groups, and Campbell Island. It would be difficult to overstate the significance of the wildlife populations of these islands. Breeding birds number in the millions – the Snares Islands alone host an estimated six million birds. Globally significant populations of many endemic, rare and endangered species are present, including the world's rarest cormorant, duck, snipe and penguin species. Marine mammals include the rare and endangered New Zealand sea lion, elephant seals and New Zealand fur seals. The vegetation of the Snares Islands, and Adam Island and Disappointment Island in the Auckland group remains substantially unmodified by people or introduced animals. The Auckland Islands have one of the richest floral assemblages of the southern ocean, with 233 species of vascular plant (six endemic), and Campbell Island is similarly diverse (Higham, 1991; Department of Conservation, 1995, 1998). For fuller descriptions of flora and fauna see UNEP-WCMC (2006a).

The islands were known and used by Maori people prior to European discovery. A substantial sealing industry peaked between 1792 and 1815, and populations of some species became extremely low. The islands have important cultural heritage, including remains of settlements and scientific expeditions. Castaways' huts, shore-based whaling stations and World War II coast-watching outposts also remain.

These islands are relatively easy to visit because of their proximity to New Zealand, and lie on a main route between Australia or New Zealand and the Ross Sea sector of Antarctica. New Zealand tourist ships carry up to 40 passengers on 10- to 20-day tours, focusing on Auckland and Campbell Islands, and cruise ships on Antarctic voyages carry 40–180 passengers, making two or three landings at different islands. Yachts too are frequent visitors (Sanson, 1994; Sanson and Dingwall, 1995; Wouters and Hall, 1995b). Dedicated tourism visits began in 1968. During the following 20 years 1300 tourists are estimated to have visited, and in the 5 years following 1998 a further 1550 tourists landed. Due to high demand in 1990/1 a seasonal limit on visitor numbers was set at 600 (Department of Conservation, 1998).

Under New Zealand law the islands are Nature Reserves – the highest form of protection – subject to a statutory Conservation Management Strategy (CMS) (Department of Conservation, 1998). Table 16.2 summarizes the key tourism management provisions of the CMS. The management approach, while permitting limited tourism to the Auckland Islands and Campbell Island, is strict and precautionary, with human activities secondary to preservation – to paraphrase the CMS, the more vulnerable or pristine an island is, the greater the justification needs to be for allowing visits. As part of that justification, the managers require that cruise tourism activities must have genuine educational or inspirational purposes. Management is based on avoidance of risk, close attention to the effects of human disturbance, and the need for a precautionary approach. The Snares, Bounty and Antipodes islands are closed to tourism, to reduce the risk of introducing alien species, particularly rodents (Department of Conservation, 1998).

Macquarie Island

Macquarie Island lies south of the Tasman Sea around 1300 km from the Antarctic continent. The general landform is of steep coastal slopes, rising to an undulating low plateau with some lakes (DPWH, 1991). The island is renowned for its wildlife, including very large breeding colonies of penguins, and albatross, petrel, cormorant, duck, rail, skua, gull and terns. Introduced wekas and cats (both now eradicated), rats and mice have affected birds, as have habitat changes due to rabbit grazing (Parks and Wildlife Service Tasmania, 2003). Marine mammals include southern elephant seals, and New Zealand, Subantarctic and Antarctic fur seals. Forty-six species of vascular plant have been recorded, with one endemic to the island. Three species of exotic pest plant are present, of which one is common and widespread (Parks and Wildlife Service Tasmania, 2003). For full descriptions of flora and fauna see Parks and Wildlife Service Tasmania (2006).

Following its discovery in 1810 the island was rapidly stripped of fur seals. Elephant seals and penguins were rendered for oil until 1919. The Australasian Antarctic Expedition 1911–1914 set up a base on the island, and a permanent station was established in 1948, managed by the Australian Government Antarctic Division. Cultural artefacts are present including remains of shore stations, sealers' huts, steam digesters and some shipwrecks (DPWH, 1990, 1991; Parks and Wildlife Service Tasmania, 2003). The first tourists visited in 1971, as part of an Antarctic cruise. Subsequent visits, occurring annually since 1992/3, usually form part of voyages to East Antarctica or voyages that include the New Zealand sub-Antarctic islands.

Table 16.2. Tourism management practices on five island groups.

	New Zealand southern islands	Macquarie Island	Heard Island	Tristan da Cunha group	South Georgia
Permit system	Yes – concessions (licences) granted also	Yes	Yes	Yes	Yes
Overnight stays	Not permitted	Not permitted	Yes – for expeditions, by permit	Not permitted	Yes – for expeditions, by permit
Environmental assessments	Operators must prepare an assessment	Not required	Not required	Not required	Not required
Overall limits	600 visitors per season, all islands	750 visitors per season	No	No	No – a limit can be imposed if required
Vessel size limit	None	200 passengers	None	None	500 passengers, except at Grytviken/King Edward Point, where larger vessels can visit. For one site, only vessels with 200 people or fewer may visit

Zoning	Yes – 'refuge' islands where tourism is permitted at some sites, and 'minimum impact' islands where tourism is not permitted – including the Snares, Bounty and Antipodes islands groups	Yes – three 'tourism management areas' established. Landings permitted at two sites, small-boat cruising at the other	Five zones ('main use', 'heritage', 'visitor access', 'wilderness' and 'restricted'). The first three permit tourism and recreational use. Three sites are approved for visits, spread around the island	Yes – visitors are permitted in the 'natural zone', coast and adjacent lowlands, on Inaccessible Island. Landing areas are restricted to two sites	Yes – authorized visitor sites are listed (32 at present). SPAs are declared
Site limits	150 visitors per year for most sites, 600 per year for three sites. Visits to shoreline sites may have more than 150 by permit. A limit of one ship visit per day applies to all sites. Daily quotas and other restrictions can be applied	60 people at once. Visits to the research station subject to approval of station leader	No more than 60 people at a time at Atlas Cove, and 30 at a time at Spit Bay and Long Beach. A limit of one landing per day at sites can be imposed	100 people ashore at a time	300 ashore at a time at Grytviken/King Edward Point – three ships may visit per day. Vessels that are not IAATO member vessels may only visit this site. 100 ashore at a time at other sites (65 at one site), with one ship at a time per site, and two ships per site per day.

Continued

Table 16.2. – *Continued*

	New Zealand southern islands	Macquarie Island	Heard Island	Tristan da Cunha group	South Georgia
Small boat cruising	Permitted for all areas, limits on distance from shore are specified for each island	Permitted at one site for wildlife viewing. Guidelines apply	By permit		
Prevention of alien species introduction	Strict measures provided for in CMS and guidelines for visits. Closure of islands and island groups to tourism	Strict measures provided for in management plan and guidelines for visits	Strict measures provided for in management plan	Strict measures provided for in management plan	Strict measures in place, including specific quarantine guidelines
Shore infrastructure	Tracks, boardwalks and signs at some sites	Boardwalks, viewing platforms and interpretive signs	None provided	None provided	Tracks, boardwalks, museum, post office and accessible buildings
Guide/visitor ratios	1:20	1:14	1:15	1:8	IAATO rules apply
Guidelines	Yes – specifying quarantine measures, avoidance of pollution and accidental wildfire, guide/visitor ratios, avoiding disturbance of wildlife	Yes – covering all aspects of visits including quarantine measures and safety	No – but management plan is prescriptive	Yes	IAATO guidelines apply

Code of conduct	Yes	Yes	Yes	Provisions in management plan	Yes – along with an information pack
Management supervision	Departmental representative accompanies all visits	Guidelines include relevant provisions Shore-based departmental staff (ranger) supervises visits	The managing authority may require an authorized official to accompany visits	Guides from Tristan da Cunha must accompany visits	All vessels must call at King Edward Point for passenger briefing by government officer. Observers placed aboard where tour leaders or masters are not experienced with South Georgia visits
Reserve category	IUCN Ia, strict Nature Reserve	IUCN Ia, strict Nature Reserve	IUCN Ia, strict Nature Reserve	IUCN Ia, strict Nature Reserve	IUCN Ia, strict Nature Reserve (SPA) and IUCN II, National Park
Quality of tourism experience	Taken into account in developing management provisions. Operators are required to provide adequate, high-quality interpretation services	Taken into account in developing management provisions including vessel size limit	Taken into account in developing management provisions	Not considered	Taken into account in developing management provisions

Continued

Table 16.2. – Continued

	New Zealand southern islands	Macquarie Island	Heard Island	Tristan da Cunha group	South Georgia
Other recreational/ adventure use	Not permitted	Not permitted	Yes – permits can be granted for special purposes such as climbing expeditions, for 'Wilderness' zone entry	Not permitted	Yes – adventure activities permitted, a regulatory regime applies to 'expeditions'
Monitoring	Site monitoring will be undertaken to determine impact levels	Yes – impact monitoring at tourism sites	Not specifically for tourism impacts	Not specified	Not specified
Fees	Levied	Levied – operators also required to accept responsibility for any search and rescue costs incurred by state government	Adequate insurance must be obtained, and operators must be willing to reimburse costs of search and rescue	Levied	Levied

CMS, Conservation Management Strategy; IUCN, International Union for Conservation of Nature and Natural Resources; SPA, Specially Protected Area; IAATO, International Association of Antarctica Tour Operators.

Tourism management policies were first introduced in 1989. Current management provisions are set out in the *Macquarie Island Nature Reserve and World Heritage Area Draft Management Plan 2003*, a statutory instrument under the Tasmanian National Parks and Reserves Management Act 2002 (Parks and Wildlife Service Tasmania, 2003). Table 16.2 summarizes the tourism management provisions, which can be varied by the managing authority as required.

The plan notes that impacts of tourism have to date been negligible, and identifies species introductions as the biggest risk associated with tourism, as well as the potential for wildlife disturbance. The management regime is strict and precautionary. The plan allows for educational tourism to promote awareness of the values of the reserve. In order to provide for management supervision, minimize risk of alien species introduction and limit access only to robust areas, tourism is limited to landings at two sites, and small-boat cruising at one site. Present guidelines limit numbers to 750 visitors each season, though in no season have they exceeded 560.

Heard Island and the McDonald Islands

Heard Island lies south of the Antarctic Convergence some 4000 km south of Australia. In area 368 km², it is dominated by a massive active volcanic cone, mostly ice-covered. The McDonald Islands, 40 km distant with an area of 2.45 km², are comparatively low-lying. They have roughly doubled in size since 1980 due to volcanic activity (Australian Government Antarctic Division, 2005). These are among the most pristine of the southern oceanic islands. Freedom from introduced animal species, and the presence of only one alien plant species, confer exceedingly high wilderness, scientific and conservation values (Australian Government Antarctic Division, 1995, 2005).

The only breeding habitat available to birds and mammals in an enormous area of ocean, the islands support very large breeding colonies of penguins, petrels and other birds (including one endemic species), and three species of seals. The severe climate, ice cover and isolation have resulted in the smallest number of vascular plant species of any sub-Antarctic island group (Australian Government Antarctic Division, 2005). For detailed descriptions of flora and fauna see Australian Government Antarctic Division (2006).

Discovered in 1853, Heard Island was subject to intensive harvesting of elephant seals, and was reoccupied sporadically for this purpose from about 1875 to 1929 (Australian Government Antarctic Division, 2005; UNEP-WCMC, 2006b). Cultural heritage sites and artefacts remaining include hut footings and ruins, barrel caches, try-works, flensing platforms, coopering sites, domestic areas and graves. Scientists visited during the late 19th and early 20th century, and the first Australian National Antarctic Research Expedition set up a station on Heard Island in 1947. The remoteness of the islands, their distance from normal tourist routes to the Antarctic, difficult access and poor weather limit possibilities for commercial tourism. Only three visits by tourist ships are recorded (in 1992, 1997 and 2002) and five small private expeditions (Australian Government Antarctic Division, 2006). The managing authorities see no likelihood of an increase in the foreseeable future (Australian Government Antarctic Division, 2005).

The islands and much of the surrounding 200-nautical-mile Exclusive Economic Zone form a vast (65,000 km²) marine reserve declared under the Environment Protection and Biodiversity Conservation Act 1999 that protects the islands and adjacent marine ecosystems. The *Heard Island and McDonald Islands Marine Reserve Management Plan* (Australian Government Antarctic Division, 2005) sets out the management regime for the islands, summarized in Table 16.2. Provisions for tourism under the management plan include controlled, low-intensity, on-site recreational visitor access, with the aim 'to manage visitor access and commercial operations in the Reserve so as to provide a safe and enjoyable experience without compromising the Reserve's values' (Australian Government Antarctic Division, 2005). In particular, Heard Island's freedom from introduced vertebrates means that visits will only be permitted under tight quarantine controls. Visits to Heard Island from non-Australian ports are strongly discouraged.

The Tristan da Cunha group

These small, steep-sided volcanic islands lie in the South Atlantic Ocean, almost midway between the southern tip of Africa and South America. Tristan da Cunha (113 km²), the largest island, forms a triangle with Inaccessible (14 km²) and Nightingale (4 km²) islands, all within sight of each other. Together they form the British Overseas Territory of Tristan da Cunha, a dependency of the United Kingdom Crown Colony of St Helena (Cooper and Ryan, 1995). Gough Island, included politically within the group, lies some 350 km to the south-east.

The islands were first discovered in 1506, and fur seals were harvested from them around 1790. First occupied in 1816, Tristan da Cunha has a permanent population of 300–350, based in the single settlement Edinburgh. Its lowland areas have been much modified by farming. Nightingale and Inaccessible islands, both steep-sided and difficult to access, remain uninhabited, though Inaccessible has been used for grazing sheep and cattle. The well-vegetated uplands of all the islands support spectacular bird breeding colonies, with threatened and endemic species and some globally significant populations represented. Subantarctic fur seals breed on the islands, and southern elephant seals occur. For full descriptions of flora and fauna see Department of Environmental Affairs and Tourism (2006a) and UNEP-WCMC (2006c). Only Inaccessible Island remains free of introduced rodents (Ryan and Glass, 2001). Nightingale Island is regularly visited by islanders for egg collecting and recreation; Inaccessible Island is visited only rarely.

The general management approach to tourism is of strict control, aiming to minimize landings on the islands to reduce risks of alien species introductions. Management plans set out criteria for tour operators, including a sound record of environmental responsibility and adequate experience in landings on exposed beaches. Parties from cruise ships may visit and land at Edinburgh by previous arrangement, though calm weather is required for handling small boats in the very restricted harbour. Visits are provided for also in plans for both Nightingale and Inaccessible islands (Table 16.2). Visitors from small cruise ships may land and walk on Nightingale, by permit, again subject to calm weather, and only in the presence of island guides. Inaccessible Island, which in 2004 was added as an extension to the Gough Island UNESCO World Heritage Site (see below), is not currently open to visitors, who may however view it from small boats offshore (Ryan and Glass, 2001).

Gough Island, a mountainous outlier with steep coastal cliffs and a high, dissected plateau, supports a low diversity of floral species including 12 endemics. Very large colonies of breeding seabirds occupy all available ground, including 48% of the world's northern rockhopper penguins, around 25% of the world's sooty albatrosses, and virtually all pairs of a northern race of the wandering albatross (Cooper and Ryan, 1994). Introduced mice have become predatory on some of the nesting seabirds. The island was first charted in 1732, and soon afterwards stripped of its fur seals. Visited occasionally during the 19th century, it was the site of a scientific survey in 1955/6, and from then onwards occupied at Transvaal Bay by a Republic of South Africa weather station. Tourist ships may have visited occasionally from 1970 (Headland, 1994b). The current Management Plan for the Gough Island Wildlife Reserve, adopted by the Government of Tristan da Cunha in 1994, states that 'there is no public access to the island, although tourist vessels from time to time express an interest in visiting'. It regards tourism as an unsuitable activity for the island, but makes provision for regulation should tourism commence, suggesting the need for an environmental impact assessment before any could proceed (Cooper and Ryan, 1994).

Regarded by the World Conservation Monitoring Centre as the least disturbed or modified of the temperate oceanic islands (UNEP-WCMC, 2006c), Gough Island was in 1995 inscribed as a UNESCO World Heritage Site, and in 2004 Inaccessible Island was included in this listing. Apart from other considerations, difficulty of access makes any form of future commercial tourism extremely unlikely at either venue.

South Georgia and the South Sandwich Islands

South Georgia lies south of the Antarctic Convergence some 2000 km east of the tip of South America. Almost 200 km long and 35 km across, mountainous and over 50% ice-covered, the island supports around 25 indigenous species of vascular plants, now in competition with some 40 introduced plant species. There are 31 species of breeding birds including endemic pipits and petrels, and estimated populations of 22 million pairs of Antarctic prions and over two million breeding pairs of macaroni penguins.

Seven albatross species are recorded, populations for some of them significant proportions of global totals. There are possibly in excess of six million Antarctic fur seals, and around 110,000 breeding female elephant seals – around 54% of the world population (British Antarctic Survey, 1999, 2006). For fuller descriptions of flora and fauna see Poncet and Crosbie (2005) and Project Atlantis (2006). For further discussion of tourism issues on South Georgia see Chapters 15 and 17, in this volume.

The South Sandwich Islands, a chain of 11 volcanic islands lying 500–600 km east-south-east of South Georgia, are of great geological and biological interest, but visited only rarely, and only by the most adventurous cruise-ship operators. Several of the islands are actively volcanic, with few possibilities for small-boat landings except in the calmest weather.

Hunting for fur seals, later elephant seals, began shortly after discovery of the island in 1775 and continued spasmodically through the 19th and early 20th centuries. Land-based whaling began in 1904; the last whaling station closed in 1965. Reindeer were introduced by the whalers for food and hunting: two genetically distinct populations occupy a significant proportion of ice-free land on the island and have a substantial impact on native vegetation. Brown rats have a severe impact on the ground-nesting birds, and house mice are also present (British Antarctic Survey, 1999, 2006). A permanent British administration operated from 1909 to 1969 at King Edward Point, when the site became a research station under the British Antarctic Survey. The station was occupied by the Argentine military during the Falklands conflict, after which it was occupied by British troops until 2001, when a fisheries research facility was constructed. Sealing, whaling and science have left South Georgia with a substantial heritage of campsites, abandoned whaling stations, scientific and administrative buildings, a church and a cemetery which includes the grave of the British explorer Sir Ernest Shackleton. A former station manager's residence at Grytviken, the main whaling centre, has been turned into a museum. A signposted heritage trail around the whaling station and cemeteries is maintained by two resident curators. A post office at King Edward Point sells philatelic items and accepts mail (British Antarctic Survey, 2006).

Forming part of the British Overseas Territory of South Georgia and the South Sandwich Islands, South Georgia is administered by a commissioner based in Stanley, Falkland Islands. An Environmental Management Plan published in 2000, and complementary *South Georgia Plan for Progress: Managing the Environment 2006–2010* (British Antarctic Survey, 2006), include substantial provisions for tourism (Table 16.2).

Tourism to the island began with the visit of a cruise ship in 1970. In the following 20 years some 3000 tourists visited. Since then there has been a rapid increase (threefold between 1996 and 2006); in 2005/6 49 cruises arrived carrying 5436 tourists (British Antarctic Survey, 2006). Cruise-ship visits involve three or four days spent at South Georgia, usually as part of an Antarctic itinerary, although some 'island-only' cruises also occur, testifying to the island's attractions. Yacht visits are also increasingly common, with 26 visits in 2005/6. Commercial adventure tourism is popular, including camping and mountaineering (for example replicating Shackleton's crossing of the island). King Edward Point (which has a resident Government Officer) and Grytviken are located in the same bay, and this area is a focus of tourism visitation.

Tourism management at South Georgia reflects a more liberal approach than at other southern oceanic islands, demonstrated for example in the number of sites provided for tourism use, the lack of a ceiling on numbers, the acceptance of large vessels, provision of attractive on-shore facilities at a pre-existing administrative centre, and encouragement of land-based activities ashore. This approach is possible on a large island with a wide choice of safe landing sites, leaving ample provision for conservation areas where landings are not permitted. The more generous management approach is perhaps best reflected in the policy statement that reasonable proposals for sustainable land-based tourism will be considered.

An interesting management feature is the preferential access given to companies that are members of the International Association of Antarctica Tour Operators (IAATO), which is felt to provide some assurance of experience and sound operational standards (British Antarctic Survey, 2006). South Georgia also accepts IAATO's guidelines as standards that must be adhered to by non-IAATO vessels at the one site where they may visit. A second interesting feature is the government's ability to charge operating companies substantial harbour and landing fees, generating revenue to pay for the effective administration of this small but developing industry.

Tourism Management on Other Island Groups

The Falkland Islands, an Overseas Territory of the UK, is the only southern island group with a permanent population (about 2000 in 2006) and a growing tourism industry based on a substantial harbour, international airport and status as a gateway port for Antarctica. For fuller details of the islands, their flora and fauna and developing tourism management *see* Summers (2005), Bertram *et al.* (Chapter 8, this volume) and Falkland Islands Development Corporation (2006).

Four remote French island groups in the southern Indian Ocean (Ile Crozet, Ile Amsterdam, Ile Saint-Paul and Iles Kerguelen) receive tourists mainly as visitors aboard a government re-supply and research vessel. All but Ile Saint-Paul have large research stations, where visitors can walk and gain an appreciation both of the islands (with rich and abundant wildlife) and of the research programmes in operation. Around four voyages of around 20 to 30 days are offered annually, with a maximum of 15 people per voyage (Rubin, 2005; Terres Australes et Antarctiques Françaises, 2006). For further details of management and conservation strategies on these islands *see* Terres Australes et Antarctiques Françaises (2006).

Marion and Prince Edward Islands, twin volcanic islands in the Indian Ocean 1800 km southeast of South Africa, are currently devoted almost entirely to research. For further details *see* Department of Environmental Affairs and Tourism (2006b). The Government of South Africa, which administers the group, has considered possibilities for tourism: a recent environmental impact assessment (Department of Environmental Affairs and Tourism, 2000) made recommendations for managing commercial tourism should it be approved, concluding for example that Marion Island is more suited to the landing of small tourist parties than large. Tourism is not currently regarded as appropriate for either of the two islands.

Bouvetøya, possibly the world's most isolated island, is claimed by Norway and protected from all visits not only by Norwegian law, but also by the extreme difficulty of landing in any but the calmest weather.

Comparisons and Conclusion

Several points of interest arise from consideration of tourism on these southern oceanic islands:

1. Despite their remoteness from the rest of the world, all the islands listed have to some degree been touched by tourism.

2. All are subject to management by sovereign powers – management that includes planning for conservation and, within that framework, planning for controlled tourism.

3. Management is generally conservative, precautionary and protective, taking due account of the islands' extremely important natural values and status as nature reserves.

4. Of general concern to managers are possibilities for: (i) introduction of alien species; and (ii) disturbance to wildlife. Management measures tend to be aimed at reducing these particular risks, focusing on controlling access, strict quarantine procedures, selection of sites where tourism can be conducted without significant impacts, and closure to tourism of particular islands where risks are considered too high.

5. Conversely, most managing authorities recognize educational and related benefits in permitting tightly controlled tourism to some islands, closing others completely to casual or recreational visitors, in some instances despite requests for more liberal visitor policies.

6. For the islands that can be visited, sites approved for tourism use are often restricted to the few where safe landing is possible, or where managers assess that good viewing experiences may be achieved with minimal environmental impacts.

7. For similar reasons limitations on number of visits or visitors allowed per season, where imposed at all, tend to be arbitrary and generally low. These approaches reflect different conditions, different levels of tourism use, and different levels of precaution adopted by the managing authority.

8. Though sensitive to human impacts, as can be seen from the damage incurred from past human activities, the southern oceanic islands have been subjected to tourism, in some cases for many years, with no marked additional degradation. The lack of measurable tourism impacts testifies to the adequacy of the management systems in place and the ecological integrity of small, properly managed ship-borne operations.

9. Management of South Georgia is more permissive than that on other islands, allowing relatively high levels of visitation and access to many authorized sites. This different approach takes full advantage of the island's size, topography and relatively disturbed natural condition, as well as its attractiveness for tourism and recreational activity deriving from its location and historical associations.

10. Only Heard Island and South Georgia are managed to provide opportunities for adventure recreation.

11. Only South Georgia appears to be operating on a scale that attracts substantial revenues – i.e. to be making money on a scale that bears some relation to management effort and costs.

12. In all cases, managers prescribe regulations through permit systems, but accept the need for self-regulation of activities through codes of conduct that make clear to visitors what requirements are in place, advise on appropriate behaviour in different circumstances, and advise on ways to comply with requirements. Most of the management systems also use guidelines to inform operators about more detailed management and administrative procedures, and operational requirements.

The management systems covering all these islands show marked similarities, and are to some degree based on each other. While there are differences that can be attributed to different physical and environmental contexts, the presence or not of managing authorities on site, and differing levels of (and demand for) tourism activity, the overall similarity of management provisions suggests that a model of 'best-practice' management has emerged for these areas – a model that is worthy of consideration in many remote wilderness areas and locations subject to increasing pressures of tourism.

References

Australian Government Antarctic Division (1995) *Heard Island Wilderness Reserve Management Plan*. Australian Government Antarctic Division, Department of the Environment, Sport and Territories, Canberra.

Australian Government Antarctic Division (2005) *Heard Island and McDonald Islands Marine Reserve Management Plan*. Australian Government Antarctic Division on behalf of Director of National Parks, Department of the Environment and Heritage, Kingston, Tasmania.

Australian Government Antarctic Division (2006) *Heard Island and McDonald Islands*. Australian Government Antarctic Division, Kingston, Tasmania; available at http://www.heardisland.aq (accessed September 2006).

British Antarctic Survey (1999) *Environmental Management Plan for South Georgia: Public Consultation Paper*. British Antarctic Survey, Cambridge, UK.

British Antarctic Survey (2006) *South Georgia Plan for Progress: Managing the Environment 2006–2010*. British Antarctic Survey, Cambridge, UK.

Cessford, G. and Dingwall, P. (1996) *Tourist Visitors and Their Experiences at New Zealand Subantarctic Islands*. Science and Research Series No. 96. Department of Conservation, Wellington.

Cessford, G. and Dingwall, P. (1998) Research on shipborne tourism to the Ross Sea region and the New Zealand sub-Antarctic islands. *Polar Record* 34(189), 99–106.

Cooper, J. and Ryan, P. (1994) *Management Plan for the Gough Island Wildlife Reserve*. Government of Tristan da Cunha, Edinburgh, Tristan da Cuhna.

Cooper, J. and Ryan, P. (1995) Conservation status of Gough Island. In: Dingwall, P. (ed.) *Progress in the Conservation of the Subantarctic Islands*. International Union for Conservation of Nature and Natural Resources, Gland, Switzerland, pp. 71–84.

Davis, P.B. (1995) Antarctic visitor behaviour: are guidelines enough? *Polar Record* 31(178), 327–334.

Department of Conservation (1995) *Draft Conservation Management Strategy: Subantarctic Islands*. Southland Conservancy Conservation Management Planning Series No. 6. Department of Conservation, Invercargill, New Zealand.

Department of Conservation (1998) *Conservation Management Strategy: Subantarctic Islands 1998–2008*. Southland Conservancy Conservation Management Planning Series No. 10. Department of Conservation, Invercargill, New Zealand.

Department of Environmental Affairs and Tourism (2000) *Environmental Impact Assessment of Tourism on Marion Island*. Compiled by Heydenrych, R. and Jackson, S., Prince Edward Islands Management Committee. Department of Environmental Affairs and Tourism, Pretoria.

Department of Environmental Affairs and Tourism (2006a) *Gough Island*. Department of Environmental Affairs and Tourism, Pretoria; available at http://home.intekom.com/gough/ (accessed September 2006).

Department of Environmental Affairs and Tourism (2006b) *SANAP @ Marion Island*. Department of Environmental Affairs and Tourism, Pretoria; available at http://marion.sanap.org.za/ (accessed September 2006).

Dingwall, P. (1995) Legal, institutional and management planning considerations in subantarctic island conservation. In: Dingwall, P. (ed.) *Progress in the Conservation of the Subantarctic Islands*. International Union for Conservation of Nature and Natural Resources, Gland, Switzerland, pp. 169–174.

DPWH (1990) *One of the Wonder Spots of the World: Macquarie Island Nature Reserve*. Department of Parks, Wildlife and Heritage, Hobart, Tasmania.

DPWH (1991) *Macquarie Island Nature Reserve Management Plan 1991*. Department of Parks, Wildlife and Heritage, Hobart, Tasmania.

Falkland Islands Development Corporation (2006) *Falkland Islands*. Falkland Islands Development Corporation, Stanley, Falkland Islands; available at http://www.falklandislands.com (accessed September 2006).

Headland, R. (1994a) Historic sites and monuments. In: Lewis Smith, R.I., Walton, D.W.H. and Dingwall, P.R. (eds) *Developing the Antarctic Protected Area System: Proceedings of the SCAR/IUCN Workshop on Antarctic Protected Areas*. Scientific Committee on Antarctic Research/International Union for Conservation of Nature and Natural Resources, Cambridge, UK, pp. 85–93.

Headland, R. (1994b) Historical development of Antarctic tourism. *Annals of Tourism Research* 21(2), 269–280.

Headland, R. (1996) *Summary of the Peri-Antarctic Islands*. Scott Polar Research Institute, Cambridge, UK; available at http://www.spri.cam.ac.uk/bob/periant.html (accessed February 1998).

Higham, T. (ed.) (1991) *New Zealand's Subantarctic Islands: A Guidebook*. Department of Conservation, Invercargill, New Zealand.

Parks and Wildlife Service Tasmania (2003) *Macquarie Island Nature Reserve and World Heritage Area Draft Management Plan 2003*. Department of Tourism, Parks, Heritage and the Arts, Parks and Wildlife Service Tasmania, Hobart, Tasmania.

Parks and Wildlife Service Tasmania (2006) *Macquarie Island World Heritage Area*. Parks and Wildlife Service Tasmania, Hobart, Tasmania; available at http://www.parks.tas.gov.au/macquarie/ (accessed September 2006).

Poncet, S. and Crosby, K. (2005) *A Visitor's Guide to South Georgia*. Wildguides Ltd, Maidenhead, UK.

Project Atlantis (2006) *South Georgia Island: online environmental resources*. South Georgia Official website; available at http://www.sgisland.org/ (accessed September 2006).

Rubin, J. (ed.) (1996) *Antarctica: A Lonely Planet Travel Survival Kit*. Lonely Planet Publications, Hawthorn, Victoria.

Rubin, J. (2005) *Antarctica*. Lonely Planet Publications, Melbourne, Victoria.

Ryan, P. and Glass, J. (2001) *Management Plan: Inaccessible Island Nature Reserve*. Government of Tristan da Cunha, Edinburgh, Tristan da Cuhna.

Sanson, L. (1994) An ecotourism case study in sub-Antarctic islands. *Annals of Tourism Research* 21(2), 344–354.

Sanson, L. and Dingwall, P. (1995) Conservation status of New Zealand's subantarctic islands. In: Dingwall, P. (ed.) *Progress in the Conservation of the Subantarctic Islands*. International Union for Conservation of Nature and Natural Resources, Gland, Switzerland, pp. 85–105.

Stonehouse, B. (1994) Ecotourism in Antarctica. In: Cater, E. and Lowman, G. (eds) *Ecotourism: A Sustainable Option?* John Wiley & Sons, Chichester, UK, pp. 195–212.

Stonehouse, B. and Crosbie, K. (1995) Tourist impacts and management in the Antarctic peninsula area. In: Hall, C.M. and Johnston, M.E. (eds) *Polar Tourism: Tourism in the Arctic and Antarctic Regions*. John Wiley & Sons, Chichester, UK, pp. 217–233.

Stonehouse, B. (ed.) (2002) *Encyclopaedia of Antarctica and the Southern Oceans*. John Wiley & Sons, Chichester, UK.

Summers, D. (2005) *A Visitor's Guide to the Falkland Islands*. Falklands Conservation, Finchley, UK.

Terres Australes et Antarctiques Françaises (2006) *Terres Australes et Antarctiques Françaises*; web page available at http://www.taaf.fr (accessed September 2006).

Tracey, P. (2001) Managing Antarctic tourism. PhD thesis, Institute of Antarctic and Southern Ocean Studies, University of Tasmania, Hobart, Tasmania.

UNEP-WCMC (2006a) *Protected Areas and World Heritage: New Zealand Subantarctic Islands*. United Nations Environment Programme World Conservation Monitoring Centre, Cambridge, UK; available at http://www.unep-wcmc.org/sites/wh/subantar.htm (accessed September 2006).

UNEP-WCMC (2006b) *Protected Areas and World Heritage: Heard Island and McDonald Islands*. United Nations Environment Programme World Conservation Monitoring Centre, Cambridge, UK; available at http://www.unep-wcmc.org/sites/wh/himi.html (accessed September 2006).

UNEP-WCMC (2006c) *Protected Areas and World Heritage: Gough Island Wildlife Reserve*. United Nations Environment Programme World Conservation Monitoring Centre, Cambridge, UK; available at http://www.unep-wcmc.org/sites/wh/gough.html (accessed September 2006).

Wouters, M. and Hall, C.M. (1995a) Managing tourism in the sub-Antarctic islands. In: Hall, C.M. and Johnston, M.E. (eds) *Polar Tourism: Tourism in the Arctic and Antarctic Regions*. John Wiley & Sons, Chichester, UK, pp. 258–276.

Wouters, M. and Hall, C.M. (1995b) Tourism and New Zealand's sub-Antarctic islands. In: Hall,
     C.M. and Johnston, M.E. (eds) *Polar Tourism: Tourism in the Arctic and Antarctic
     Regions*. John Wiley & Sons, Chichester, UK, pp. 277–295.

17 Tourism Management for Antarctica

ESTHER BERTRAM[1] AND BERNARD STONEHOUSE[2]

[1]Royal Holloway, University of London, Egham, Surrey, UK; [2]Scott Polar Research Institute, University of Cambridge, Lensfield Road, Cambridge CB2 1ER, UK

Introduction

> Current regulations governing tourism under the Antarctic Treaty, though considerable, are inadequate and insufficiently integrated with other measures for regulating human access to Antarctica and use of its resources.
>
> (IUCN, 1992: 7)

Some 15 years after this judgement, despite the expansion and development of Antarctic tourism (Chapter 9, this volume), management of the industry under the Antarctic Treaty System (ATS) continues to lack certain critical elements. The following points apply:

1. Tourism in Antarctica, notably ship-borne tourism to sites along the coast of Antarctic Peninsula, continues to expand and diversify at accelerating rates (Chapter 9, this volume).

2. There is no overall strategy, based on clearly stated goals and objectives, within the ATS specifically to manage or control Antarctic tourism, either the industry as a whole or its operations in the field.

3. Antarctic tourist operations are managed under regulations within the 1991 Protocol to the Antarctic Treaty, which were designed mainly to cover activities at scientific stations, and are applicable only with difficulties to the quite different activities of passengers at landing sites. (A fifth Annex to the Protocol, dedicated to Antarctic tourism management, was proposed at the 17th Antarctic Treaty Consultative Meeting (ATCM), but not adopted.)

4. Tourist landings are permitted under 'programmatic' (i.e. generalized) Environmental Impact Assessments (EIAs) that take no account of individual differences between landing sites. Tour operators are required to 'monitor' the effects of their activities, but have no regulations, guidelines or advice on how to monitor. Nor, in the absence of management plans for individual sites, is it clear what they are expected to monitor for.

5. There are no provisions for site monitoring by experienced ecologists who are independent of the industry, or for reporting sites at which changes due to tourist pressures may have occurred, or for the investigation or remediation of such changes.

6. There are no provisions for temporarily closing or resting sites where such changes are shown to have occurred.

7. There are neither mechanisms for enforcement, nor resources available to implement enforcement, of regulations relating to tourism under the Antarctic Treaty.

This chapter examines relevant aspects of management systems that have arisen in other areas that share some of the features of Antarctica, and discusses whether management measures in general use elsewhere might be applied to Antarctic tourism.

Tourism Regulation under the Antarctic Treaty System

Companies and individual tourists belonging to states that are parties to the Antarctic Treaty are responsible to the laws of their own countries, prescribed in Acts that specify each country's responsibilities under the Treaty. Most in fact participate under US, British, German, Australian or New Zealand law. That many operate from ships registered in states that are not party to the Treaty, or that some visitors are from states that have no Treaty responsibilities, have not so far raised management problems.

However, no participating states have so far provided effective, long-term forms of rangering service, inspection or supervision, with law-enforcement capability, to ensure that requirements under their various Antarctic Acts are met. This suggests an assumption that operators and tourists, without policing or enforcement, obey the rules sufficiently to ensure the continuing well-being of the Antarctic environment.

Environmental impact assessment procedures outlined in Annex I of the Protocol remain the sole gatekeepers for Antarctic access (Hemmings and Roura, 2003: 21). Tourism activities are permitted under Initial Environmental Evaluation (IEE) procedures, with an inherent assumption that any activity permitted imposes 'less than a minor or transitory impact'. Although infrequent site assessments have been performed, there have been no regularly scheduled and scientifically consistent monitoring studies to show whether impacts are indeed less than minor or transitory, or whether the more rigorous Comprehensive Environmental Evaluation (CEE) procedure would be more appropriate, either overall or for any particular activity or site.

Day-to-day activities of operators and tourists in Antarctica are based on guidelines and codes of conduct agreed between the International Association of Antarctica Tour Operators (IAATO); for further details of IAATO see Bertram (2007) and Landau and Splettstoesser (2007), both in this volume. In practice this follows from a complete absence of effective law enforcement. As Stonehouse (1998: 55) comments:

Though several Treaty nations now provide conservation law and regulations covering Antarctica, providing for penalties in case of infringement, there are no rangers or inspectors empowered to supervise sites and to see that regulations are observed, and no guides other than the ship-borne guides provided by the tour companies. It is very unlikely that legal charges against passengers who infringe regulations, even seriously, could be carried effectively through the courts of their native countries. In these circumstances good guidelines, and the goodwill on which they rely, remain Antarctica's only practical, on-the-spot defence against despoliation.

Tour operators maintain that current guidelines and codes of conduct for Antarctica are adequate, but it is unclear that these forms of self-regulation address all issues arising from tourist activity (Enzenbacher, 1992: 261). As noted by Wouters (cited in Johnston and Hall, 1995: 304), the current spirit of cooperation amongst tour operators should be encouraged, but may need to be supplemented by an enforcement mechanism.

Tourism Management in Similar Regions Elsewhere

How is tourism managed in other polar and sub-polar regions? Could methods used elsewhere instruct Antarctic tourism management? Here we review tourism management in three such areas – South Georgia, Svalbard and Glacier Bay – which unlike Antarctica are regulated under national sovereignty. Table 17.1 summarizes some similarities and differences in management at the three localities, compared with management of tourism in Antarctic Peninsula. Data for this section are based mainly on studies by Hall and Wouters (1994), Hall and Johnston (1995), Wouters and Hall (1995), Nuttall (1998) and Tracey (2001).

South Georgia

A glaciated island some 200 km long in the Atlantic sector of the Southern Ocean, South Georgia is administered from the Falkland Islands as part of the British Overseas Territory of South Georgia and the South Sandwich Islands. Local administration and management are effected by a Marine Officer based at King Edward Point, the government centre in Cumberland Bay. Visiting ships are usually required to call at King Edward Point to allow clearance by the Customs and Immigration Officer. South Georgia is often included in the itineraries of Antarctic cruise ships. Five ships carrying some 500 landing passengers visited the island in 1992/3: 10 years later some 45 ships brought approximately 3500 passengers (Scott and Poncet, 2003: 23). A guidebook has recently been published (Poncet and Crosbie, 2005).

Table 17.1. Management characteristics and methods in four polar and sub-polar localities.

	South Georgia[a]	Svalbard[b]	Glacier Bay[c]	Antarctic Peninsula
Governing body	Government of South Georgia and the South Sandwich Islands	Government of Norway	United States National Park Service	Antarctic Treaty System
Area (approximate)	3750 km²	62,000 km²	13,052 km²	70,000 km²
Surface area ice-covered	60%	70%	25%	95%
Are area designations consistent with IUCN management categories?	Yes – areas match IUCN Categories 1a and II, and National Park	Yes – some are equivalent to IUCN Category I Nature Reserve	Yes – US National Park, IUCN Category II	No – continent is a 'Special Conservation Area', with SMAs and SPAs
No. of visitors per year	3,600	~45,000	~400,000	~20,000
Are there upper limits on annual visitor numbers?	Yes – but not currently in use	Yes – but not currently in use	Yes – daily and seasonal limits	No
Are there upper limits on numbers of ships?	No	Yes – for particular popular areas	Yes – daily and seasonal limits	No
Are there specific tourism management plans?	Yes	Yes	Yes	No
Are EIAs required for ship-borne visits?	No	No	Operators assess impacts of activities on park values through permit system	Programmatic EIA covers all activities
Are additional permits needed for ship-borne visits?	No	Yes – to certain areas	Yes	No
Are areas zoned for visitor use?	Yes – open, protected, environmentally sensitive	Yes – nature reserve, national park, outdoor recreation and excursion zones. Tourism allowed in all	Yes – closure of areas to traffic for periods. Types of recreational activities are clearly designated	No
Are ship movements in the area restricted?	Yes	No	Yes	No

Are landing points restricted?	Yes – to permitted areas	No	No landings allowed	No
Are passenger landing fees payable?	Yes – £50 per passenger	Not currently	Yes – $5 per passenger	No
Are there management personnel in the area?	Yes – at King Edward Point	Yes – rangers	Yes – rangers	No
Are management personnel placed aboard?	No	No	Yes	No
Are there limits on numbers allowed ashore at one time?	Not currently	Not currently	Cruise-ship visitors may not land	IAATO-based guideline of 100 is generally observed
Are codes of conduct provided?	Yes	Yes	Yes	IAATO-based guidelines
Are post-visit reports required?	Yes	No	No	Yes
Are tourism impacts measured?	No	No	Yes	No

IUCN, International Union for the Conservation of Nature and Natural Resources; SMA, Specially Managed Area; SPA, Specially Protected Area; EIA, Environmental Impact Assessment; IAATO, International Association of Antarctica Tour Operators.
[a]Source: McIntosh and Walton (2000).
[b]Source: Ministry of Environment (1994).
[c]Source: NPS (1995).

Visitor activities are subject to a Management Plan (McIntosh and Walton, 2000: 59), drawn up in 2002 and reviewed on a 5-year cycle: for a recent overview see Walton (2003: 13). The plan includes management objectives for tourism, for which several areas with designation similar to Category II of the International Union for the Conservation of Nature and Natural Resources (IUCN) have been set aside. Cruise-ship operators require permits and must indicate the sites they will visit beforehand. Non-IAATO operators can visit only Grytviken; IAATO members may apply for permits to visit other sites as well. Permitting enables regulation both of types of vessels visiting and numbers of visitors. IAATO tour operators are required to fill in a post-visit form similar to that used in the Antarctic. Relatively few places are totally restricted to tourists: Areas of Special Tourist Interest (ASTIs) were proposed in a 1975 Conservation Ordinance, but have not been designated. No guides or supervision are normally provided for visiting ships.

For ship-borne tourists the island's attractions include sheltered fjords, spectacular alpine scenery, extensive beaches and islets, penguin colonies, seals and flying birds. Abandoned whaling stations are currently in a dangerous state of dereliction and closed to visitors, but parts of Grytviken whaling station have recently been cleared for visitors, and the whaling museum is a popular attraction. Camping, climbing and other forms of land-based tourism are possible.

The importance of monitoring is recognized in the management plan (McIntosh and Walton, 2000: 59):

> Appropriate environmental monitoring programs will be developed at the most vulnerable sites, after the Government has collected site-specific base line data for the island's fauna and flora and natural environment. These data will form the baseline against which any environmental changes will be measured and will highlight particular sensitivities at each site.

No details are given on what is to be monitored or how needs would be assessed, but baseline studies at some of the most visited sites are currently being undertaken. If the monitoring programmes produce evidence of significant impacts, the South Georgia Government can limit visits to some sites at certain times of the year.

For further comments on possibilities for management on South Georgia see Snyder and Stonehouse (Chapter 15, this volume).

Svalbard

Part of Norwegian sovereign territory 1000 km north of the Norwegian mainland, Svalbard is an archipelago of eight islands with no indigenous population but several permanent settlements including Longyearbyen, the administrative centre, and outlying coal mines (Viken, 1995: 75). Tour operators seeking to visit must notify the Governor's office of travel arrangements, including proposed landing areas, and provide proof of insurance (Ministry of Justice, 1993). Protected areas, including three national parks on Spitsbergen (the main island) and a number of plant and bird sanctuaries, cover more than 50% of the land. All the other islands are nature reserves (IUCN Category 1a), under strict protection (Ministry of Justice, 1993).

A management plan for tourism and recreation, covering all areas up to 4 nautical miles offshore and all types of recreation, protects the islands to preserve their wilderness character. A zoning system allows specific areas for tourism – the Outdoor Recreation Area and the Excursion Area – the latter designated around the current settlements. Prior notice of visits to these areas is not required. The management plan suggests the need for improved visit statistics, site monitoring, possible restrictions on cruise tourism, possible limits on vessel numbers and possible fee imposition. A monitoring scheme is postulated, indicating that it should have simple, measurable parameters, good routines for measurement and follow-up, and that it should be carried out by those responsible for daily management. It does not appear to have been implemented.

The authors' own experience of landing at Longyearbyen and Magdalena Bay from a British cruise liner revealed some of the shortcomings of landings that are neither prepared for and regulated by ship-borne guides nor overseen by shore-based rangers. The 400 or more passengers landing on a warm, sunny day in Magdalena Bay – an area of great natural beauty and historic and wildlife interest – were greeted by the ship's band already ashore and invited to wander freely while a barbecue was prepared. All resident wildlife had disappeared, except for a flock of herring gulls that gathered expectantly around the barbecue and a colony of Arctic terns that (unreasonably, in the view of some of the visitors) attacked those who unwittingly approached their nests among the dunes. A ranger's hut ashore was unoccupied; the small group of rangers available on Svalbard were engaged elsewhere. The historic grave sites were fenced off and permanent notices explained their interest, but few visitors knew enough about the site even to seek them out.

The situation was epitomized by two female British passengers overheard on arrival. One indicated that she would not go ashore because there was nothing to see. The other reminded her that she would miss the barbecue by staying aboard. This is a reminder that many visitors fail to value a natural area outside their normal experience unless specific information about it is provided; and that ship-borne naturalists and historians, when present and given opportunities to speak, play crucial roles in providing information both afloat and ashore.

A notable feature of several landing points in Svalbard is the degree to which the terrain shows evidence of use by man. This is not surprising, in view of the numbers (~45,000) visiting each year, but well-worn paths among dunes behind beaches – which are only just starting to appear in Antarctica – are prominent in heavily visited areas.

Glacier Bay, Alaska

Glacier Bay National Park, in southern Alaska, holds World Heritage status as part of a contiguous group of US and Canadian parks, and is also a UNESCO Biosphere Reserve. One of 16 Alaskan parks managed by the US National Park Service (NPS), it offers a wide range of recreational activities to visitors arriving by sea or air. There are no permanent residents: the only settlement is the Park Headquarters, though Gustavus, a small local community of 285 (1990 census), lies just outside the park boundary. Tourist accommodation is available at Glacier Bay Lodge and at locally owned bed-and-breakfasts. Over 353,000 tourists visited in 2004, over 90% of them by cruise ships (NPS, 2005). There is a commercial airport 10 miles from park headquarters.

The park has a General Management Plan, a Development Concept Plan and a Cultural Resources Plan which stipulate that the natural values and cultural resources of the region should be protected, as should the visitor experience of solitude (even if it is experienced alongside fellow travellers): respect for the human experience is no less important than for the natural environment. The permissible built environment as reflected in the Development Concept Plan is intended to support both park management and visitor satisfaction. Some 85% of the Park area is administered as wilderness, including five marine areas called 'wilderness water'. Land areas are managed under a Wilderness Visitor Use Management Plan.

From 1984 a permit system regulated the entry of cruise ships and smaller tour boats, the primary motivation being the protection of humpback whales that feed in Glacier Bay and surrounding waters. In 1992 a Vessel Management Plan and Environmental Assessment for Vessels in Glacier Bay was prepared, assessing impacts of the industry on wildlife and the wilderness experience. The result was a cap on total numbers of vessels visiting each season. Some areas are closed to ships during wildlife nesting and breeding seasons. Ships entering the region for cruising are required to take official rangers aboard to interpret the experience and ensure that regulations are observed.

A monitoring system exists for the park in conjunction with the Vessel Management Plan, investigating ship numbers, marine mammal–vessel interactions and identification of sensitive resources. This monitors for the objectives stated in the plan (NPS, 1995), which are to:

> allow ecological processes to continue unimpaired by visitor use. Protect marine and terrestrial vegetation from adverse effects of visitor use. Identify areas that have special sensitivities for wildlife, solitude or other values and develop methods for protecting these special sensitivities.

Glacier Bay receives the most cruise-ship visitors of the areas summarized in this review, and its designation as a national park means that tourism must be catered for within its overall management plan. Monitoring appears – at least on paper – to be in line with the objectives and could, therefore, act as an effective method for identifying whether the management framework is proving effective. Significantly, ship-borne tourists are not allowed to make landings: the Glacier Bay experience is therefore directly comparable with large-ship (liner) cruising in Antarctic waters. However, absence of landings has not exempted Glacier Bay from environmental damage and threats to personal safety. Cruise ships have foundered upon rocks and experienced a variety of operational difficulties – events that have required emergency measures both to minimize environmental damage and to evacuate stranded tourists and crews.

Comparisons and contrasts

These examples all have relevance for the management of Antarctic tourism:

1. South Georgia, Svalbard and Glacier Bay under sovereignty have *designated areas* consistent with IUCN worldwide classifications; Antarctica under the ATS has not.

2. Only Antarctica has *no provision for setting upper limits to numbers of annual visitors or ships*.

3. Venues under sovereignty publish *management plans* with stated *management objectives*, which include *specific provision for tourism* as part of their overall strategies. All designate areas zoned for visitor use. Antarctica has no management plan beyond provisions derivable from the Protocol, no tourist zones, and the ATS has specifically rejected the need for special tourism management strategies (see 'Introduction', point 3).

4. Glacier Bay sets *rigorous requirements for ship-borne visits*; Antarctica requires only *non-specific programmatic EIAs*.

5. Glacier Bay has a separate vessel management plan which places restrictions on activities for the benefit of wildlife and the visitor experience, including speed restrictions, mitigation requirements and fees for access. *Cruise liner operators submitting EIAs for visiting Antarctica make a point of noting that, in the absence of more positive regulation, they intend to operate as though in Glacier Bay* (Stonehouse and Brigham, 2000: 347).

6. Governments in South Georgia and Svalbard *restrict ship movements* to avoid congestion at popular sites. In Antarctica this is managed by IAATO, based on operators filing day-by-day itineraries at the start of each season.

7. South Georgia and Svalbard *restrict landing points* to approved zones. Antarctic operators are free to land anywhere except in scheduled reserves and (with permission) at research stations.

8. *Management personnel* with responsibilities for tourism are present in Svalbard and Glacier Bay, but not in South Georgia or Antarctica. Anomalously (possibly without ATS sanction, but in the general view laudably) the New Zealand Antarctic Heritage Trust refuses visitors access to historic huts in Victoria Land unless their representative is present.

9. Only in Glacier Bay are *observers placed aboard cruise ships* to ensure adherence to regulations. Superficially restrictive, this results in a more flexible form of management, as the observers are authorized to allow changes to itineraries in line with conditions at particular sites, despite permitting restrictions.

10. There is provision for monitoring and *measuring tourist impacts* in all four venues, but only in Glacier Bay does it appear to be taken seriously. This reflects the influence of strictly enforced laws and regulations from the jurisdictional perspective, and the responsibility that can be shown by those manning a very large and profitable operation, with clearly stated objectives and adequate funding.

Though the tourism management systems of the venues under sovereignty are more comprehensive and clearly enunciated than those operating in Antarctica, they do not inevitably effect better management. The authors' experiences on Svalbard demonstrated that, in a venue subject to nominally sound management procedures but with a poorly manned rangering service, ill-briefed passengers going ashore at a sensitive site may gain little or nothing from the experience, while contributing substantially to wear-and-tear. South Georgia's emerging programme of tourism management shows promise but, in the absence of even a token dedicated ranger service, seems likely to lack either teeth or feedback. Management in Antarctica has been left almost entirely to the industry, IAATO providing the only significant level of regulation. Indeed the same level of management responsibility appears in IAATO-member ships when they operate in the Arctic or in the sub-Antarctic outside the Treaty area – a fact which South Georgia authorities take into account when issuing permits.

Arctic ship-borne tourism has recently benefited from Antarctic experience. Hundreds of thousands of visitors to the northern circumpolar regions each year sample the mythology of the Arctic and northern wilderness (Johnston, 1995: 28). Numbers appear set to grow considerably due to easy land and air access, and the region would benefit from finding ways to allow such levels of visits to remain sustainable (Johnston and Hall, 1995: 309). In 1997 the World Wildlife Fund Arctic Programme produced a report, *10 Principles for Arctic Tourism* (WWF Arctic Programme, 1997), stressing in particular four points: (i) support for the preservation of wilderness and biodiversity; (ii) respect for local cultures; (iii) the importance of trained staff; and (iv) the need for tourism to be educational. It prepared also a Code of Conduct for Tour Operators in the Arctic and a Code of Conduct for Arctic Tourists, based to a significant degree on the IAATO guidelines and codes of conduct. These included recommendations that tour operators:

- Actively support conservation initiatives and ensure tourists are aware of them.
- Encourage tourists to donate time or money to local conservation initiatives.
- Ensure artefacts are not removed.
- Encourage a staff-to-tourist ratio of 1:15 to 1:20 for ship-borne tourism. (WWF Arctic Programme, 1997)

The success of such a system depends on the appropriateness of the codes for different geographical sites, as well as the commitment of those involved to incorporate these principles into their operational structures (Johnston, 1997). As noted by Davis (1995b: 332–333), codes are most effective when implemented as part of a larger management strategy.

Management Options for the Antarctic

Svalbard, Glacier Bay, and to some extent, South Georgia use standard management planning systems, which typically include: (i) a logical process for management with a clear, traceable rationale for decision making; (ii) decisions based on scientific, technical analysis; (iii) a designation for protection; (iv) an overall management plan with clear management objectives; (v) zoning including designated tourist areas; (vi) management objectives for landing sites developed from the overall management plan; and (vii) a permitting system that regulates numbers. Within this framework several management strategies are available and have indeed been suggested as applicable to Antarctica.

Carrying capacity

The carrying capacity of a site may be interpreted variously as: (i) a level of use that can be sustained to a point beyond which environmental damage occurs; or (ii) a level of crowding that tourists will accept. Either way it varies according to season, user behaviour and environment (Ceballos-Lascuráin, 1996: 131). It is generally not considered effective for visitor management (Williams, 1998: 116), particularly when based on simple limitation of numbers. As Kuss *et al.* (1990: 163) note in their review of literature concerning visitor disturbance of wildlife:

> The number of people using an area has a smaller role in human–wildlife relationships than characteristics such as frequency of use ... type ... and behaviour.

Similarly Naveen *et al.* (2001: 122), who have observed many Antarctic landing sites, state that:

> Overall numbers of visitors and tour ships are less important than: where visitors make landings; how many people go ashore during zodiac landings ...; and how frequently zodiac landings occur.

We are unaware of any published estimate of carrying capacity for any Antarctic landing site. The number of 100 ashore at a time, sometimes regarded as an indication of carrying capacity, is more likely to be based on: (i) the practicalities of getting shore parties back aboard in case of emergency; and (ii) maintaining a favourable ratio of staff to passengers ashore. Carrying capacity is not considered in any of the management plans of the comparable venues discussed above.

Limits of acceptable change

Glacier Bay management uses elements of a 'limits of acceptable change' (LAC) management planning strategy, which evaluates how much change is acceptable in an environment and manages accordingly. Designed originally for designated Wilderness Areas managed by the US Forest Service, LAC techniques since their inception have demonstrated considerable flexibility and resiliency. LAC planning is based on the following.

1. Background review and evaluation of conditions.
2. Selection of likely changes and indicators for change.
3. Inventory of resource and social conditions.
4. Specification of quality standards for indicators.
5. Prescription of desired conditions within zone(s) of development.
6. Agreement of management action to maintain quality.
7. Implementation monitoring and review. (Adapted from Hall and Page, 2002: 138)

Emphasis is on the conditions desired for the area, rather than how much use the area can tolerate (Stankey et al., 1985). When the conditions meet the limit of acceptable change, then remedial actions need to be taken, usually in the form of limiting use of the region.

Wilderness Areas are so designated to protect their 'wilderness character'. Article 3, Subsection 1 of the Protocol similarly defines '… protection of the Antarctic environment and dependent and associated ecosystems and the intrinsic value of Antarctica, including its wilderness and aesthetic values …' as '… fundamental considerations in the planning and conduct of all activities within the Antarctic Treaty area'. In practical terms Wilderness Areas share with polar regions an almost complete lack of infrastructure and tenuous emergency response capabilities, and management measures that prove effective for one might well be applicable to the other.

The benefits of LAC planning are its measurable objectives and allowance for a diversity of visitor experiences; also quantifiable standards, flexibility and allowance for public involvement. Stankey et al. (1985) explain that only a few of these systems have been implemented because they were developed by natural area managers without tourism industry involvement. Davis (1995a) has suggested that a tourism management system based on an LAC framework could work in the Antarctic, in which only activities in keeping with the wilderness qualities of the area should be permitted. However, Bertram (2005: 63–66) concludes that the wilderness concept has not been sufficiently defined in Treaty-related instruments, and instead proposes ecological integrity based on Crosbie (1998) as a management objective for monitoring sites – a systems approach within an LAC framework which could then be used to decide on desired conditions to be maintained at each site.

Multiple resource management planning

Snyder (2003) and Snyder and Stonehouse (Chapter 15, this volume) have proposed for South Georgia a Multiple Resource Management Plan (MRMP), which we here consider for possible application to Antarctic sites. The MRMP, containing elements of limits of acceptable change, aims to establish human activities that are compatible with the environmental quality, cultural integrity and economic feasibility of the region. It creates objectives to: (i) conserve the area's natural and cultural resources; (ii) protect the health, safety and enjoyment of the tourist; (iii) diminish the potential liabilities for the tour operators; and (iv) support the long-term economic feasibility of responsible resource management in the area, and includes monitoring to assess sustainability. An outline of the MRMP process appears in Fig. 17.1.

The first phase of an MRMP identifies the environmental, socio-cultural and economic aspects of the region, including an inventory of natural attractions of interest to tourists, and the sensitivity or vulnerability of the area including natural dynamics. Snyder (2003: 3–4) proposes that for South Georgia economic factors would assess the willingness of private investors, entrepreneurs and businesses to support ecotourism projects. The ATS lacks the powers of a sovereign government to supervise or benefit from commercial operations, but as tour operators are predominantly IAATO members, IAATO could in his view conceivably continue in its role of enforcing ecologically sensitive operations, as well as supervising a management and monitoring fund derived from a conservation tax on tourists (see 'Antarctic site monitoring programmes' below).

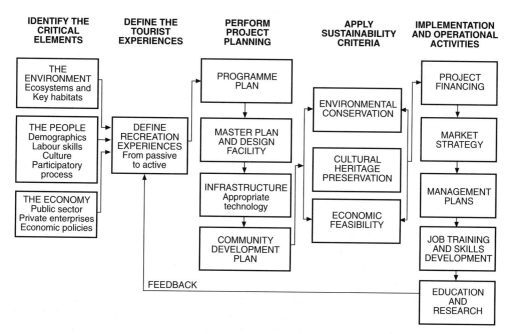

Fig. 17.1. Multiple resource management planning. (Source: Snyder, 2003.)

The second phase defines the 'tourist experience', to determine the types of recreational activities that can be offered and their associated support system. This also takes into account visitor satisfaction, which is often absent in other approaches to ensure sustainable tourism (Snyder, personal communication, 2004). The third phase creates a Tourism Management Plan, including how the activities proposed will fit with the environmental and socio-cultural elements of the continent. Applied to Antarctica, this plan would provide the foundation on which an overall management plan for the Antarctica would be drawn up. The fourth and last phase determines sustainability, querying for example to what extent the region can tolerate visitation: geographical distribution of visitors and types of recreation. This would include rare species that may need to be monitored, as well as level of erosion and water pollution.

This system uses the basic concept of management by objectives and the zoning proposed in the Environmental Management Plan for South Georgia. It presents an approach similar to LAC, allowing feedback mechanisms to constantly re-evaluate the system, and includes a range of different values in its systems methodology.

Management Plans for Landing Sites

The development of management plans is a key element of integrated planning processes (CONCOM, 1986; cited in Keage *et al.*, 1989: 310):

> [Management plans] are aimed at on-site decision-makers and site users, but have additional value as an educational means of pursuing wider conservation objectives and practical implementation of a conservation ethic and should normally be an explicit component of a site plan. They should generally convey the planning principles that have been developed to ensure that conservation objectives are achieved.

Antarctic tourism management would benefit from an overall management plan for the continent, from which management objectives and plans for individual sites could be developed. Stankey (1982: 151) lists five steps necessary for creating a management plan for an area:

1. Collecting baseline data.
2. Creating a planning framework which provides set standards and action plans.
3. Preparing for the consequences of actions.
4. Developing a management philosophy.
5. Establishing a range of environmental classes in natural areas.

For tourism, establishing environmental classes could include setting aside areas robust enough to allow frequent visitation. Provisions for protection of areas within the Treaty area are currently focused on scientific objectives rather than environmental issues. There is an additional requirement for the active management of landing sites, possibly involving a system of general management plans for all sites, and special plans for those that are considered most vulnerable to change.

Antarctic Specially Protected Areas, Antarctic Specially Managed Areas and Areas of Special Tourist Interest

Annex V of the Protocol provides for Antarctic Specially Protected Areas (ASPAs) and Antarctic Specially Managed Areas (ASMAs).

ASPAs, set aside for long-term scientific studies, can be entered only by permits; as permits are not issued for tourist visits, ASPAs are closed to tourists.

ASMAs, according to Annex V, Article 4.1, include areas where:

> activities are being conducted or may in the future be conducted ... to assist in the planning and coordination of activities, avoid possible conflicts, improve cooperation between Parties or minimise environmental impacts.

These were designed particularly to control station areas, where they aim to minimize the environmental impacts of station activities.

ASTIs, originally designated at the 8th ATCM (1975), were subject to controversy within Treaty deliberations and no such areas have ever been designated. Representatives may have argued that such areas where 'tourist activities could be systematically assessed' (Roberts, 1977: 101) would become 'honey pots' – areas where impacts are concentrated (IUCN, 1991: 35). There are already a number of sites that could constitute designated 'tourist sites' as they are visited by most cruise ships each year (Chapter 9, this volume). Possible benefits of ASTIs have recently been reviewed in a report (France, 2006) tabled at the 21st ATCM.

Walton (1994: 95) proposed that, in the absence of ASTIs, tourist sites could be designated as ASMAs and provided with management plans which recognize tourism as the principal activity, with selected ASPAs designated as control areas against which to measure impact. Investigation of impacts would allow framing of clear management rules for each site and an assessment of the degree of impact permitted before reducing tourist pressure. Kriwoken and Rootes (2000: 147) also suggest that these areas would allow for regional assessment, and reducing cumulative effects of ship-borne tourism. However, the ASMA category was designed to meet the need to manage research stations and their immediate environs, and requires on-site management. Only one variant has so far found favour – an ASMA covering several stations, scientific localities and tourist landing sites in Admiralty Bay, South Shetland Islands (Richardson, 1999: 9). On-site management of landing sites, other than at stations, is not considered as an option in this plan, and would not currently be feasible for most individual tourist landing sites.

If the ASTI concept is not favoured by the ATS, could tourist visits be restricted only to certain sites, to avoid general degradation of all the sites currently in use? This would prove unpopular with tour operators, expedition leaders and ships' captains, who favour being able to choose from the full range of sites, as and when opportunities allow. It is likely also that ATCMs would be unwilling to attempt imposing restrictions of movement on any Antarctic stakeholder, other than by measures already argued and formally agreed.

Levels of usage

Of the 270 or more landing sites identified in US National Science Foundation statistics (IAATO, 2001) as having been visited since 1995/6, only a few are visited often enough to be considered at risk of human-induced changes (Stonehouse and Bertram, 2001; see also Chapter 9, this volume). In 2001, for example, over half of the recognized sites remained unvisited. Among the most popular sites, only nine received more than 40 visits per season (on average two or three visits per week), and only 16 sites were visited more than once weekly. Though overall visits have continued to increase since 2000/1 (see also Chapter 9, this volume), this analysis suggests that site conservation is likely to be needed at only a small number of sites that are either particularly sensitive to human disturbance or among the most popular, or possibly both.

Monitoring for Change in Landing Sites

> The Australian Antarctic Division should immediately develop management
> regimes for environmental protection and establish procedures for the planning,
> administration and monitoring of tourist activities within the AAT.

This statement appears among the conclusions of an early report, *Tourism in Antarctica* (Commonwealth of Australia, 1989), which Herr and Davis (1996: 348) cite as among the first to consider possible impacts of tourism, their monitoring and mitigation.

'Monitoring' in an environmental context is a method of detecting and measuring changes by collecting time-series of data for defined purposes, and observing trends in selected variables (Walton *et al.*, 2001: 33). Perceived as vital to all aspects of conservation (Abbott and Benninghoff, 1990), it is used particularly to identify alterations to ecosystems caused by human activity, beyond those caused by natural variations.

Spellerberg (1991: 186) regards monitoring as the preferred basis for environmental impact assessments, stressing that simplicity, reliability and stability are essential for both the selection of parameters and for data collection. He summarizes the value of ecological and biological monitoring as means of: (i) establishing whether ecosystems and populations are being managed and conserved effectively; (ii) assessing the best use of the land; (iii) indicating the state of the environment; and (iv) advancing knowledge of the dynamics of the ecosystem. Figure 17.2, based on Spellerberg's model of steps to be used in an environmental monitoring programme, illustrates his view that effective monitoring demands a series of review stages to assess the appropriateness of the objectives, variables and methods, and ultimately to determine whether or not specified management objectives have been met.

Antarctic site monitoring programmes

A Scientific Committee on Antarctic Research/Council of Managers of National Antarctic Programs workshop (SCAR/COMNAP, 1996: 9), charged with considering monitoring of sites within the Antarctic Protected Area System (though not tourist landing sites), defined three objectives:

1. To protect the Antarctic's scientific value.
2. To help in the continuous improvement of Antarctic environmental management.
3. To meet legal requirements of the Protocol and national legislation.

To achieve these objectives the workshop recommended four stages in developing a monitoring programme:

1. Developing management objectives and specific informational needs.
2. Developing a testable hypothesis specific to the site to be monitored, and implementing a pilot study to ensure that the proposed design is feasible.
3. Implementing a full study using the most appropriate technologies, methodologies, statistical designs and data management techniques.

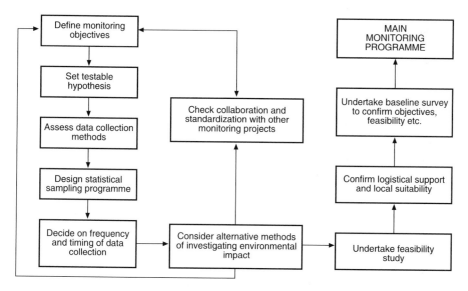

Fig. 17.2. Steps to be used in the design of an environmental monitoring programme. (Adapted from Spellerberg, 1991: 182.)

4. Assessing data on a regular basis, and developing specific management recommendations as a result.

A workshop convened in 2000 by the US National Science Foundation and Environmental Protection Agency and IAATO considered the 'Assessment of possible cumulative environmental impacts of commercial ship based tourism in the Antarctic Peninsula area'. It concluded that there was a need for site monitoring, coordination with related long-term monitoring programmes, tour planning and expedition long-term planning and evaluation, but failed to recommend how cumulative impacts could be identified (Hofman and Jatko, 2000: 78–81).

Despite many statements of intent and agreements in principle, Antarctic landing sites are not currently being monitored to any standards resembling those discussed above. The most probable reason is that effective monitoring to worthwhile standards is time-consuming work for professional biologists, and thus likely to prove expensive. There is no money available for it: the industry pays no taxes for the privilege of using Antarctica, and the Treaty makes no provision for collecting revenues towards the costs of effective management.

A secondary reason is that site-specific management objectives – considered by Spellerberg and other management ecologists as essential components of monitoring – have not been defined. While responsibility for the undefined monitoring specified by the Protocol is accepted by IAATO as a condition of their permitting, neither collectively nor individually are tour operators likely to consider monitoring of landing sites to full professional standards. Nor would such monitoring be possible in the absence of defined management objectives, which must surely be a responsibility of those who issue permits.

Thus monitoring is a legal requirement for permitting under the Protocol, but whether by design or default, standards of monitoring have not been defined. While monitoring to full professional standards is, or can become, a complex and expensive long-term process, token monitoring without management objectives invites derision. What level of monitoring is required to ensure the lasting integrity of landing sites?

The Ecosystem Monitoring Programme (CEMP) adopted by the Convention on the Conservation of Antarctic Marine Living Resources (CCAMLR), which approaches standard ecological principles of monitoring more closely than anything so far proposed for tourism, suggests that bird populations and breeding success parameters should be measured for a minimum of 10 years for trends to be established (CCAMLR, 1992). Walton and Shears (1993: 84) regard management and storage of data as paramount to the success of a monitoring programme, as well as the ability to duplicate this monitoring across sites. Systems for monitoring change at individual landing sites would need to be incorporated into a larger management planning system, as discussed above for other similar wilderness areas. This could include the development of management plans for Antarctica as a whole, within which tourism would be integrated, with the potential for zoning areas, permitting and fees for visitation. Revenue so generated might determine amounts to be spent and numbers of sites monitored, but the overall management planning system, of which such monitoring would be a part, would need to be implemented within an ATS-recognized regime.

Minbashian (1997) and Crosbie (1998), both ecologists with considerable ship-board experience in Antarctic tourism, have proposed much simpler systems of monitoring that would fully meet the requirements of the COMNAP workshop cited above, and incidentally add to our knowledge of the dynamics of Antarctic coastal ecosystems. However, neither elaborate nor simple developments in monitoring appear currently to be under consideration.

Instead the Treaty is focusing on a minimal objective of designing 'guidelines' for visitor behaviour at the most popular sites. Under Resolution 5 (Site Guidelines for Visitors) arising from the 20th ATCM (2005), guidelines were developed for four sites. An Intersessional Contact Group on Site Guidelines (ICG) for Visitors to Antarctica of the ATS Committee for Environmental Protection has revised these guidelines, and also drafted new guidelines for a further seven sites which now await approval (United Kingdom, 2006: 3). This process, undertaken by the ICG using the CEP online Discussion Forum, involved three rounds of discussion including all key stakeholders, as well as information from the IAATO site-specific guidelines (IAATO, 2003). In addition site visits took place in February 2006 with representatives of Australia, Argentina, Norway, the UK, IAATO and Oceanites (United Kingdom, 2006).

In demonstrating care for landing sites on an individual basis, and in recognizing the need for increased management at the most used sites, such guidelines are a positive and long-awaited development. However, in a view expressed by the Antarctic and Southern Ocean Coalition (ASOC, 2006: 6) and generally shared by management ecologists, guidelines remain a reactive rather than proactive management measure, which do not develop techniques for active management of sites or add to the development of ongoing monitoring or assessment of these sites.

Meanwhile Antarctic tourism is following a predictable path of development and diversification, and landing sites are being visited by more and more tourists each year. From our own observations and those of many colleagues, the first signs of degradation are starting to appear at heavily visited sites, in the form of worn tracks and trampling. While these seldom persist over winter, they are indications of more serious and lasting changes. They are most readily detectable, not by 'monitoring', but by simple, straightforward observation on the part of naturalists who visit the sites regularly, and would no doubt be interested to pass on their observations to any body capable of effecting remediation.

Recovery and remediation

Bölter and Stonehouse (2001: 402) tabulate a vocabulary of terms relating to disturbance and recovery of soil ecosystems, based on a range of authors, that is particularly relevant to polar environments. Recovery is the process by which a soil or ecosystem achieves biological and physical stability following disturbance; remediation is human-induced processes that encourage recovery. Recovery may be complete if the ecosystem is re-established in its original state, or functional if it achieves stability at a level different from the original.

Suppose that, after a period of four or five years, either monitoring or simple observation confirmed the degradation of a particular Antarctic landing site: what recovery processes would be available, and what remediation would be practicable?

In the words of Bölter and Stonehouse (2001: 403):

> Remediation and ultimate restoration of Antarctic landscapes to functional levels of recovery are entirely possible, so long as expectations remain modest and patience is exercised. ... Recovery that in temperate regions might be expected within a human life span may take two or three times longer in Antarctica. Remediation is costly, competing with scientific research for time and funding, and governments have so far proved reluctant to spend lavishly in restoring disturbed or damaged sites.

The forms of damage most likely to appear on Antarctic landing sites are: (i) reduction in size of bird breeding colonies; and (ii) man-made paths over gravel or vegetated surfaces. Both present ambiguities: bird colonies respond also to environmental changes that are not tourist-induced (Patterson *et al.*, 2003) and vegetation is constantly damaged by seals, birds and weather in the absence of tourists.

However, if damage is perceived, and there is reason to believe it is due to tourist visits, what remedies are available? Penguins, giant petrels and skuas cannot be coerced into rejoining depleted colonies, and working parties are unlikely ever to be employed in the probably thankless task of replacing moss or turf. As Stonehouse and Bertram (2001; cited in Bertram, 2005: 310) have pointed out:

> Should environmental damage to a site become apparent, for which tourist visits are deemed to be responsible, the only practicable remedies will be to restrict access to the part of the site that is at risk, or curtail visits for several seasons to encourage natural restitution.

The most pressing need is therefore for a simple system of reporting signs of wear and tear, followed by inspection, assessment and a recommendation for restriction of visits. Such a system falls far short of professional monitoring, but is also less likely to be dismissed, avoided or simply ignored on grounds of expense or practicability.

Immediate, temporary restrictions of visits to sites, unlikely to be effected through ATCMs, currently lie in the hands of IAATO, on a voluntary basis. IAATO has in the past afforded protection to more than one site by circulating information and requesting a voluntary abstention from visits by its members.

Should longer protection be required with Protocol sanction, an ATCM could declare it off-limits as an ASPA, with the purpose, in the words of Article 3 of the Protocol:

to protect outstanding environmental, scientific, historic aesthetic or wilderness values, any combination of these values, or ongoing or planned scientific research.

The required research, expressed in the management plan required under Protocol regulations, would consist of a well-defined monitoring programme with the management objective of measuring the site's recovery. ASPAs are designated for indefinite periods, but management plans require reviewing at least every 5 years. It would thus be possible to retain a site within ASPA status for as long as it takes to reach a complete or at least satisfactory level of functional recovery, and to release it thereafter.

There remains the legal requirement of permit holders to 'monitor', which is currently neglected for want of understanding, methodology or motivation. Minbashian (1997) and Crosbie (1998) have suggested uncomplicated and direct forms of monitoring, based on a simple, even 'programmatical' management plan that is capable of rapid adaptation to the needs of particular sites. A case for closure could be met by combined IAATO and Treaty action to ensure the rapid safeguarding of a site at risk.

Summary and Conclusions

Provisions for environmental protection under the Protocol to the Antarctic Treaty work well for the scientific activities for which they were designed, but in relation to tourism fall far short of practices in other polar and sub-polar areas. Marshalled by IAATO, the tourist industry has a sound record of self-regulation. However, as numbers of ships and tourists continue to increase, there is a growing need for management of landing sites, especially those that are among the most popular and frequently visited. Monitoring, though required under the Protocol, in a technical sense may be an over-elaboration of what landing sites need initially for their protection against over-use. What is required is a simple system of damage-reporting, inspection, and if necessary temporary closure – all measures well within existing capabilities of IAATO and the ATS.

References

Abbott, B. and Benninghoff, W.S. (1990) Orientation of environmental change studies to the conservation of Antarctic ecosystems. In: Kerry, K. and Hempel, G. (eds) *Antarctic Ecosystems: Ecological Change and Conservation.* Springer-Verlag, Berlin, pp. 394–403.

ASOC (2006) Managing Antarctic tourism: a critical review of site-specific guidelines. *Agenda Item 7, Antarctic Treaty Consultative Meeting XXIX*, Edinburgh, 12–23 June. Antarctic and Southern Ocean Coalition, Washington, DC.

Bertram, E. (2005) Tourists, gateway ports and the regulation of shipborne tourism in wilderness regions: the case of Antarctica. PhD thesis, Royal Holloway, University of London, London.

Bölter, M. and Stonehouse, B. (2002) Uses, preservation, and protection of Antarctic coastal regions. In: Beyer, L. and Bölter, M. (eds) *Geoecology of Antarctic Ice-free Coastal Landscapes.* Springer-Verlag, Berlin, pp. 393–407.

CCAMLR (1992) *Ecosystem Monitoring Programme Standard Methods* A6.1. Convention on the Conservation of Antarctic Marine Living Resources, Hobart, Tasmania.

Ceballos-Lascuráin, H. (1996) *Tourism, Ecotourism and Protected Areas: The State of Nature Based Tourism around the World and Guidelines for Its Development.* International Union for the Conservation of Nature and Natural Resources, Gland, Switzerland.

Commonwealth of Australia (1989) *Tourism in Antarctica: Report of the House of Representatives Standing Committee on Environment, Recreation and the Arts.* Australian Government Publishing Service, Canberra.

CONCOM (1988) *Plans of Management for Protected Areas: A Report by the CONCOM Working Group on Management of National Parks.* Department of Arts, Heritage and Government, Canberra.

Crosbie, K. (1998) Monitoring and management of tourist activities in the Maritime Antarctic. PhD thesis, Scott Polar Research Institute, Cambridge, UK.

Davis, P. (1995a) Wilderness and visitor management and Antarctic tourism. PhD thesis, Scott Polar Research Institute, Cambridge, UK.

Davis, P. (1995b) Antarctic visitor behaviour: are guidelines enough? *Polar Record* 31(178), 327–334.

Enzenbacher, D.J. (1992) Antarctic tourism and environmental concerns. *Marine Pollution Bulletin* 25(9–12), 258–265.

France (2006) Establishment of Areas of Special Tourist Interest. *Agenda Item 12 (WP18), Antarctic Treaty Consultative Meeting XXIX*, Edinburgh, 12–23 June.

Hall, C.M. and Johnston, M. (eds) (1995) *Polar Tourism: Tourism in the Arctic and Antarctic Regions.* John Wiley & Sons, Chichester, UK.

Hall, C.M. and Page, S.T. (2002) *The Geography of Tourism and Recreation: Environment, Place and Space.* Routledge, London.

Hall, C.M. and Wouters, M. (1994) Managing nature tourism in the sub-Antarctic. *Annals of Tourism Research* 21(2), 355–374.

Hemmings, A. and Roura, R. (2003) A square peg in a round hole: fitting impact assessment under the Antarctic Environmental Protocol to Antarctic tourism. *Impact Assessment and Project Appraisal* 21(1), 13–24.

Herr, R.A. and Davis, B.W. (1996) ATS decision making and change: the role of domestic politics in Australia. In: Stokke, O.S. and Vidas, D. (eds) *Governing the Antarctic.* Cambridge University Press, Cambridge, UK, pp. 331–360.

Hofman, R.J. and Jatko, J. (eds) (2000) *Assessment of Possible Cumulative Environmental Impacts of Commercial Ship Based Tourism in the Antarctic Peninsula Area.* Proceedings of a Workshop held in La Jolla, California, 7–9 June. National Science Foundation, Washington, DC.

IAATO (2001) Report of the International Association of Antarctica Tour Operators 2000–1. *Agenda Item 5b, Antarctic Treaty Consultative Meeting XXIV*, St Petersburg, Russian Federation, 9–20 July. International Association of Antarctica Tour Operators, Basalt, Colorado.

IAATO (2003) IAATO site specific guidelines 2003 in the Antarctic Peninsula. *Agenda Item 10, Antarctic Treaty Consultative Meeting XXVI*, Madrid, 9–20 June. International Association of Antarctica Tour Operators, Basalt, Colorado.

IUCN (1991) *A Strategy for Antarctic Conservation.* International Union for the Conservation of Nature and Natural Resources, Gland, Switzerland.

IUCN (1992) Tourism in Antarctica. *IP 18, Antarctic Treaty Consultative Meeting XVII*, Venice, Italy, 11–20 November. International Union for the Conservation of Nature and Natural Resources, Gland, Switzerland.

Johnston, M. (1995) Patterns and issues in Arctic and sub-Arctic tourism. In: Hall, C.M. and Johnston, M. (eds) *Polar Tourism: Tourism in the Arctic and Antarctic Regions.* John Wiley & Sons, Chichester, UK, pp. 27–42.

Johnston, M. (1997) Polar tourism regulation strategies: controlling visitors through codes of conduct and legislation. *Polar Record* 33(184), 13–20.

Johnston, M. and Hall, C.M. (1995) Visitor management and the future of tourism in polar regions. In: Hall, C.M. and Johnston, M. (eds) *Polar Tourism: Tourism in the Arctic and Antarctic Regions.* John Wiley & Sons, Chichester, UK, pp. 297–313.

Keage, P.L., Hay, P.R. and Russell, J.A. (1989) Improving Antarctic management plans. *Polar Record* 25(155), 309–314.

Kriwoken, L. and Rootes, D. (2000) Tourism on ice: environmental impact assessment of Antarctic tourism. *Impact Assessment and Project Appraisal* 18(2), 138–150.

Kuss, F.R., Graefe, A.R. and Vaske, J.J. (1990) *Visitor Impact Management: A Review of Research.* National Parks and Conservation Association, Washington, DC.

McIntosh, E. and Walton, D. (2000) *Environmental Management Plan for South Georgia.* British Antarctic Survey, Cambridge, UK.

Minbashian, J. (1997) Biological integrity: an approach to monitoring human disturbance in the Antarctic Peninsula region. MPhil thesis, Scott Polar Research Institute, Cambridge, UK.

Ministry of Environment (1994) *Management Plan for Tourism and Outdoor Recreation in Svalbard.* Ministry of Environment, Oslo.

Ministry of Justice (1993) *Regulations Relating to Tourism and Other Travel in Svalbard.* Ministry of Justice, Oslo.

Nuttall, M. (1998) *Protecting the Arctic. Indigenous Peoples and Cultural Survival.* Harwood Academic Publishers, Amsterdam.

Naveen, R., Forrest, S.C., Dagit, R.G., Blight, L.K., Trivelpiece, W.Z. and Trivelpiece, S.G. (2001) Zodiac landings by tourist ships in the Antarctic Peninsula region 1989–99. *Polar Record* 37(201), 121–132.

NPS (1995) *Glacier Bay National Park and Preserve Vessel Management Plan and Environmental Assessment.* US Department of the Interior, National Park Service, Denver Service Centre, Denver, Colorado.

NPS (2005) *Statistical Abstracts prepared by the National Park Service Social Science Programme.* Public Use Statistics Office, Denver, Colorado.

Patterson, D., Easter-Pilcher, A. and Fraser, W. (2003) The effects of human activity and environmental variability on long-term changes in Adélie penguin populations at Palmer Station, Antarctica. In: Huiskes, A.H.L., Gieskes, W.W.C., Rozema, J., Schorno, R.M.L., van der Vies, S.M. and Wolff, W.J. (eds) *Antarctic Biology in a Global Context.* Backhuys, Leiden, The Netherlands, pp. 301–307.

Poncet, S. and Crosbie, K. (2005) *A Visitor's Guide to South Georgia.* Wildguides, Old Basing, UK.

Richardson, M.G. (1999) *Regulating Tourism in the Antarctic: Issues of Environment and Jurisdiction.* Antarctic Project Report 2/99. Fridtjof Nansen Institute, Lysaker, Finland, pp. 1–19.

Roberts, B.B. (1977) Conservation in the Antarctic. *Philosophical Transactions of the Royal Society* 279, 97–104.

SCAR/COMNAP (1996) *Monitoring for Environmental Impacts from Science and Operations in Antarctica.* Scientific Committee on Antarctic Research, Cambridge, UK.

Scott, J. and Poncet, S. (2003) *South Georgia Environmental Mapping Report.* Government of South Georgia and the South Sandwich Islands, Stanley.

Snyder, J.M. (2003) A systems approach to sustainable resource management in South Georgia: applying the approach to Grytviken. Paper presented at *The Future of South Georgia, South Georgia Association Conference*. Strategic Studies Inc., Centennial, Colorado.

Spellerberg, I.F. (1991) *Monitoring Ecological Change*. Cambridge University Press, Cambridge, UK.

Stankey, G.H. (1982) The role of management in wilderness and natural-area preservation. *Environmental Conservation* 9(2), 149–155.

Stankey, G.H., Cole, D.N., Lucas, R.C., Peterson, M.E. and Frissell, S. (1985) *The Limits of Acceptable Change System for Wilderness Planning*. Research Paper INT-176. US Department of Agriculture Forest Service, Intermountain Forest and Range Experiment Station, Ogden, Utah.

Stonehouse, B. (1998) Polar shipborne tourism: do guidelines and codes of conduct work? In: Humphreys, B., Pederson, A., Prokosch, A.Ø., Smith, P. and Stonehouse, B. (eds) Linking Tourism and Conservation in the Arctic. Proceedings from Workshops on 20–22 January 1996 and 7–10 March 1997 in Longyearbyen, Svalbard. *Norsk Polarinstitutt Meddelelser* 159, 49–58.

Stonehouse, B. and Bertram, E. (2001) Monitoring and assessment of impacts at Antarctic tourist landing sites. Paper presented at *Antarctic Biology in a Global Context, VIII SCAR International Biology Symposium*, Amsterdam, 27 August–1 September. Abstract S6003.

Stonehouse, B. and Brigham, L. (2000) The cruise of MS *Rotterdam* in Antarctic waters. *Polar Record* 36(199), 347–349.

Tracey, P. (2001) Managing Antarctic tourism. PhD thesis, Institute of Antarctic and Southern Ocean Studies, University of Tasmania, Hobart, Tasmania.

United Kingdom (2006) Report of the CEP Intersessional Contact Group on site guidelines for visitors to Antarctica. *CEP Agenda Item 7, Antarctic Treaty Consultative Meeting XXIX*, Edinburgh, 12–23 June.

Viken, A. (1995) Tourism experiences in the Arctic – the Svalbard case. In: Hall, C.M. and Johnston, M. (eds) *Polar Tourism: Tourism in the Arctic and Antarctic Regions*. John Wiley & Sons, Chichester, UK, pp. 73–84.

Walton, D. (1994) Summary and conclusions. In: Lewis-Smith, R.I., Walton, D.H. and Dingwall, P.R. (eds) *Developing the Antarctic Protected Area System*. Proceedings of the SCAR/IUCN Workshop on Antarctic Protected Areas, 29 June–2 July 1992, Cambridge, UK. International Union for the Conservation of Nature and Natural Resources, Cambridge, UK, pp. 15–26.

Walton, D. (2003) *Environmental Management Plan for South Georgia. The Future of South Georgia: A Programme for the Next 10 Years*. South Georgia Association, Cambridge, UK.

Walton, D.W. and Shears, J. (1993) The need for environmental monitoring in Antarctica: baselines, environmental impact assessments, accidents and footprints. *International Journal of Analytical Chemistry* 55, 77–90.

Walton, D.W., Scarponi, G. and Cescon, P. (2001) A scientific framework for environmental monitoring in Antarctica. In: Caroli, S., Cescon, P. and Walton, D.W. (eds) *Environmental Contamination in Antarctica: A Challenge in Analytical Chemistry*. Elsevier Science, Oxford, UK, pp. 33–53.

Wouters, M. and Hall, C.M. (1995) Managing tourism in the sub-Antarctic islands. In: Hall, C.M. and Johnston, M. (eds) *Polar Tourism: Tourism in the Arctic and Antarctic Regions*. John Wiley & Sons, Chichester, UK, pp. 277–295.

WWF Arctic Programme (1997) WWF Arctic tourism project: linking tourism and conservation in the Arctic. *WWF Arctic Bulletin* 4.97. WWF Arctic Programme, Oslo.
Williams, S. (1998) *Tourism Geography*. Routledge, London.

Index

Abercrombie & Fitch 16
Aboriginal Tourism Canada 86
access
   allemansratt (freedom of access) 115
   difficulties 7–9
   enhanced by climate change 40
Adélie penguins 218
Adventure Network International/Antarctic
      Logistics and Expeditions
      (ANI/ALE, formerly Adventure Network
      International) 173–175
adventure tourism
   characteristics and management challenges
      64–65
   incidents causing concern 176–180
   increased control and management 16
   land-based 173–175
   other forms 175–176
   policy and regulations in the Antarctic 171,
      180–183
      position of IAATO 171
   ship-based 172–173
   significance and scope 171–172
   South Georgia 252
Alaska
   Anaktuvuk Pass 72–75
   economic impact of tourism 117–119
   Gold Rush 23–24
   hunting and fishing statistics 59
   lessons learned 81–82
   nature tourism 63
   Northwestern Alaska 75–78

   origins of Alaskan tourism 20–23, 71–72
   personal value of wildlife 119
   restored building in Skagway 67(illust)
   tourist numbers 54, 118
   Yakutat 78–80
   see also Glacier Bay, Alaska
Alaska Commercial Company 21
Alaska Native Council 65
Alaska's Recreational Riches (Alaska
      Development Board 1946) 117–118
allemansratt 115
Amundsen, Roald 27
Anaktuvuk Pass, Alaska 72–75
angling *see* hunting and fishing
animals *see* wildlife
Antarctic: maps 34, 150
   Antarctic and Southern Ocean Coalition
      138, 142–143, 214
Antarctic (bulletin, New Zealand Antarctic
      Society) 211
Antarctic Convergence 33
Antarctic Logistics Centre International 176
Antarctic Non-government Activity Newsletter
      170
Antarctic Science, Tourism and Conservation
      Act (1996, USA) 219–220
Antarctic Treaty System 13, 131
   Consultative Meetings (ATCMs) 215, 299
   environmental impact assessments 286
   Environmental Protocol 1996 165–166,
      171, 214–215, 234, 299–300
   incorporation of port state jurisdiction
      143–144

Antarctic Treaty System *continued*
    management of tourism 206
        elements lacking 285–286, 287, 303
    prohibition of mineral extraction 36
    and South Africa 136–137
    survey of treaty and its implications
        213–214
Appleton's Guide Book to Alaska and the
    Northwest (Scidmore) 21
Arctic Climate Impact Assessment (Arctic
    Council) 38–39
Arctic Council 38, 39
Arctic haze 38
Arctic International Programme World Wildlife
    Fund 243–244
Arctic Marine Strategic Plan (Arctic Council)
    39
Arctic region 33(map)
Arctic Trail 116
Areas of Special Tourist Interest 299–300
Argentina
    see Ushuaia, Argentina
Asker, Phil 190
ASOC *see* Antarctic and Southern Ocean
    Coalition
Atlantic Monthly, The 19
Australia
    Antarctic overflights 190–193
    Economic Exclusion Zones 138
    Heard Island and McDonald Islands
        270–274(tab) 275–276
    Hobart *see* Hobart, Australia
Auyuittuq National Park Reserve 90
aviation
    adventure tourism 173
    air shuttles 127, 132, 134, 138, 142
    commercial air transport 57–58
    comparison with ship-borne tourism
        157–158
    flight/cruise operations 155
    International Antarctic Centre, Canterbury
        Airport, NZ 140
    overflights 20, 158
        from Australia 190–193
        author's experiences 193–194
        early initiatives 188–190
        passenger survey 194–195
        safety 194
        from South America 193
        tourist numbers 201(tab)
    significance for tourism 26–27

surveys of activities 216
tourist numbers 123(tab) 158(tab) 160,
    161(tab)

Baffin Island *see under* Nunavut
Bahia Paradiso: sinking 218
barriers to entry
    access 7–9
    cost 11–12
    environmental conditions 9–11
    jurisdictional constraints 12–13
    language 105–106
    time 12
bears, hunting of 60, 243
behaviour, tourist 238–239
Bennett, Thomas 17
Borchgrevink Coast, Victoria Land 192(illust)
boundaries: of polar regions 32–35
Bradford, William 17, 107

Canada
    benefits of aboriginal tourism 86–87
    economic impact of tourism 102–104
    Five Finger Rapids 24(illust)
    impact of Gold Rush on Arctic tourism
        23–24
    Kamestatin Lake 84–85
    Parks Canada 242
    sport fishing 60–61
capacity, carrying: of Antarctic landing sites
    295–296
    see also Nunavut
Cape Town, South Africa
    current status 137
    location and distance 125(map), 126(tab)
    origin and development 127, 128(tab)
    pre-WW2 harbour for whaling fleets 136
Carroll, Captain James 20–21, 22
Chaillu, Paul du 19
Chile
    Puerto Williams 134
    Punta Arenas *see* Punta Arenas, Chile
Christchurch/Lyttelton, New Zealand
    current status 140–142
    economic significance of national
        expeditions 141(tab)
    location and distance 125(map), 126(tab)
    origin and development 127, 128(tab), 139
circles, polar 32, 33–34

climate: as barrier to entry 8
Cloutier, Sheila Watt 44
codes of conduct *see* guidelines and
    regulations
Collis, Mrs Septima 238–239
community-based tourism 85–86
conservation, environmental *see under*
    environment
Convention for the Conservation of Antarctic
    Marine Living Resources 214–215,
    302
Convention on the Regulation of Antarctic
    Mineral Resource Activities 215
Cook, Captain James 248
Cook, Dr Frederick 19
Council of Managers of National Antarctic
    Programs (COMNAP) 171
Croydon Travel Melbourne 190–191
cruises *see* ship-borne tourism
Crystal Symphony (cruise ship) 153(illust)
culture
    culture and heritage tourism 65–67
    and management of tourism 235–237,
        253–254
Cunard Princess 5(illust)
Cuverville Island, Danco Coast 217

definitions: polar regions 32–35
Denmark: input into Greenland tourism 108
Disenchantment Bay, Alaska 78, 79
Drygalski Fjord, South Georgia 249(illust)

Economic Exclusion Zones 138
economy and finance
    cost of travel as barrier to access 11–12
    economic impact of national expeditions
        141(tab)
    economic impact of tourism
        aboriginal tourism 86–87
        Alaska 117–119
        Antarctic overflights 191
        Canada 102–104
        cruise ships 53–57
        culture and heritage tourism 65
        Finland 104–106
        to gateway ports 129(tab), 133(tab),
            135(tab)
        Greenland 106–109
        hunting and fishing 58–62

        Iceland 109–111
        nature tourism 63
        Norway 111–112
        Russia 113–115
        Sweden 115–117
    economic limitations to Antarctic tourism
        204–205
    implications for resource management
        255–256
Ecosystem Monitoring Programme 302
ecosystems: define polar regions 33
ecotourism: Alaska 75
elitism 238–239
emergencies 151
Endeavour (cruise ship) 152(illust)
environment
    as barrier to entry 9–11
    cause and effect determination 235
    changes
        due to climatic warming 38–39
        due to human activities 35–38, 44
        effects on wildlife 40–43
        role of tourists 45–46
    conservation in Antarctic 254–255
        Environmental Protocol *see under*
            Antarctic Treaty System
        future trends 216
        IUCN report 215–216
        landing operations 217–219
        limits on Antarctic tourism 206–207
        monitoring for change 300–305
        recovery and remediation 303–305
    New Zealand Conservation Management
        Strategy 270–274(tab)
equipment 16
expeditions, private 174
Explorer's Journal, The 211

Falkland Islands *see* South Georgia
    Stanley *see* Stanley, Falkland Islands
Finland 104–106
fishing *see* hunting and fishing
flights *see* aviation
fur trapping 36

Galapagos 204
Gateway Antarctic, University of Canterbury,
    NZ 140

gateway ports, Antarctic
   air-borne tourism 142
   definition 124
   early tourism 28, 128–129
   economic significance of Antarctic 129,
         133(tab), 135(tab), 141(tab)
   flight/cruise operations 155
   journey times 125–126
   locations and distances 125(map), 126(tab)
   origins and development 126–128
   port state jurisdiction 142–144
   South African and Australasian sectors
         136–142
   South American sector 130–136
   tourist numbers 126(fig)
gentoo penguins 217, 218
Glacier Bay, Alaska 117, 238–239,
         288–289(tab), 292–293
   first tourists 20–21, 22
Gold rush 23–24, 117
Gough Island 277
Greenland 54, 106–109
Grytviken, South Georgia 248–249, 252(illust)
   access restrictions 253–254
guidebooks 19
   Appleton's Guide Book to Alaska and the
         Northwest (Scidmore) 21
guidelines and regulations
   discussions and research 219–220
   Environmental Protocol 1996 see under
         Antarctic Treaty System
   first code of conduct 214
   IAATO 15, 153–154, 206
   World Wildlife Fund 243–244
guides, licensed 244

Halfmoon Island, South Shetland Islands 217
Hannah Point, Livingston Island, South
         Shetland Islands 217–218
Harper's Weekly 18
Harriman Alaska Expedition 22–23
Hayes, Dr. Isaac I. 17–18, 107
Hayes/Bradford cruise 17–19
haze, Arctic 38
Heard Island and McDonald Islands
         270–274(tab), 275–276
Hedgepeth, Joel 212
helicopters 157–158
Herbert, Sir Wally 26
Hillary, Sir Edmund 29–30

Hobart, Australia
   current status 137–139
   location and distance 125(map), 126(tab)
   origin and development 127, 128(tab), 137
hunting and fishing
   allocation and management issues 76–77
   campaign against illegal commercial fishing
         138
   change in attitudes and practice 16
   economic significance 58–62
   by indigenous peoples 78–79
Hurley, Frank 27–28

IAATO see International Association of
         Antarctic Tour Operators
ice
   as barrier to entry 7, 8
   decrease in Arctic pack ice 37
   importance for Inuit travel 45
Ice Breaker (magazine, Hobart) 138
Ice Hotel, Jukkasjarvi 117
icebreakers: use as tourist cruise ships 25–26,
         53(illust)
Iceland
   economic impact of tourism 109–111
   sport fishing 59
   tourist numbers 54, 109
Ilulissat Icefjord 108
indigenous populations
   characteristics 90
   and cultural/heritage tourism 65–66
   effects of climate change 44–45
   impact of tourism 46
   need to understand market 74–75
infrastructure, Antarctic 234–235
Initial Environmental Evaluations 166
Instituto Fuegino de Turismo 132
International Association of Antarctic Tour
         Operators (IAATO) 129, 150–151,
         198–199
   guidelines and regulations 153–154, 156,
         157, 206, 234
   landing site database 217
   membership categories 199–200
   position on adventure tourism 171,
         172–173
   preferential access to South Georgia 279
   statistics on tourism 164, 190, 193,
         201–202
   tourism management highlights 202–203

International Geophysical Year (IGY) 29, 136, 151
International Union for the Conservation of Nature and Natural Resources 215–216
International Whaling Commission 249
Intourist 25, 113
islands, Southern Oceanic
   Heard Island and McDonald Islands 275–276
   impacts of tourism 267
   links to Antarctic tourism 267–268
   location and characteristics 264–265
   other groups 279–280
   tourist experience 265–266
   tourist management practices 270–274(tab)
   Tristan da Cunha group 276–277
   see also Macquarie Island; South Georgia
isotherms 32, 34–35

James Caird (lifeboat) 28
Journey to Spitzbergen, A (J. Lainige) 15
jurisdictions, polar: constraints on tourism 12–13

Kamchatka 242–243
Kamestatin Lake, Canada 84–85
Kipling, Rudyard 56, 239
Knox, George 211–212
Kotzebue, Alaska 75–76, 77(illust)
Kungsleden trail 116
Kwa'ashk'i Kwaan clan 78–79

Lainige, John 15
Lamb, Hubert 37
Land of the Midnight Sun, The (du Chaillu) 19
landing sites, Antarctic 158–160, 164–165, 217–219
   carrying capacity 295–296
   management plan development 298–300
   monitoring for change 300–303
   recovery and remediation 303–305
language: as barrier 105–106
legislation, national 206
liability, corporate and personal 255–256
licensing: of guides 244
Limits of Acceptable Change 232, 296

Lindblad, Lars Eric 30, 45, 250
Lindblad Explorer (cruise ship) 26
'Lindblad pattern' 45–46, 152, 153
logistics: and limitations to Antarctic tourism 205–206
Lyttelton see Christchurch/Lyttelton, New Zealand

Macquarie Island 138, 269
   tourist management practices 270–274(tab) 275
Madrid Protocol 191
management: of Antarctic wilderness and tourism 219–220, 231–232
   comparisons 270–274(tab), 288–289(tab), 293–295
   from cultural perspective 235–237
   environmental cause and effect 235
   landing site management plans 298–300
   potential options 295–298
   self-regulation 240–244
   South Georgia see under South Georgia
   through Antarctic Treaty System see under Antarctic Treaty System
   tourist behaviour 237–239
   see also national parks; protected areas
mapping 7, 8
   during International Geophysical Year (IGY) 29
   tourism value maps 90, 91(illust)
Marine Mammal Protecton Act (1972, USA) 79
Maritime Antarctic 149
mass tourism 52–53
Matumeak, Warren 44–45
Mawson, Douglas 28–29
McDonald Islands see Heard Island and McDonald Islands
Merriam, C. Hart 22–23
minerals: extraction prohibited in Antarctic 36
monitoring: of landing sites 300–305
Mount Erebus disaster 190
Muir, John 20, 21, 22
multiple resource management 256–260, 297–298
museums 73, 74, 76, 92, 129, 140, 249, 252

NANA Museum of the Arts, Kotzebue, Alaska 76

Nansen, Fridtjof 16
National Antarctic Programs (NAPs) 171,
        175–176
National Geographic Magazine 21
national parks 25, 77, 96–97
    Auyuittuq National Park Reserve 90
    economic significance 103–104
    Finland 62, 105
    impact of climate change 43
    Parks Canada 242
    Sweden 116
    see also protected areas
*National Survey of Fishing, Hunting and
        Wildlife Associated Recreation* (USA)
        59, 63
nature tourism 62–64, 110
Naveen, Ron 218
New Zealand
    Christchurch/Lyttelton *see*
            Christchurch/Lyttelton, New Zealand
    southern islands 268–269, 270–274(tab)
New Zealand Antarctic Society 211
Nobile, General Umberto 25
Northeast Greenland National Park 62
Northern Alaska Tour Company 73–74
Northwest Alaska Regional Corporation
        (NANA) 75–76
Northwest Passage 104
Northwestern Alaska 75–78
Norway
    economic impact of tourism 54, 111–112
    Svalbard 288–289(tab), 290–291
Nunamiut 72, 73
    development of tourist services 74–75
Nunavut 65
    Baker Lake traditional camp 101(illust)
    cruises 99
    future prospects 100
    national and territorial parks 96–97
    Nunavut Tourism (organization) 95–96
    origins of tourism 85–86
    Pangnirtung, Baffin Island
        location 89(map)
        Marshall Macklin Monaghan Office
            88(illust)
        pilot project on community tourism
            87–94
        planned visitors' centre 93(illust)
        tourism value maps 90, 91(illust)
    personal experiences 94
    strategic planning 97–98

Oceanites 218
Oulanka National Park 62, 105
overflights *see under* aviation

Paanajarvi National Park 62
Pacific Coast Steamship Company 20, 21,
        22
Pampa (ship) 28
Panama Canal 130
Pangnirtung *see under* Nunavut
Panther (steamship) 18
Parks Canada 242
paths, trekking and cycling 116
penguins 217, 218, 250(illust)
perceptions
    of hosts 74–75
    of tourists
        as barrier to access 10–11
        and climate change 43–44
        of personal safety 63
photography 27–28
polar bears 41, 42(illust), 94, 95(illust)
Polar Front 33, 34
pollution 37–38, 44
Port Stanley *see* Stanley, Falkland Islands
port state jurisdiction 142–144
Porter, George 45
ports, gateway *see* gateway ports, Antarctic
Primus Oil Stove 16
Project Antarctic Conservation 215–216,
        217
protected areas 233, 241–243, 299–300
    site monitoring 301–305
    see also national parks
psychology: and limitations to Antarctic
        tourism 205
Puerto Williams, Chile 134
Punta Arenas, Chile
    current status 133–134
    flight/cruise operations 155
    location and distance 125(map), 126(tab)
    origin and development 127, 128(tab), 130
    tourist numbers 126(fig)
Putin, President Vladimir 114

railways 24–25
    White Pass & Yukon Railroad Company
        23–24
    Whitehorse Railroad, Yukon 68(illust)

regulations: and limitations to Antarctic tourism
    206
research: on Antarctic tourism
    1960s to 1990 211–215
    1990s
        future trends 216
        IUCN report 215–216
        landing operations 217–219
        management of tourism 219–220
    Environmental Protocol 214–215
    tourist industry 210–211
rivers 37–38
Roosevelt, Theodore (US President) 15
Ross Dependency 136
Russian Federation
    economic impact of tourism 113–115
    hunting and fishing 61–62
    protected areas 242–243
    use of Finland as gateway 106
    use of nuclear icebreakers for cruises 25–26
    see also Soviet Union

safety
    Antarctic overflights 194
    factor in destination selection 63, 163
    passengers liners in Antarctic 155
Scandinavia
    economic impact of tourism 111–112,
        115–117
    sport fishing 59–60
Scidmore, Eliza Ruhamah 21, 238
Scientific Committee for Antarctic Research
    (SCAR) 211–212
scientists
    attitudes to tourists 212
    contrasted to tourists 219
    role in tourism 22
Scott, Robert Falcon 27
seals 248, 269
    effect of cruise ships 79–80
    hunting 36
    in lives of indigenous peoples 78–79
    risks to tourists 10
self-regulation 240–244
Sensemeier, Ray 78, 79
Shackleton, Sir Ernest 27, 249
ship-borne tourism
    Antarctic
        adventure tourism 172–173
        comparison with air-borne tourism
            157–158

control of ships through port state
    jurisdiction 142–143
expedition ships and small sailing vessels
    152–153
first attempts 28, 29–30, 151
flight/cruise operations 155
IAATO rules 151, 152, 153–154, 166,
    167
Initial Environmental Evaluations 166
landing sites 158–160, 164–165
larger ships with landings 153–154, 162
'Lindblad pattern' 45–46, 152, 153
passenger liners 154–155, 216
Southern Oceanic islands see islands,
    Southern Oceanic
tourist numbers 156–157, 201(tab)
Arctic
    effect of ships on seals 79–80
    exploratory cruises 26, 30
    Greenland 108
    Harriman Alaska Expedition 22–23
    Hayes/Bradford (1869) 17–19
    proactive collaboration with communities
        99
    Soviet and Russian initiatives 25–26,
        53(illust)
    economic impact 53–57
    largest mass tourism activity 4–5
    tourist numbers 123(tab), 151–152, 153,
        155, 160, 161(tab)
Simon Paneak Memorial Museum, Anaktuvuk
    Pass, Alaska 73, 75
Skagway, Alaska 67(illust)
Smith, Dick 188
South Africa
    Antarctic interests 136–137
    Cape Town see Cape Town, South Africa
South Georgia 277–279
    growth of tourism 250–251
    heritage and natural resources 248–249
    location and map 247
    management of tourism
        economic and financial issues 255–256
        environment, culture and wildlife
            253–255
        multiple resource management 256–260
    multiple resource management 297–298
    patterns of tourism 252–253
    tourist management practices 270–274(tab),
        288–289(tab), 290
South Pole: expeditions 27–28

South Sandwich Islands 278
South Shetland Islands 217–218
Southern Ocean Rim States (SORS) 124
Soviet Union 25, 113
    environmental damage 37–38
    see also Russian Federation
sport
    role in establishment of tourism 15–16
    see also hunting and fishing
Stanley, Falkland Islands 124
    current status 134–136
    economic significance of tourism 135(tab)
    origin and development 127, 128(tab), 130
    tourist numbers 126(fig)
steamships
    Panther 18
    role of companies in developing tourism
        24–25
Survey of Recreational Fishing in Canada
        (2000) 60
surveys, topographic and cadastral 7
Svalbard 288–289(tab), 290–291
Sweden 115–117

Tasmania see Hobart, Australia
Thomas Cook and Sons 27
Tourism Satellite Accounts 105
tourists
    Antarctic
        definition 149
        nationalities 129, 162–163
        numbers 54, 149, 151–152, 153, 155,
            156–157, 160, 161(tab), 163
    Arctic
        lady tourist of 1880s 20(illust)
        numbers 3, 54, 66
        spending 53, 59, 60–61, 63
    management of behaviour and numbers
        237–239
    perceptions of polar regions see perceptions
    and pollution 44
    term of opprobium 147
tours, guided 19–20
transport
    Alaska 118
    impact of new modes 8–9
    role of transport providers in developing
        tourism 24–27
    Sweden 115
    see also aviation; railways; ship-borne tourism

treaties, international 233–234
tree-line 33, 34
Tristan da Cuhna 270–274(tab), 276–277

UK: backs port state jurisdiction 143–144
United States Antarctic Journal 211
Ushuaia, Argentina
    current status 131–132
    early tourism 28
    economic significance of tourism 133(tab)
    location and distance 125(map), 126(tab)
    origin and development 127, 128(tab)
    tourist numbers 126(fig)

Vancouver: gateway for Arctic tourism 104
Vicuña, Orrego 142–143
visas 114
Vize, Prof. V. Yu. 25

Warbelow's Air Ventures 73, 74
warming, global
    consequences for tourism 40–45
    effects 38–39
whaling 36, 248, 249
    environmental contamination 253
White Pass & Yukon Railroad Company
        23–24
Whitehorse Railroad, Yukon, Canada 68(illust)
wilderness
    definition and expectations 77–78
    impact of aviation 26–27
Wilderness Act, USA 77
wildlife
    effects of tourists 79–80, 217–218
    impact of climate change 40–43
    loss of habitat
        due to river diversion 37
    New Zealand southern islands 268
    of personal importance to Alaskans 119
    pollutants in Antarctic organisms 38
    South Georgia 254–255, 277–278
    see also nature tourism
World Wildlife Fund 243–244

Yakutat, Alaska 78–80